VOLUME ONE HUNDRED AND TWENTY NINE

ADVANCES IN
CANCER RESEARCH

Hsp90 in Cancer: Beyond the Usual Suspects

VOLUME ONE HUNDRED AND TWENTY NINE

Advances in
CANCER RESEARCH

Hsp90 in Cancer: Beyond the Usual Suspects

Edited by

JENNIFER ISAACS
Cell and Molecular Pharmacology
Medical University of South Carolina
Charleston, SC, USA

LUKE WHITESELL
Whitehead Institute
Cambridge, MA, USA

ELSEVIER

AMSTERDAM • BOSTON • HEIDELBERG • LONDON
NEW YORK • OXFORD • PARIS • SAN DIEGO
SAN FRANCISCO • SINGAPORE • SYDNEY • TOKYO
Academic Press is an imprint of Elsevier

Academic Press is an imprint of Elsevier
50 Hampshire Street, 5th Floor, Cambridge, MA 02139, USA
525 B Street, Suite 1800, San Diego, CA 92101-4495, USA
The Boulevard, Langford Lane, Kidlington, Oxford OX5 1GB, UK
125 London Wall, London, EC2Y 5AS, UK

First edition 2016

ISBN: 978-0-12-802290-0
ISSN: 0065-230X

For information on all Academic Press publications
visit our website at store.elsevier.com

DEDICATION

The editors dedicate this volume to Susan Lindquist, a founder of the field of HSP90 research who continues to push it forward and inspire scientists from diverse disciplines to do the same.

CONTENTS

CONTRIBUTORS

Ephraim Ansa-Addo
Hollings Cancer Center, Department of Microbiology and Immunology, Medical University of South Carolina, Charleston, South Carolina, USA

Brian S.J. Blagg
Department of Medicinal Chemistry, The University of Kansas, Lawrence, Kansas, USA

Stuart K. Calderwood
Department of Radiation Oncology, Beth Israel Deaconess Medical Center, Harvard Medical School, Boston, Massachusetts, USA

Diana Dunn
Department of Urology; Department of Biochemistry and Molecular Biology; Cancer Research Institute, SUNY Upstate Medical University, Syracuse, New York, USA

Pablo C. Echeverría
Département de Biologie Cellulaire, Université de Genève, Sciences III, Geneva, Switzerland

Gaurav Garg
Department of Medicinal Chemistry, The University of Kansas, Lawrence, Kansas, USA

Michael W. Graner
Department of Neurosurgery, University of Colorado Denver, Aurora, Colorado, USA

Feng Hong
Hollings Cancer Center, Department of Microbiology and Immunology, Medical University of South Carolina, Charleston, South Carolina, USA

Jennifer S. Isaacs
Department of Cell and Molecular Pharmacology, Medical University of South Carolina, Hollings Cancer Center, Charleston, South Carolina, USA

Sami Jamal
Department of Urology; Cancer Research Institute, SUNY Upstate Medical University, Syracuse, New York, USA

Daniel Jarosz
Chemical & Systems Biology; Developmental Biology, Stanford University School of Medicine, Stanford, California, USA

Daniel G. Jay
Department of Developmental, Molecular, and Chemical Biology, School of Medicine, Tufts University, Boston, Massachusetts, USA

Anuj Khandelwal
Department of Medicinal Chemistry, The University of Kansas, Lawrence, Kansas, USA

Zihai Li
Hollings Cancer Center, Department of Microbiology and Immunology, Medical University of South Carolina, Charleston, South Carolina, USA

Jonelle B. Miller
Department of Urology; Cancer Research Institute, SUNY Upstate Medical University, Syracuse, New York, USA

Mehdi Mollapour
Department of Urology; Department of Biochemistry and Molecular Biology; Cancer Research Institute, SUNY Upstate Medical University, Syracuse, New York, USA

Len Neckers
Urologic Oncology Branch, Center for Cancer Research, National Cancer Institute, Bethesda, Maryland, USA

Didier Picard
Département de Biologie Cellulaire, Université de Genève, Sciences III, Geneva, Switzerland

Evangelia Vartholomaiou
Département de Biologie Cellulaire, Université de Genève, Sciences III, Geneva, Switzerland

Daniel Senh Wong
Graduate Program in Cellular and Molecular Physiology, Sackler School of Graduate Biomedical Sciences, Tufts University, Boston, Massachusetts, USA

Mark R. Woodford
Department of Urology; Cancer Research Institute, SUNY Upstate Medical University, Syracuse, New York, USA

Bill X. Wu
Hollings Cancer Center, Department of Microbiology and Immunology, Medical University of South Carolina, Charleston, South Carolina, USA

Yongliang Zhang
Hollings Cancer Center, Department of Microbiology and Immunology, Medical University of South Carolina, Charleston, South Carolina, USA

PREFACE

Over 20 years ago, potent and specific natural products were first discovered as direct inhibitors of the protein folding function of HSP90. This initial discovery was first greeted with considerable skepticism, followed by a tidal wave of enthusiasm for the targeting of HSP90 as an anticancer strategy. Much of this interest was driven by the potential of HSP90 inhibition to accomplish what other molecularly targeted anticancer therapies do not: the simultaneous disruption of multiple signaling pathways critical to tumor cell growth and survival. Such a combinatorial attack on the oncoproteins chaperoned by HSP90 was embraced by the community as a "rational approach" to addressing the heterogeneity and complexity characteristic of the most common human cancers. Small biotechnology companies and subsequently large pharmaceutical concerns saw promising commercial potential. Plunging into the discovery and clinical development of second-generation, synthetic HSP90 inhibitors of various chemical classes, industry has now delivered excellent drugs with optimized pharmacology, reduced toxicity, and promising effects in animal models. Unfortunately, the clinical activity of these HSP90 inhibitors, alone or in combination, has been disappointingly modest in most cancer trials reported to date. At this point, it seems the honeymoon is over and the infatuation of pharma with HSP90 has waned. If the partnership is to be rescued and effective treatments delivered, it is now up to the scientific community to dig deeper into the complex biology of HSP90, rethink conventional wisdom, and advance beyond overly simplistic models that fail to capture reality, either in the laboratory or in the clinic.

In the nine chapters of this volume, we have sought to abandon the well-trodden "usual suspects" of most HSP90 reviews and survey understudied, sometimes counterintuitive, sometimes controversial topics. The unifying theme is their relevance to the roles of HSP90 in cancer biology and how these roles might be most effectively exploited to treat cancers. The volume starts with a thoughtful bioinformatic effort from the Picard laboratory to mine the literature for lesser known clients of HSP90. The impairment of these unusual suspects which include tumor suppressors, chromatin-remodeling factors, and regulators of metabolism is likely to have previously ignored impact on the net effect of HSP90 inhibition upon cancer cells. We then move on to a chapter from the Mollapour lab describing how specific cancer-associated posttranslational modifications to HSP90 alter its chaperoning function and sensitivity to small-molecule inhibitors. This chapter addresses the provocative concept that clients are not only assisted in their folding by HSP90 but once properly folded can also "return the favor" by modifying HSP90 itself in a feedback/feedforward phenomenon. Such effects require much more study to fully understand how they impact the activity of HSP90 inhibition against cancers. To explore this concept further from a chemical biological perspective, our next chapter from the Blagg laboratory provides a tour-de-force summary of the ever-expanding array of small molecules reported to alter HSP90 function in nonclassical ways. These act either through direct binding to sites outside the N-terminal ATPase pocket of HSP90 or through effects on co-chaperones, modifying factors or even other components of the protein homeostasis network in which HSP90 plays an integral role. Our next two chapters deal with the greatly underappreciated functions of HSP90 within the nucleus. Traditionally thought of exclusively as a chaperone for cytosolic proteins, HSP90 actually plays important roles in the control of transcription as detailed in the chapter by Calderwood and Neckers. These effects on transcription are mediated not just by direct effects on the basal transcriptional machinery but also include far-reaching effects on epigenetic modifiers and chromatin structure as laid out in the chapter by Isaacs. Given the emerging appreciation that epigenetic changes have profound effects upon cancer development and drug responses, it is well worth noting that many of the clinical effects of Hsp90 inhibitors may be due to their impact upon the epigenetic machinery. Continuing the theme of noncytosolic functions for HSP90, a chapter from the Jay laboratory addresses emerging, understudied roles for extracellular HSP90 in modifying the tumor microenvironment and altering the metastatic potential of cancers. Implications for the development of cell-impermeant inhibitors of HSP90 as antimetastatic agents provide much food for thought. Continuing consideration of noncytosolic HSP90 functions, the next chapter from the Li laboratory provides a comprehensive overview of the roles in cancer biology of GRP94, the endoplasmic reticulum-resident homolog of HSP90. The often-neglected but likely important effects of conventional HSP90 inhibitors on GRP94 function are discussed. The potential utility of GRP94-selective inhibitors as anticancer and immunomodulatory agents is also examined in light of GRP94's prominent function in the processing and presentation of cancer-associated neoantigens. To expand on the theme of immune function, the next chapter by Graner tackles the complex role of HSP90-family chaperones in tumor immunology more globally. Here the complex, often conflicted literature in this area is distilled to come up with a model in which the immunological consequences of HSP90 inhibition are highly dose dependent. Maximally tolerated HSP90 inhibition leads to profound suppression of both innate and adaptive immune functions, while more modest inhibition can actually stimulate antigen processing and anticancer immune responses, activities that have been near completely ignored in the clinical development of inhibitors.

Finally, to close the volume, a chapter by Jarosz steps back to provide a very broad whole organism perspective. It summarizes how the basic biochemical functions of HSP90 have been found to act at genetic and epigenetic levels to support the adaptability and evolution of organisms ranging fungi to plants, fruit flies, and fish. Increasing evidence indicates that by modulating the phenotypic expression of genetic variation and sculpting the architecture of entire networks in cancers, HSP90 plays a similar evolutionary role in malignant progression by supporting the ability of tumors to adapt to new environments and therapeutic interventions. Indeed, the best way to exploit HSP90 as a target in advanced cancers may well be in combination with other active agents to limit evolvability and the rapid emergence of drug resistance, which is arguably the greatest current barrier to curing many cancers. It is hoped that by addressing the fundamental issues raised within this volume, the ongoing clinical development of HSP90 inhibitors will become less empiric, more rational, and more successful. After 20 years, multidisciplinary efforts to understand HSP90 have let us get the crab by the claw, but it is going to take a lot more work to put it in the box.

JENNIFER ISAACS
LUKE WHITESELL
January 2016

CHAPTER ONE

Unusual Suspects in the Twilight Zone Between the Hsp90 Interactome and Carcinogenesis

Evangelia Vartholomaiou[1], Pablo C. Echeverría[1], Didier Picard[2]
Département de Biologie Cellulaire, Université de Genève, Sciences III, Geneva, Switzerland
[2]Corresponding author: e-mail address: didier.picard@unige.ch

Contents

Abstract

The molecular chaperone Hsp90 has attracted a lot of interest in cancer research ever since cancer cells were found to be more sensitive to Hsp90 inhibition than normal cells. Why that is has remained a matter of debate and is still unclear. In addition to increased Hsp90 dependence for some mutant cancer proteins and modifications of the Hsp90 machinery itself, a number of other characteristics of cancer cells probably contribute to

[1] These authors contributed equally to this work.

Advances in Cancer Research, Volume 129
ISSN 0065-230X
http://dx.doi.org/10.1016/bs.acr.2015.08.001

this phenomenon; these include aneuploidy and overall increased numbers and levels of defective and mutant proteins, which all contribute to perturbed proteostasis. Work over the last two decades has demonstrated that many cancer-related proteins are Hsp90 clients, and yet only few of them have been extensively investigated, selected either on the basis of their obvious function as cancer drivers or because they proved to be convenient biomarkers for monitoring the effects of Hsp90 inhibitors. The purpose of our review is to go beyond these "usual suspects." We established a workflow to select poorly studied proteins that are related to cancer processes and qualify as Hsp90 clients. By discussing and taking a fresh look at these "unusual suspects," we hope to stimulate others to revisit them as novel therapeutic targets or diagnostic markers.

1. INTRODUCTION

1.1 Hsp90 and Cancer

The molecular chaperone Hsp90 is a highly abundant cytosolic protein, whose primary function is to assist client proteins in their maturation and in maintaining their stability. It achieves this task as part of multi-chaperone complexes in association with several co-chaperones. Early on, it was speculated that the Hsp90 molecular chaperone machine may assist up to 10% of all cytosolic proteins at some stage of their life cycle (Nathan, Vos, & Lindquist, 1997). Recent studies confirmed the high number, variety, and complexity of these interactions (Echeverria, Bernthaler, Dupuis, Mayer, & Picard, 2011; Taipale et al., 2012). Inhibitor studies support the view that the intrinsic ATPase activity of Hsp90 and the associated conformational changes are essential for its function as a major hub of protein–protein interaction networks (Echeverria & Picard, 2014; Fierro-Monti et al., 2013). The Hsp90-dependent proteome participates in many key cellular processes that are related to the development and homeostasis of normal as well as cancer cells. Considering the ever-growing number of Hsp90-regulated proteins, their identity sometimes buried in an ocean of supplementary tables and information, many can go unnoticed to a readership hungry for new Hsp90 clients involved in specific cellular processes. In particular, in the context of cancer, there is a need to understand the impact of different kinds of Hsp90 inhibitors, of which several are in clinical trials (reviewed in Neckers & Workman, 2012), and to identify new relevant Hsp90 targets, notably those that specific types of tumors may be addicted to.

The goal of this review is to shed light on some of the proteins of the Hsp90 interactome (Hsp90Int) that, to this date, have not been extensively studied as Hsp90-interacting factors and yet are important for cancer. These are proteins that one could think of as the "unusual suspects" contributing to the Hsp90 client/cancer connection.

1.2 Extracting Poorly Studied Cancer-Related Hsp90 Clients from the Hsp90 Interaction Database

For the purposes of this review, we worked out a workflow (Fig. 1) to select proteins that are known (or predicted) to be Hsp90 interactors and to be involved or studied as key players in the initiation or progression of cancers, but for which the connection has not yet been substantially explored. We made use of the comprehensive databases of the Picard lab: our list of Hsp90 interactors, which is based on high-quality literature mining (http://www.picard.ch/downloads/Hsp90interactors.pdf), and our database of the Hsp90Int (from Echeverria et al., 2011, publicly available at http://www.picard.ch/Hsp90Int). The latter also includes interactors of Hsp90 co-chaperones and captures the range of interactors experimentally verified or predicted from other organisms. Following the decision tree depicted in Fig. 1, we collected the data previously mentioned and imported it into open-source software for the management of networks (Cytoscape)

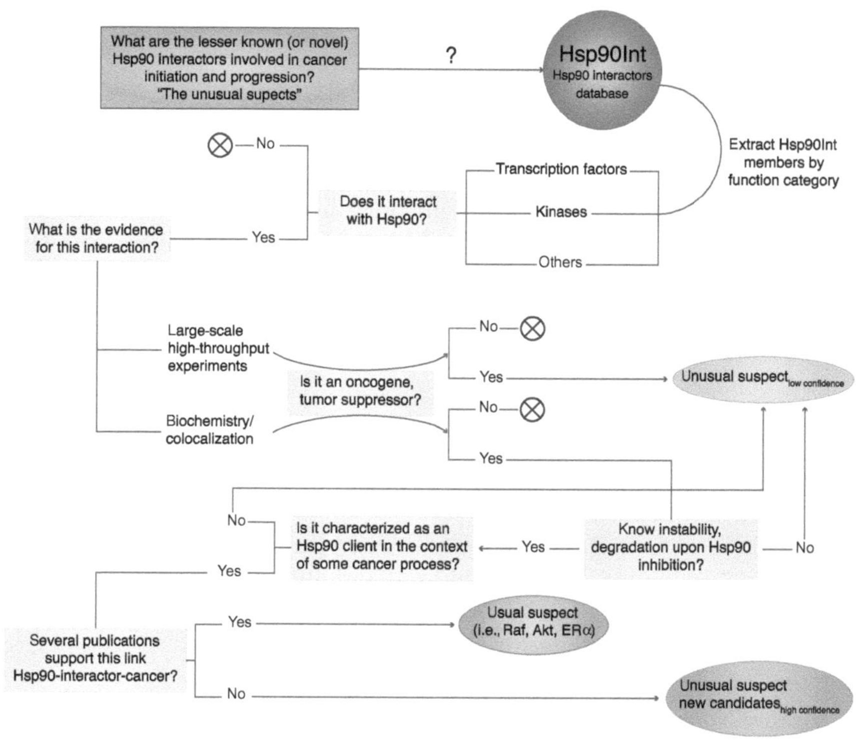

Figure 1 Workflow for the identification of cancer-related Hsp90 interactors as "unusual suspects." (See the color plate.)

(Shannon et al., 2003), for further analysis. We fed this Hsp90 network with metadata corresponding to information about the oncogene or tumor suppressor status of each component of the network using lists from a recent publication by Vogelstein et al. (2013) and an online resource from the Bushman laboratory (http://www.bushmanlab.org/links/genelists). Hsp90Int already contains metadata concerning the experimental systems and publication IDs. The exploration of these data (following the workflow of Fig. 1) returned a substantial number (57) of "unusual suspects" that we discuss in the following sections. When its interaction with Hsp90 is only known from a high-throughput screen and not yet validated biochemically in some publication, we consider the suspect of low confidence. Likewise, we put a protein in this same category if nothing is known about the functional relevance of its interaction with Hsp90 (e.g., whether it is degraded upon Hsp90 inhibition or whether its interaction with Hsp90 has known significance in cancer biology). If, on the other hand, there have been a few reports describing the "suspect" as an important Hsp90 client in some cancer-related process, we classify this "suspect" as *unusual* with high confidence. If, however, the same case is supported by dozens of publications, it becomes a "usual suspect" and is not further discussed here.

Figure 2 displays the 400 cancer-related proteins represented in the Hsp90Int database with their connections to different types of cancer (for the purposes of this figure only those included in the KEGG database). "Unusual suspects" are explicitly highlighted in blue. It is noteworthy that many of the new candidates we found with our methodology had only recently been added to the Hsp90Int by a systematic and quantitative study that surveyed most human kinases, transcription factors (TFs), and E3 ligases for interaction with Hsp90β (Taipale et al., 2012). This was done using an improved version of LUminescence-based Mammalian IntERactome (LUMIER) technology (Barrios-Rodiles et al., 2005). Even though this study was high throughput, we felt that the LUMIER assay was done in a way that yielded high-confidence interactors, notably because the levels of both bait and prey were carefully controlled.

We have previously pointed out (Fierro-Monti et al., 2013) that unexpected and undesirable effects of Hsp90 inhibitors on oncoproteins and tumor suppressors must be carefully evaluated and weighed in considering the use of such compounds to treat cancer. Indeed, there are some oncoproteins that are upregulated and some tumor suppressors that are downregulated by Hsp90 inhibitors. Such effects constitute a potential risk in targeting Hsp90 for treatment. In the following sections, we will also discuss several of these cases.

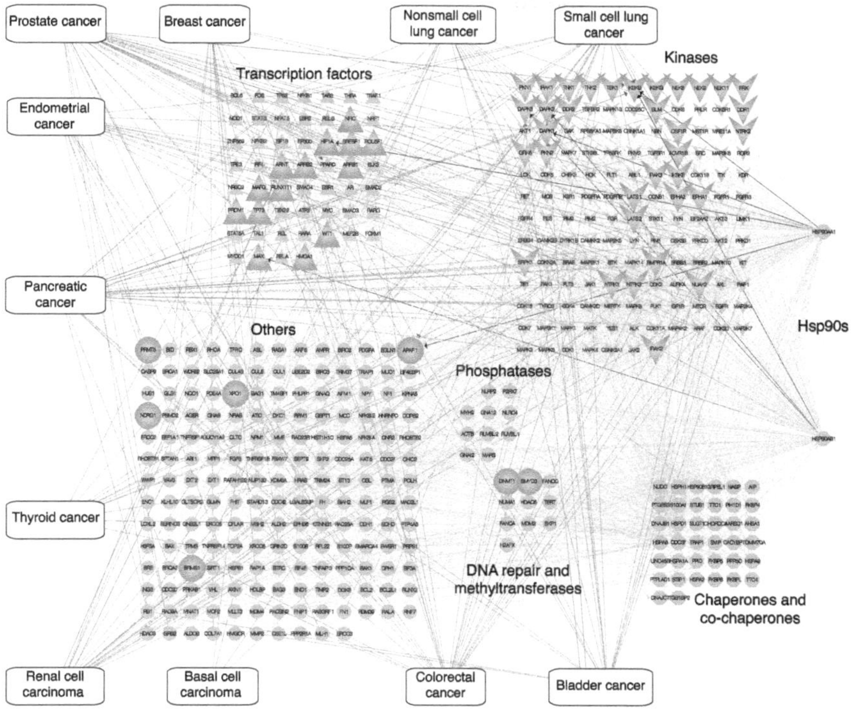

Figure 2 Cancer genes in the Hsp90Int database. All cancer genes present in the protein–protein interaction (PPI) network database Hsp90Int were extracted and clustered according to their main functions. In addition to the PPI links among them (in light gray), the connections between these proteins and different cancers are depicted (in red), and for some highlighted ones as thicker arrows. Chaperones and co-chaperones that maintain the interconnectivity of the network were kept in the scheme. The "unusual suspects" are highlighted with increased size and colored in light blue. Note that for simplicity, only cancer connections from the KEGG cancer pathways resource (http://www.genome.jp/kegg/disease/cancer.html) were used here, whereas the text draws on a more comprehensive analysis that includes other databases and literature mining. (See the color plate.)

2. TFs AND COFACTORS

Several members of the steroid receptor family and other TFs are oncogenic signal transduction proteins that rely on Hsp90 for maturation and/or stabilization. These include the estrogen receptors ERα and ERβ (Renoir, 2012) and the androgen receptor (Culig & Santer, 2014). Other TFs are the heat-shock factor 1 (HSF1), p53, SP1, NF-κB, STATs, and

BCL6. These examples of oncogenic factors and their Hsp90 connection have been widely described in the literature (for an updated review, see Khurana & Bhattacharyya, 2015). Following the protocol described in the decision tree of Fig. 1, we found several interesting, but less well-trodden examples.

2.1 TFs Enabling Metabolic Changes

Cells respond to reduced oxygen levels through the hypoxia-inducible transcription factor 1 α (HIF1A) of the bHLH family. As one of the major mediators of the hypoxic response, HIF1A is responsible for activating hypoxia-responsive genes, which are involved in several aspects of oncogenesis and malignant progression, including proliferation, altered metabolism, neoangiogenesis, invasion, metastasis, and therapy resistance (Burroughs et al., 2013; Luo, Wang, Wu, Jiang, & Wu, 2014; Mujcic, Hill, Koritzinsky, & Wouters, 2014). HIF1A is vital for cancer cell metabolism because it transactivates genes encoding the major enzymes of glycolysis, such as glucose transporters and most glycolytic enzymes, thereby supporting the Warburg phenomenon (Le et al., 2010). HIF1A acts as a heterodimer together with another TF, the aryl hydrocarbon receptor nuclear translocator (ARNT) (also known as HIF-1β). It is this complex that is involved in tumor as well as more generally embryonic development (reviewed by Harris, 2002). Hsp90 interacts with HIF1A through its PAS domain (Isaacs, Jung, & Neckers, 2004; Ueda, Xu, Morimoto, Kawabe, & Imaoka, 2008), and the Hsp90 inhibitors geldanamycin and 17-allylaminogeldanamycin (17-AAG) induce proteasomal degradation of HIF1A (Gradin et al., 1996; Isaacs et al., 2002). In the nucleus, HIF1A is released from Hsp90 because of competition by ARNT (Isaacs et al., 2004). ARNT replaces Hsp90 in stabilizing HIF1A and allows DNA binding. The TF MAFG of the bZIP family was found to interact with HIF1A and to regulate the hypoxic response of cells by sequestering HIF1A in the nucleus (Ueda et al., 2008). MAFG was also found to interact with Hsp90 (Taipale et al., 2012). Overall, it is clear that Hsp90 helps regulate the hypoxia response, and consequently, it could be a key to sustaining the processes regulated by this response that enable cancer initiation and progression. Incidentally, MAFG is known to be activated by the kinase BRAF, another Hsp90 client, to allow the recruitment of a corepressor complex to gene promoters subjected to the "CpG island methylator phenotype," a feature relevant to colorectal cancers (CRCs) (Fang, Ou, Hutchinson, & Green, 2014). Thus, MAFG is a novel oncoprotein worth characterizing in the context of its liaison with Hsp90.

The p53 homolog TAp73 has attracted increasing interest. TAp73 is stabilized in tumors by hypoxia through HIF1A-mediated repression of the specific ubiquitin ligase that targets TAp73 for degradation (Dulloo et al., 2015). TAp73 is a critical regulator of the angiogenic transcriptome and is sufficient to directly activate the expression of several angiogenic genes (Dulloo et al., 2015), resulting in tumors with increased vascularization. It has recently been reported that Hsp90 prevents TAp73 degradation by the proteasome, a process exacerbated by the inhibition of histone deacetylase HDAC1 (Zhang, Xu, & Chen, 2013). TAp73 is expressed in several isoforms, which differentially inhibit or promote carcinogenesis (reviewed in Soldevilla, Millan, Bonilla, & Dominguez, 2013). It remains to be determined if and how Hsp90 may affect the different isoforms.

These relatively new players collaborate with established Hsp90 client proteins. HIF1A associates with STAT3 for cooperative activation of HIF1A target genes in cancer cell lines (Pawlus, Wang, & Hu, 2014). HSF1 and HIF1A regulate the expression of the same gene, *FOXM1*, under different conditions. In turn, the protein FOXM1 promotes tumor cell proliferation (Xia et al., 2009) and is required for cell cycle progression (Dai et al., 2013). Since it has been found to interact with Hsp90β (Taipale et al., 2012), FOXM1 also qualifies as an unusual suspect in its own right for the purposes of this review.

Changes in lipid metabolism are another key factor for cancer development. Sterol regulatory element-binding protein 1 (SREBP-1) is the master TF of the bHLH family that controls lipid homeostasis. It serves as a critical link between oncogenic signaling and tumor metabolism, supporting the EGFR/PI3K/Akt signaling pathway that promotes growth and survival in glioblastoma and potentially other cancer types (Guo, Bell, Mischel, & Chakravarti, 2014). There is only one paper indicating that SREBP-1 interacts with Hsp90 (Taipale et al., 2012), but this makes it an excellent candidate for further studies on how Hsp90 supports metabolic changes in cancer.

2.2 TFs in Leukemia

A couple of examples illustrate how Hsp90 is a crucial factor in myeloid leukemia, potentially through some still poorly studied Hsp90 TF clients. The aberrant overexpression of the TF Wilms tumor 1 (WT1), a member of the zinc finger C2H2 family, in myeloid leukemia plays an important role in blast cell survival and resistance to chemotherapy. High expression of WT1 is associated with relapse and shortened disease-free survival in patients

(Bansal et al., 2010). Hsp90 associates with and stabilizes WT1. Hsp90 inhibitors such as 17-AAG and STA-9090 target WT1 for degradation (Bansal et al., 2010), supporting the notion that WT1 is a client protein and a relevant target in myeloid leukemia. A similar case is observed with RUNX1-ETO (RUNX1T1), a fused TF resulting from a balanced translocation. This event is one of the most common chromosomal translocations found in patients with acute myeloid leukemia (AML), and a critical driver of leukemogenesis (Peterson & Zhang, 2004). Its interaction with Hsp90 has only been described in one paper (Komori, Sueoka, Fujiki, Ishii, & Kozu, 1999), and the destabilization of RUNX1T1 by Hsp90 inhibitors was reported years later (Yang, Thompson, Brandt, & Hiebert, 2007). The importance of several other Hsp90 client proteins (for example, Flt3, c-Kit, Akt, BCL6, STAT3) for AML has been extensively studied; such work provided a scientific rationale for clinical trials to treat this disease with Hsp90 inhibitors (Lancet et al., 2010).

2.3 Unusual Suspects Among Hsp90 TF Clients in Multiple Other Cancer Types

The β-arrestins (ARRB) are cofactors that regulate transcription. It has been shown in animal models that ARRB expression affects tumor initiation time, growth rate, vascularization, survival under hypoxic conditions, invasiveness, and metastatic potential. Studies in human cancer patients have demonstrated that dysregulation of ARRB expression, localization, or phosphorylation is associated with more aggressive cancer phenotypes and poor outcome in malignancies involving the breast, lung, prostate, brain, and hematological system (Sobolesky & Moussa, 2013). The interaction of ARRB1 and 2 with Hsp90β was found by a global proteomic search for β-arrestin-interacting proteins (Xiao et al., 2007). It would be interesting to know to what extent Hsp90 regulates their stability and function and how relevant this is in the context of several tumor types.

The POU family of proteins is a very important group of homeobox-containing TFs involved in development. Apart from their function in developmental processes, they have been found to be associated with different types of cancer (reviewed by Purkayastha & Roy, 2015). OCT4 (also known as POU5F1) is reportedly involved in germ cell tumors as well as cancers of the prostate, brain, and bladder; its overexpression leads to increased cellular proliferation, invasiveness, colony formation, anchorage-independent growth, and tumorigenicity of xenografts. OCT4 was proposed as an Hsp90 client protein that is protected from degradation by Hsp90 (Bradley,

Bieberich, Mivechi, Tangpisuthipongsa, & Wang, 2012). In addition, Hsp90 also participates in the processing or maturation of OCT4 mRNA by mechanisms yet to be defined (Bradley et al., 2012), but which make OCT4 highly dependent on Hsp90 and a potentially interesting target in multiple cancer types.

NFIC is a member of the CTF/NFI family of TFs that interact with Hsp90 (Taipale et al., 2012). NFIC is involved in regulation of the brain fatty acid-binding protein and glial fibrillary acidic protein gene expression in malignant glioma cell lines (Brun et al., 2009). Furthermore, NFIC is upregulated in gastric cancer (Jiang, Yang, Lu, & Ma, 2014). NFIC is another example of a poorly characterized Hsp90 interactor that might at the same time constitute a useful anticancer target. Another interesting case is the protein high-mobility group A1 (HMGA1). This protein is a member of the HMGI/HMGY family of TFs that play an important role in promoting cell proliferation and motility, epithelial–mesenchymal transition, and maintenance of stemness (Cleynen & Van de Ven, 2008). Overexpression of the HMGA1 gene in human breast cancer has been demonstrated, with a positive correlation between HMGA1 protein levels and the metastatic phenotype of human breast cancer cell lines (Liu, Guerra-Vladusic, Kurakata, Lupu, & Kohwi-Shigematsu, 1999). Similarly, higher protein levels of HMGA1 in human breast cancer tissues are significantly associated with breast cancer progression and poor prognosis (Huang, Huang, Dai, & Yang, 2015). In addition, high levels of HMGA1 drive metabolic alterations that contribute to CRC pathogenesis through effects on fatty acid synthesis (Williams et al., 2015). Other than the finding that HMGA1 interacts with Hsp90 (Taipale et al., 2012), not much else is known about a potential role for Hsp90 in regulating its function.

2.4 TFs as Tumor Suppressors

As mentioned above, since Hsp90 supports an essential network of tumor suppressors, Hsp90 inhibition could also lead to a reduction of these proteins. An example of an Hsp90-interacting tumor suppressor is Max (Taipale et al., 2012). Max antagonizes Myc-dependent cell transformation (reviewed by Cascon & Robledo, 2012). As a consequence, the destabilization of Max could promote tumor development. Another example is the Hsp90 interactor PRDM1 (Taipale et al., 2012). Also known as B lymphocyte-induced maturation protein 1 (BLIMP1), this TF prevents B-cell lymphomas. It is a transcriptional repressor that is essential for the proper differentiation of germinal center (GC) B-cells to plasma cells. The loss of BLIMP1 in GC B-cells could contribute to pathogenesis (Vrzalikova, Woodman, & Murray, 2012).

3. KINASES

Among the wide range of Hsp90 clients, kinases probably comprise the category with the most members. Indeed, Hsp90 was discovered in 1981 as a protein coprecipitating with the tyrosine kinase v-Src (Brugge, Erikson, & Erikson, 1981). Since then, a plethora of kinases have been found to interact with Hsp90. Many of these interactions have been very well characterized in the context of cancer, including those with ErbB2, Akt, Raf, and different cyclin-dependent kinases (CDKs). However, there are other kinase clients of Hsp90, which have received less attention. Applying our workflow (Fig. 1), we were able to pinpoint kinases associated with a variety of functions and pathways. Kinases that are not discussed in the following sections are summarized in Table 1.

3.1 Receptor Kinases

Some of the best-characterized Hsp90 clients are membrane-associated receptors. ErbB2 and the IGF1 receptor are well-known examples that are involved in cancer cell proliferation and survival. However, there are several less well-known cases.

ACVR1B/ALK4 together with ACVR2A or ACVR2B forms a receptor complex for activin, a member of the transforming growth factor-β superfamily. Upon ligand binding, ACVR1B becomes phosphorylated and activated, and in turn phosphorylates and activates Smad proteins (reviewed by Harrison et al., 2004). ACVR1B/ALK4 is a strong interactor of Hsp90 (Taipale et al., 2012). Somatic mutations of ACVR1B have been identified in pancreatic cancer where the gene was described as a tumor suppressor (Su et al., 2001). Tumor-specific truncated forms of ACVR1B that lack the kinase domain are expressed in human pituitary adenomas (Alexander, Bikkal, Zervas, Laws, & Klibanski, 1996); in this disease, they were reported to have a dominant-negative effect (Zhou et al., 2000), which blocked the antiproliferative function of activin.

Closely related to ACVR1B is the bone morphogenetic protein receptor type 1A (BMPR1A), also an Hsp90 client (Haupt et al., 2012;

Table 1 Additional Kinases That Were Identified as "Unusual Suspects" (and Are Not Further Discussed in the Text)

Kinase	Function, Pathway	Hsp90 Interaction	Cancer Relevance
SRPK1	Splicing, phosphorylates SR splicing factors	Taipale et al. (2012) and Zhong, Ding, Adams, Ghosh, and Fu (2009)	High levels in prostate (Zhong et al., 2009), breast, colon, and pancreas cancers (Hayes, Carrigan, & Miller, 2007), and hepatocellular carcinoma (Zhou et al., 2013)
IRAK family: IRAK1 IRAK2 IRAK3	Toll-like receptor signaling; interleukin-1 receptor signaling	De Nardo et al. (2005) and Taipale et al. (2012)	Overexpressed and activated in lymphoid and myeloid malignancies reviewed in Rhyasen and Starczynowski (2015)
FRK	Positively regulates PTEN	Taipale et al. (2012)	Tumor suppressor induces growth arrest (Craven, Cance, & Liu, 1995; Meyer et al., 2003). Less migratory and invasive cells upon overexpression (Shi et al., 2015)
LATS1 LATS2	Hippo pathway; phosphorylation of the oncogenic YAP/TAZ	Huntoon et al. (2010)	LATS1 KO female mice develop ovarian stromal cell tumors and soft tissue sarcomas. Low levels of LATS1 and LATS2 mRNA in human breast cancer correlated with large tumor size, high metastatic incidence to the lymph nodes, and absence of estrogen and progesterone receptor (Chan et al., 2011)

Continued

Table 1 Additional Kinases That Were Identified as "Unusual Suspects" (and Are Not Further Discussed in the Text)—cont'd

Kinase	Function, Pathway	Hsp90 Interaction	Cancer Relevance
NUAK2	AMPK-related kinase; energy homeostasis	Al-Hakim et al. (2005) and Taipale et al. (2012)	Contributes to viability, migration, and metastatic potential of cancer cells (Sun, Gao, Chien, Li, & Zhao, 2013). Expression of NUAK2 is positively associated with metastasis and poor clinical outcome in melanoma patients (Namiki et al., 2011, 2015)
PKN family: PKN1 PKN2	Bind Rac and Rho GTPases	Taipale et al. (2012)	Important for cell migration (Raftopoulou & Hall, 2004). PKN1 expressed in breast cancer and PKN2 expressed in bladder tumor cells. Both are important for the migration and invasion of bladder tumor cells (Lachmann et al., 2011)
TNK1	Interacts with phospholipase Cγ (PLC-γ); involved in cell death, it enhances TNFα-induced apoptosis by inhibiting NF-κB activation	Taipale et al. (2012)	Mice that lack TNK1 (and its splice variant KOS1) develop spontaneous tumors. Human and murine B-cell lymphomas show loss of TNK1/KOS1 (May et al., 2010)
ACK1/ TNK2	Promotes the polyubiquitination and degradation of the tumor suppressor Wwox	Taipale et al. (2012)	Oncogenic properties (Mahajan, Whang, Mohler, & Earp, 2005). Overexpression and overactivation of TNK2 associated with highly malignant prostate, breast, pancreatic, and lung cancers (Mahajan & Mahajan, 2015)

Continued

Table 1 Additional Kinases That Were Identified as "Unusual Suspects" (and Are Not Further Discussed in the Text)—cont'd

Kinase	Function, Pathway	Hsp90 Interaction	Cancer Relevance
GRK6	Phosphorylates G protein-coupled receptors	Tiedemann et al. (2010)	High levels in CRC (Tiedemann et al., 2010). Missense mutations in gastric (Forbes et al., 2006) and breast (Stephens et al., 2005) carcinomas

Taipale et al., 2012). It is a transmembrane serine/threonine kinase, which upon activation by its ligand phosphorylates Smad proteins (reviewed by Hardwick, Kodach, Offerhaus, & van den Brink, 2008). It has been described as a tumor suppressor since it is involved in the suppression of the Wnt-β-catenin pathway (He et al., 2004). Mutations of *BMPR1A* are associated with juvenile polyposis (Howe et al., 2001), and patients with this disorder are at high risk of developing CRC (Brosens et al., 2007).

The Eph receptors bind ligands that are anchored to membranes and play a very important role in the nervous system. They regulate many cellular processes including migration and cell-substrate adhesion (reviewed by Lisabeth, Falivelli, & Pasquale, 2013; Pasquale, 2005). The role of Eph receptors in cancer is complex: they can either suppress or promote carcinogenesis depending on context. Some have been identified as potential anticancer targets, such as EphA2 in ovarian cancer (Landen, Kinch, & Sood, 2005) and EphB2 in CRC (Mao et al., 2004). Both EphA1 and EphA2 interact with Hsp90 (Annamalai, Liu, Gopal, & Isaacs, 2009; Taipale et al., 2012). The role of EphA2 in cancer is extensively studied and well described. It is overexpressed in a variety of cancers such as prostate, ovary, brain, and breast cancers where it correlates with poor prognosis and has been targeted by different strategies in preclinical studies (reviewed by Tandon, Vemula, & Mittal, 2011). In contrast, the role of EphA1 in cancer is less clear with only few studies focusing on this receptor. In clear cell renal cell carcinoma, the absence of EphA1 is considered a favorable prognostic factor (Toma et al., 2014), and in pancreatic ductal adenocarcinoma the expression of EphA1 correlates with tumor size and histopathological grade (Giaginis et al., 2010). In CRC, epigenetic silencing of receptor expression correlated with poor prognosis even though expression of the receptor was heterogeneous among different patients as compared to normal tissue (Herath, Doecke, Spanevello, Leggett, & Boyd, 2009).

CSF1R, a recently identified Hsp90 interactor (Taipale et al., 2012), is a tyrosine kinase and the receptor for macrophage colony-stimulating factor 1, which regulates the proliferation and differentiation of macrophages and osteoclasts. Upon binding of its ligand, the receptor becomes activated and autophosphorylated, then phosphorylates downstream molecules, and triggers a cascade of events that eventually lead to cell survival, proliferation, and differentiation (reviewed by Pixley & Stanley, 2004). The oncogenic potential of CSF1R has received a lot of attention in the context of tumor-associated macrophages (TAMs). The role of TAMs in cancer remains controversial because they are associated with more favorable prognosis in pancreatic cancer but not in thyroid and lung cancer (reviewed by Qian & Pollard, 2010). Inhibition of TAM functions by the use of CSF1R inhibitors is also under investigation as a way to improve current treatment approaches, most notably in breast cancer (reviewed by Sullivan & Pixley, 2014).

The family of neurotrophin receptor tyrosine kinases (NTRKs) is thought to play an important role in carcinogenesis. It comprises the three members NTRK1, NTRK2, and NTRK3, all of which are interactors of Hsp90 (Bernstein, Russell, Wong, Fishelevich, & Smith, 2001; Taipale et al., 2012). Activation of these receptors triggers the Ras/Rap-MAPK, PI3K-Akt, and PLCγ-PKC cascades, all playing important roles in cell proliferation and survival pathways (reviewed by Arevalo & Wu, 2006). A study conducted in neuroblastoma patients showed that expression of NTRK1 correlates with favorable risk factors and outcome, whereas NTRK2 correlates with risk factors but not with outcome (Light et al., 2012). Fusions of NTRKs with other kinases have been detected in different solid tumors such as thyroid carcinoma, lung adenocarcinoma, colon adenocarcinoma, and head and neck squamous cell carcinoma (Stransky, Cerami, Schalm, Kim, & Lengauer, 2014). Mutations in NTRK3 are found in CRC (Bardelli et al., 2003) and in lung cancer (Davies et al., 2005). However, the contribution of these mutations to tumorigenesis has yet to be examined experimentally.

The two members of the discoidin domain receptor family, DDR1 and DDR2, are receptor tyrosine kinases (RTKs) that bind collagen. DDRs support communication between the cell and the extracellular matrix, thereby regulating migration and invasion, two important features of cancer cells. Their activation is different compared to the majority of RTKs because upon binding of ligand, their autophosphorylation is slow (Shrivastava et al., 1997; Vogel, Gish, Alves, & Pawson, 1997) and oligomerization and internalization of the receptor to early endosomes take place (Mihai, Chotani, Elton, & Agarwal, 2009). Their interaction with Hsp90 was identified and validated in a study of global effects on the proteome in cancer cell lines upon inhibition of Hsp90 (Wu, Moghaddas Gholami, & Kuster, 2012). Their expression in human cancer tissues does not follow a common pattern; they are down- or upregulated in different types of cancer or in different stages of the same cancer type (reviewed by Valiathan, Marco, Leitinger, Kleer, & Fridman, 2012).

3.2 Kinases Involved in Mitosis

Protein kinases regulating mitosis are among the best known clients of Hsp90: CDKs, Plk, Aurora B, and ChK1. The NEK kinase family is much less well recognized. It has 11 members with major roles in cell division and checkpoint control. NEK2 has a role in centrosome splitting, and NEK6, 7, and 9 are involved in regulating spindle dynamics (reviewed by Moniz, Dutt, Haider, & Stambolic, 2011). Of all members of this family, only NEK8, NEK9, and NEK11 interact with Hsp90 (Taipale et al., 2012). Their association with carcinogenesis is not yet well described. Cancer-associated mutations have been identified in several members of the family, but their functional effects remain elusive. NEK8 is overexpressed in primary breast tumors (Bowers & Boylan, 2004), and mutations have been detected in liver, ovarian, and stomach tumors. In ovarian cancer, NEK9 and NEK11 and, in lung and brain cancers, NEK11 are mutated as well (reviewed by Moniz et al., 2011).

3.3 NF-κB-Independent Roles of the IKK Complex

The inhibitor of nuclear factor κ-B kinase (IKK) complex is a central regulator of the NF-κB signaling cascade. The interaction between Hsp90/CDC37 and the IKK complex is important for its maturation and activation (Bouwmeester et al., 2004; Chen, Cao, & Goeddel, 2002). The complex role of NF-κB in cancer and inflammation has been well described over the years (reviewed by Hoesel & Schmid, 2013). The IKK complex phosphorylates the inhibitors of NF-κB, which results in their ubiquitination and proteasomal degradation; their degradation in turn allows NF-κB to translocate to the nucleus and to upregulate the transcription of genes that are involved in cell proliferation, survival, the immune response, and inflammation. Because the IKK complex is an upstream regulator of NF-κB, its importance in cancer is obvious. However, IKKα and IKKβ also have

NF-κB-independent roles. IKKα phosphorylates additional substrates, which are linked with positive regulation of cell proliferation such as cyclin D1. The IKKα nuclear fraction has also been linked with the metastatic potential of prostate carcinomas. IKKβ phosphorylates Dok1, which positively regulates cell motility, and the tumor suppressor p53, which leads to its proteasomal degradation. Furthermore, IKKβ activates the mTOR pathway, which ultimately stimulates angiogenesis through induction of VEGF (reviewed by Chariot, 2009). There is evidence that the noncanonical IKK-related kinase IKKε is an oncoprotein. IKKε has high homology to the two IKK subunits IKKα and IKKβ, but its role in the regulation of NF-κB is unclear (reviewed by Hacker & Karin, 2006). Breast and ovarian cancer cell lines as well as ovarian primary tumors have elevated levels of IKKε. IKKε has been suggested to function as an oncoprotein in clear cell renal cell carcinoma, prostate cancer, and esophageal squamous cell carcinoma (reviewed by Verhelst, Verstrepen, Carpentier, & Beyaert, 2013). A second noncanonical IKK-related kinase, TBK1, is also a client of Hsp90 (Bouwmeester et al., 2004; Taipale et al., 2012). It is overexpressed in lung, breast, and colon cancers, participates in the RAS-activated transformation pathway, is required in KRAS-dependent cells, and has been found to induce angiogenesis (reviewed by Shen & Hahn, 2011).

3.4 Kinases Associated with Cell Death Signaling

Proteins involved in cell death pathways are of major importance in cancer. Normally acting as tumor suppressors, they are often found to be downregulated and/or mutated in cancer.

The death-associated protein kinases (DAPK1, DAPK2, DAPK3) are serine/threonine kinases, which play an important role in cell death pathways, apoptosis, and autophagy (reviewed by Bialik & Kimchi, 2006). Being clients of Hsp90 (Citri et al., 2006; Zhang, Nephew, & Gallagher, 2007), some have been shown to be degraded upon inhibition of Hsp90 (Citri et al., 2006). As proapoptotic kinases and tumor suppressors, their regulation is perturbed in cancer. Dysregulation happens mostly at the level of their expression, since it has been shown that hypermethylation of the promoter of the genes leads to downregulation of their transcription. However, expression of nonfunctional DAPK has been detected in lung and renal cell carcinomas, implying a contribution of posttranslational regulation in some cases (reviewed by Michie, McCaig, Nakagawa, & Vukovic, 2010).

The receptor-interacting protein (RIP) kinase family is comprised of seven members, but only RIP1 and RIP2 have been found to interact with Hsp90 (Bouwmeester et al., 2004; Lewis et al., 2000). Members of this family have major roles in death receptor signaling and immune responses (reviewed by Declercq, Vanden Berghe, & Vandenabeele, 2009; Festjens, Vanden Berghe, Cornelis, & Vandenabeele, 2007). Upon intracellular or extracellular stimulation, RIP1 and RIP2 can activate several MAPKs and NF-κB, thus being part of an antiapoptotic cascade of events. However, when the antiapoptotic effect is insufficient, they participate in pathways driving apoptotic or necrotic cell death (reviewed by Zhang, Lin, & Han, 2010). Their involvement in cancer is not yet well understood. One study of RIP2 in triple-negative breast cancer cell lines reported that RIP2 activity is associated with increased migration and invasion, probably by a mechanism that involves NF-κB and c-Jun (Singel et al., 2014).

4. OTHER IMPORTANT HSP90 INTERACTORS

In addition to TFs and kinases, Hsp90 also interacts with proteins in other functional classes that play important roles in carcinogenesis.

4.1 Methyltransferases

The epigenetic changes caused by methylation play a significant role regulating the transcription of many genes, potentially contributing to a variety of disorders. Both histone and DNA methyltransferases have been identified as Hsp90 interactors, and several of these are discussed below.

SMYD3 is a histone methyltransferase, which interacts with Hsp90 and whose activity is increased as a result of this interaction (Hamamoto et al., 2004). The expression of SMYD3 is high in CRC and in hepatocellular carcinomas where it displays oncogenic properties (Hamamoto et al., 2004). Inhibition of the interaction of SMYD3 with Hsp90 causes mislocalization of SMYD3 and decreased cell proliferation (Brown et al., 2015). SMYD3 methylates VEGFR1, increasing its activity, and MAP3K2, which activates the RAS-RAF-MEK-ERK cascade (reviewed by Hamamoto, Saloura, & Nakamura, 2015). Both of these nonhistone targets of SMYD3 have critical roles in tumorigenesis.

PRMT5 is a histone arginine methyltransferase, described as a repressor of transcription, which binds Hsp90 (Maloney et al., 2007). In transformed cells, the levels of *PRMT5* expression are elevated. RNAi-mediated knockdown of *PRMT5* leads to slower cell growth. The repressive function of PRMT5 in transcription seems to be important for the epithelial–mesenchymal transition and therefore for migration and invasion as well (reviewed by Yang & Bedford, 2013). *PRMT5* overexpression is observed in many types of cancer including colon, gastric, bladder, ovarian, and non-small cell lung cancers; in the latter two instances, its overexpression is associated with shorter survival. However, evidence from cell culture experiments suggests that the role of PRMT5 may vary depending on the cell context (reviewed by Stopa, Krebs, & Shechter, 2015).

DNMT1 is a DNA methyltransferase that interacts with Hsp90 (Zhou, Agoston, Atadja, Nelson, & Davidson, 2008). The transcription of *DNMT1* is upregulated by the oncogenic Ras-c-Jun signaling pathway and suppressed by the tumor suppressors p53 and Rb. Its expression is also regulated by several microRNAs (reviewed by Lin & Wang, 2014). The upregulation of DNA methyltransferases can lead to hypermethylation of CpG islands in regulatory sequences of tumor suppressor genes, resulting in their silencing and the promotion of tumorigenesis. It has been reported that *DNMT1* is overexpressed in many types of cancer, notably leukemia and colon, kidney, prostate, ovarian, breast, and liver cancers (reviewed by Subramaniam, Thombre, Dhar, & Anant, 2014).

4.2 Helicases, Apoptotic Factors, and More

BLM belongs to the RecQ family of DNA helicases with important roles in DNA replication, recombination, and repair. *In vitro* experiments have demonstrated that Hsp90 interacts with BLM, which results in the modulation of its helicase activity, particularly with DNA substrates that resemble telomeres (Bhattacharyya et al., 2009). Mutations in BLM lead to a rare genetic disease, called Bloom syndrome; among other problems, patients are predisposed to various types of cancer characterized by early onset and the occurrence of multiple primary tumors (reviewed by Croteau, Popuri, Opresko, & Bohr, 2014). Depletion or mutation of murine BLM leads to chromosomal instability (reviewed by Croteau et al., 2014), one of the hallmarks of cancer (Hanahan & Weinberg, 2011). In normal human tissues, only proliferating cells are BLM positive and display nuclear localization of the protein. However, BLM levels are elevated, especially in tumors of lymphoid and epithelial origins (Turley, Wu, Canamero, Gatter, & Hickson, 2001).

The N-Myc downstream-regulated gene 1 (*NDRG1*) is a member of a family of four genes. As its name implies, its expression is dowregulated by the oncoproteins c-Myc and N-Myc. An effort to define the interactome of NDRG1 identified Hsp90 (Tu, Yan, Hood, & Lin, 2007). NDRG1 seems to have an important but complicated role in carcinogenesis. In most cancer types, it is expressed at lower levels than in normal tissues, notably in glioma and cancers of the colon, prostate, breast, and esophagus. A positive correlation between NDRG1 expression and survival in glioma, breast, and colorectal cancers has also been reported (reviewed by Melotte et al., 2010). NDRG1 may inhibit the growth of the primary tumor, suppress metastasis, and regulate angiogenesis. This has led to its classification as a tumor suppressor (reviewed by Bae et al., 2013). On the other hand, *NDRG1* is overexpressed in hepatocellular carcinomas, cutaneous and oral squamous carcinomas, and cervical and renal cancers (reviewed by Melotte et al., 2010). It is obvious that the role of NDRG1 in cancer is not uniform, but depends on context. Its potential as a therapeutic target in tumors, where NDRG1 may have an oncogenic role, is currently under investigation (reviewed by Bae et al., 2013).

Breast cancer metastasis suppressor 1 (BRMS1) is a metastasis suppressor that exerts little effect on the growth of primary tumors. It interacts with the SIN3:HDAC complex, which regulates transcription by repressive remodeling of the chromatin (reviewed by Hurst & Welch, 2011a). BRMS1 binds Hsp90 and this interaction leads to its stabilization, which may be important for its suppressive function (Hurst et al., 2006). The inhibitory role of BRMS1 for metastasis has been described not only for breast cancer but also for melanoma, ovarian, and nonsmall cell lung cancers (reviewed by Hurst & Welch, 2011b). Interestingly, the effect of BRMS1 in malignant melanoma seems to depend on its localization, in that cytoplasmic BRMS1 inhibits melanoma progression, whereas nuclear localization is reported to promote invasion (Slipicevic et al., 2012).

The apoptotic proteasome-activation factor 1 (APAF-1) is a cytoplasmic protein with a major role in the intrinsic pathway of apoptosis. In normal cells, it resides in the cytoplasm as a monomer, but when stimulated by cytochrome *C* release from mitochondria, it assembles the apoptosome and activates procaspases (reviewed by Yuan & Akey, 2013). APAF-1 interacts with both Hsp70, which regulates it negatively, and Hsp90 (Saleh, Srinivasula, Balkir, Robbins, & Alnemri, 2000). The interaction between Hsp90 and APAF-1 is reported to inhibit the formation of the active APAF-1 complex (Pandey et al., 2000). As a proapoptotic regulator, APAF-1 is considered a tumor suppressor. Loss of APAF-1 has been reported in melanoma, glioblastoma, ovarian, and leukemic cell lines; in some instances, drug resistance has been associated with downregulation of apoptosome components including APAF-1 (reviewed by Ledgerwood & Morison, 2009).

Exportin-1 (XPO1 or CRM1) participates in the transport of RNA and proteins from the nucleus to the cytoplasm. It interacts with Hsp90 (Falsone, Gesslbauer, Tirk, Piccinini, & Kungl, 2005) as part of a larger complex containing Aha1, importin-4, and importin-α6 (Echeverria et al., 2011; Sun, Hartson, & Matts, 2015). Exportin-1 mediates the export of several cancer-related proteins. Among them are many tumor suppressors such as p53, APC, Rb, and FOXO (reviewed by Turner, Dawson, & Sullivan, 2012). The cytoplasmic localization of these proteins potentially results in their degradation and disrupts the transcriptional programs that they regulate, leading to reduced apoptosis and increased cell growth signals. Increased levels of exportin-1 are found in ovarian, pancreatic, and cervical cancers, osteosarcoma and glioma (Turner et al., 2012), mantle cell lymphoma, and multiple myeloma (reviewed by Senapedis, Baloglu, & Landesman, 2014). Its overexpression correlates with bigger tumor size, increased incidence of metastasis, and poor prognosis (Turner et al., 2012). Inhibitors of exportin-1 are being extensively studied with several compounds already in clinical trials, with some showing promising early results (Senapedis et al., 2014).

5. CONCLUDING REMARKS

The number of Hsp90 interactors has grown dramatically over the past few years. Some of these interactors have been well characterized, including their contributions to cancer. In this chapter, we have made an effort to identify some of the least well-known interactors of Hsp90 that are involved in carcinogenesis. We were able to identify 57 "unusual suspects" with high confidence by following the procedure schematized in Fig. 1. The range of pathways and processes in which these "suspects" participate is very wide: TFs involved in the control of metabolism, in developmental processes, and in the maintenance of stemness; kinases involved in signal transduction, mitosis, and cell death pathways; and proteins involved in the epigenetic control of gene expression by methylation. These are but a few prominent examples emphasizing the importance of Hsp90 for a wide range of cellular processes.

Among the diverse clientele of Hsp90, there are many proteins, both usual and unusual suspects, which are involved in carcinogenesis. Inspired by the recently revised hallmarks of cancer (Hanahan & Weinberg, 2011), we were able to pair most of our "unusual suspects" with at least one hallmark (Fig. 3). Their important role in carcinogenesis is highlighted

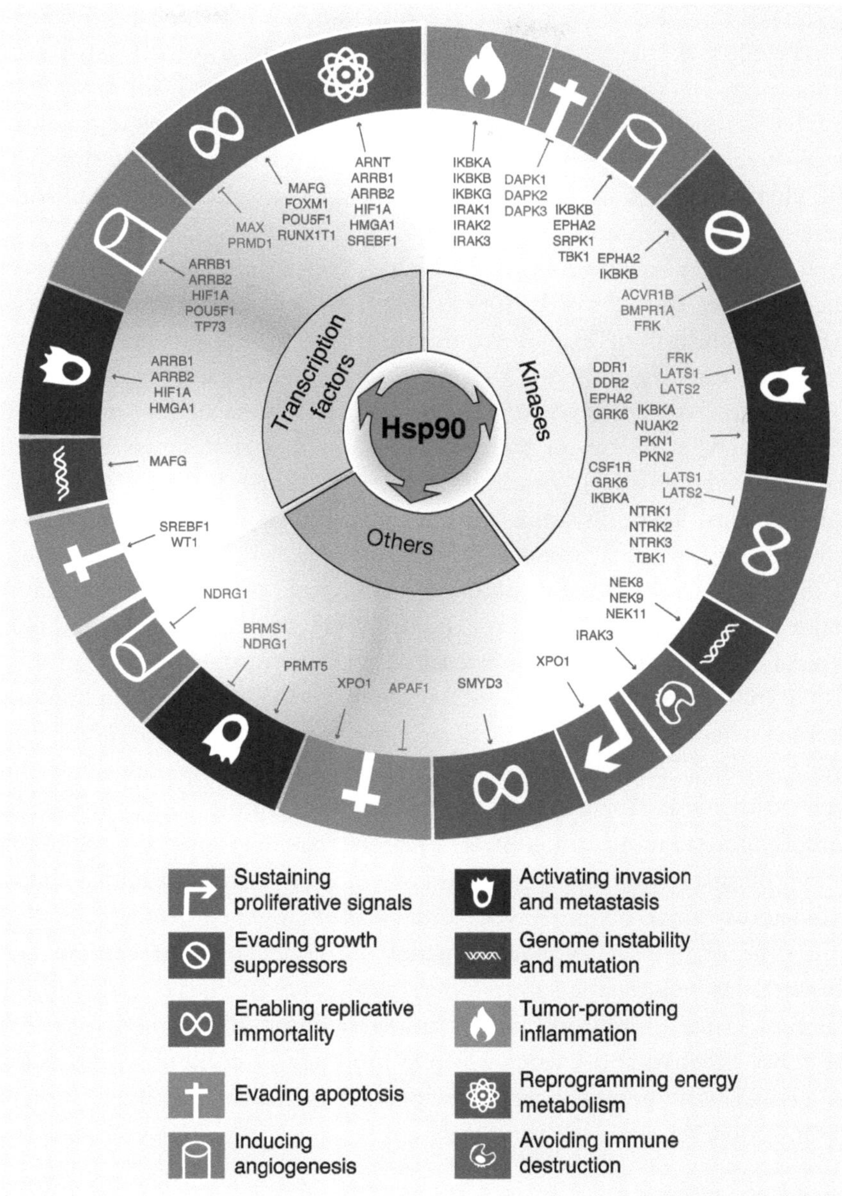

Figure 3 "Unusual suspects" as supporting spokes for a wheel of cancer hallmarks. "Unusual suspects" identified in this review (tumor suppressors in blue and oncoproteins in red) were mapped to the different hallmarks of cancer. Hsp90 takes a central position, stabilizing and/or activating networks of cancer facilitators as well as protectors against cancer. *Adapted from Hanahan and Weinberg (2011), Copyright (2011), with permission from Elsevier.* (See the color plate.)

by the fact that despite their relatively small number, our "unusual suspects" could easily form the spokes of a wheel of cancer hallmarks with Hsp90 at its center (Fig. 3). The hallmarks of "sustained angiogenesis," "limitless replicative potential," and "tissue invasion and metastasis" were well represented in all three of our main classifications of Hsp90 interactors. Only time will tell whether there is any significance to the fact that other cancer hallmarks were underrepresented.

It might have been expected that we would find a range of oncoproteins (shown in red in Fig. 3), whose depletion upon pharmacological inhibition of Hsp90 would exert desirable anticancer activity. However, our new suite of "unusual suspects" also features a number of tumor suppressors (shown in blue in Fig. 3). These tumor suppressors are involved in the cancer hallmarks "sustained angiogenesis," "limitless replicative potential," "evading apoptosis," "insensitivity to growth signals," and "tissue invasion and metastasis" (Fig. 3). As we previously pointed out for "usual suspects" (Fierro-Monti et al., 2013), impairment of these tumor-suppressive factors by inhibition of Hsp90 could play a major role in limiting the efficacy of Hsp90-targeted therapeutics and help explain the rather disappointing results seen so far in the clinical testing of Hsp90 inhibitors. The fundamental problem is that, to Hsp90, a client is a client and hence, from an Hsp90 perspective, oncoproteins and tumor suppressors look alike. While it might not be possible to completely avoid this type of collateral damage in the use of Hsp90 inhibitors, more careful and comprehensive monitoring of an enlarged panel of markers including oncoproteins, tumor suppressors, and even cancer-unrelated Hsp90 clients could allow one to be more aware of it and perhaps even help to reduce it. This type of "Hsp90 biomarker panel" could be used to explore and ultimately to exploit the subtle differences that exist between chemically different Hsp90 inhibitors. In the future, it might allow one to identify the most effective and selective Hsp90 inhibitor or combination therapy for a particular type of cancer, or even to "personalize" specific treatment for an individual patient's tumor.

ACKNOWLEDGMENTS

Hsp90-related work in the laboratory of Didier Picard is supported by the Swiss National Science Foundation and the Canton de Genève.

REFERENCES

Alexander, J. M., Bikkal, H. A., Zervas, N. T., Laws, E. R., Jr., & Klibanski, A. (1996). Tumor-specific expression and alternate splicing of messenger ribonucleic acid encoding activin/transforming growth factor-β receptors in human pituitary adenomas. *Journal of Clinical Endocrinology and Metabolism*, *81*, 783–790.

Al-Hakim, A. K., Goransson, O., Deak, M., Toth, R., Campbell, D. G., Morrice, N. A., et al. (2005). 14-3-3 cooperates with LKB1 to regulate the activity and localization of QSK and SIK. *Journal of Cell Science*, *118*, 5661–5673.

Annamalai, B., Liu, X., Gopal, U., & Isaacs, J. S. (2009). Hsp90 is an essential regulator of EphA2 receptor stability and signaling: Implications for cancer cell migration and metastasis. *Molecular Cancer Research*, 7, 1021–1032.

Arevalo, J. C., & Wu, S. H. (2006). Neurotrophin signaling: Many exciting surprises! *Cellular and Molecular Life Sciences*, *63*, 1523–1537.

Bae, D. H., Jansson, P. J., Huang, M. L., Kovacevic, Z., Kalinowski, D., Lee, C. S., et al. (2013). The role of NDRG1 in the pathology and potential treatment of human cancers. *Journal of Clinical Pathology*, *66*, 911–917.

Bansal, H., Bansal, S., Rao, M., Foley, K. P., Sang, J., Proia, D. A., et al. (2010). Heat shock protein 90 regulates the expression of Wilms tumor 1 protein in myeloid leukemias. *Blood*, *116*, 4591–4599.

Bardelli, A., Parsons, D. W., Silliman, N., Ptak, J., Szabo, S., Saha, S., et al. (2003). Mutational analysis of the tyrosine kinome in colorectal cancers. *Science*, *300*, 949.

Barrios-Rodiles, M., Brown, K. R., Ozdamar, B., Bose, R., Liu, Z., Donovan, R. S., et al. (2005). High-throughput mapping of a dynamic signaling network in mammalian cells. *Science*, *307*, 1621–1625.

Bernstein, S. L., Russell, P., Wong, P., Fishelevich, R., & Smith, L. E. (2001). Heat shock protein 90 in retinal ganglion cells: Association with axonally transported proteins. *Visual Neuroscience*, *18*, 429–436.

Bhattacharyya, S., Keirsey, J., Russell, B., Kavecansky, J., Lillard-Wetherell, K., Tahmaseb, K., et al. (2009). Telomerase-associated protein 1, HSP90, and topoisomerase IIα associate directly with the BLM helicase in immortalized cells using ALT and modulate its helicase activity using telomeric DNA substrates. *Journal of Biological Chemistry*, *284*, 14966–14977.

Bialik, S., & Kimchi, A. (2006). The death-associated protein kinases: Structure, function, and beyond. *Annual Review of Biochemistry*, *75*, 189–210.

Bouwmeester, T., Bauch, A., Ruffner, H., Angrand, P. O., Bergamini, G., Croughton, K., et al. (2004). A physical and functional map of the human TNF-α/NF-κB signal transduction pathway. *Nature Cell Biology*, *6*, 97–105.

Bowers, A. J., & Boylan, J. F. (2004). Nek8, a NIMA family kinase member, is overexpressed in primary human breast tumors. *Gene*, *328*, 135–142.

Bradley, E., Bieberich, E., Mivechi, N. F., Tangpisuthipongsa, D., & Wang, G. (2012). Regulation of embryonic stem cell pluripotency by heat shock protein 90. *Stem Cells*, *30*, 1624–1633.

Brosens, L. A., van Hattem, A., Hylind, L. M., Iacobuzio-Donahue, C., Romans, K. E., Axilbund, J., et al. (2007). Risk of colorectal cancer in juvenile polyposis. *Gut*, *56*, 965–967.

Brown, M. A., Foreman, K., Harriss, J., Das, C., Zhu, L., Edwards, M., et al. (2015). C-terminal domain of SMYD3 serves as a unique HSP90-regulated motif in oncogenesis. *Oncotarget*, *6*, 4005–4019.

Brugge, J. S., Erikson, E., & Erikson, R. L. (1981). The specific interaction of the Rous sarcoma virus transforming protein, $pp60^{src}$, with two cellular proteins. *Cell*, *25*, 363–372.

Brun, M., Coles, J. E., Monckton, E. A., Glubrecht, D. D., Bisgrove, D., & Godbout, R. (2009). Nuclear factor I regulates brain fatty acid-binding protein and glial fibrillary acidic protein gene expression in malignant glioma cell lines. *Journal of Molecular Biology, 391*, 282–300.

Burroughs, S. K., Kaluz, S., Wang, D., Wang, K., Van Meir, E. G., & Wang, B. (2013). Hypoxia inducible factor pathway inhibitors as anticancer therapeutics. *Future Medicinal Chemistry, 5*, 553–572.

Cascon, A., & Robledo, M. (2012). MAX and MYC: A heritable breakup. *Cancer Research, 72*, 3119–3124.

Chan, S. W., Lim, C. J., Chen, L., Chong, Y. F., Huang, C., Song, H., et al. (2011). The Hippo pathway in biological control and cancer development. *Journal of Cellular Physiology, 226*, 928–939.

Chariot, A. (2009). The NF-κB-independent functions of IKK subunits in immunity and cancer. *Trends in Cell Biology, 19*, 404–413.

Chen, G., Cao, P., & Goeddel, D. V. (2002). TNF-induced recruitment and activation of the IKK complex require Cdc37 and Hsp90. *Molecular Cell, 9*, 401–410.

Citri, A., Harari, D., Shohat, G., Ramakrishnan, P., Gan, J., Lavi, S., et al. (2006). Hsp90 recognizes a common surface on client kinases. *Journal of Biological Chemistry, 281*, 14361–14369.

Cleynen, I., & Van de Ven, W. J. (2008). The HMGA proteins: A myriad of functions (review). *International Journal of Oncology, 32*, 289–305.

Craven, R. J., Cance, W. G., & Liu, E. T. (1995). The nuclear tyrosine kinase Rak associates with the retinoblastoma protein pRb. *Cancer Research, 55*, 3969–3972.

Croteau, D. L., Popuri, V., Opresko, P. L., & Bohr, V. A. (2014). Human RecQ helicases in DNA repair, recombination, and replication. *Annual Review of Biochemistry, 83*, 519–552.

Culig, Z., & Santer, F. R. (2014). Androgen receptor signaling in prostate cancer. *Cancer Metastasis Reviews, 33*, 413–427.

Dai, B., Gong, A., Jing, Z., Aldape, K. D., Kang, S. H., Sawaya, R., et al. (2013). Forkhead box M1 is regulated by heat shock factor 1 and promotes glioma cells survival under heat shock stress. *Journal of Biological Chemistry, 288*, 1634–1642.

Davies, H., Hunter, C., Smith, R., Stephens, P., Greenman, C., Bignell, G., et al. (2005). Somatic mutations of the protein kinase gene family in human lung cancer. *Cancer Research, 65*, 7591–7595.

Declercq, W., Vanden Berghe, T., & Vandenabeele, P. (2009). RIP kinases at the crossroads of cell death and survival. *Cell, 138*, 229–232.

De Nardo, D., Masendycz, P., Ho, S., Cross, M., Fleetwood, A. J., Reynolds, E. C., et al. (2005). A central role for the Hsp90.Cdc37 molecular chaperone module in interleukin-1 receptor-associated-kinase-dependent signaling by toll-like receptors. *Journal of Biological Chemistry, 280*, 9813–9822.

Dulloo, I., Phang, B. H., Othman, R., Tan, S. Y., Vijayaraghavan, A., Goh, L. K., et al. (2015). Hypoxia-inducible TAp73 supports tumorigenesis by regulating the angiogenic transcriptome. *Nature Cell Biology, 17*, 511–523.

Echeverria, P. C., Bernthaler, A., Dupuis, P., Mayer, B., & Picard, D. (2011). An interaction network predicted from public data as a discovery tool: Application to the Hsp90 molecular chaperone machine. *PLoS One, 6*, e26044.

Echeverria, P. C., & Picard, D. (2014). A global view of the proteome perturbations by Hsp90 inhibitors. In W. A. Houry (Ed.), *The molecular chaperones interaction networks in protein folding and degradation: Vol. 1* (pp. 133–149). New York: Springer.

Falsone, S. F., Gesslbauer, B., Tirk, F., Piccinini, A. M., & Kungl, A. J. (2005). A proteomic snapshot of the human heat shock protein 90 interactome. *FEBS Letters, 579*, 6350–6354.

Fang, M., Ou, J., Hutchinson, L., & Green, M. R. (2014). The BRAF oncoprotein functions through the transcriptional repressor MAFG to mediate the CpG island methylator phenotype. *Molecular Cell, 55*, 904–915.

Festjens, N., Vanden Berghe, T., Cornelis, S., & Vandenabeele, P. (2007). RIP1, a kinase on the crossroads of a cell's decision to live or die. *Cell Death and Differentiation, 14*, 400–410.

Fierro-Monti, I., Echeverria, P., Racle, J., Hernandez, C., Picard, D., & Quadroni, M. (2013). Dynamic impacts of the inhibition of the molecular chaperone Hsp90 on the T-cell proteome have implications for anti-cancer therapy. *PLoS One, 8*, e80425.

Forbes, S., Clements, J., Dawson, E., Bamford, S., Webb, T., Dogan, A., et al. (2006). Cosmic 2005. *British Journal of Cancer, 94*, 318–322.

Giaginis, C., Tsourouflis, G., Zizi-Serbetzoglou, A., Kouraklis, G., Chatzopoulou, E., Dimakopoulou, K., et al. (2010). Clinical significance of ephrin (eph)-A1, -A2, -a4, -a5 and -a7 receptors in pancreatic ductal adenocarcinoma. *Pathology Oncology Research, 16*, 267–276.

Gradin, K., McGuire, J., Wenger, R. H., Kvietikova, I., Whitelaw, M. L., Toftgard, R., et al. (1996). Functional interference between hypoxia and dioxin signal transduction pathways: Competition for recruitment of the Arnt transcription factor. *Molecular and Cellular Biology, 16*, 5221–5231.

Guo, D., Bell, E. H., Mischel, P., & Chakravarti, A. (2014). Targeting SREBP-1-driven lipid metabolism to treat cancer. *Current Pharmaceutical Design, 20*, 2619–2626.

Hacker, H., & Karin, M. (2006). Regulation and function of IKK and IKK-related kinases. *Science's STKE, 2006*, re13.

Hamamoto, R., Furukawa, Y., Morita, M., Iimura, Y., Silva, F. P., Li, M., et al. (2004). SMYD3 encodes a histone methyltransferase involved in the proliferation of cancer cells. *Nature Cell Biology, 6*, 731–740.

Hamamoto, R., Saloura, V., & Nakamura, Y. (2015). Critical roles of non-histone protein lysine methylation in human tumorigenesis. *Nature Reviews. Cancer, 15*, 110–124.

Hanahan, D., & Weinberg, R. A. (2011). Hallmarks of cancer: The next generation. *Cell, 144*, 646–674.

Hardwick, J. C., Kodach, L. L., Offerhaus, G. J., & van den Brink, G. R. (2008). Bone morphogenetic protein signalling in colorectal cancer. *Nature Reviews Cancer, 8*, 806–812.

Harris, A. L. (2002). Hypoxia—A key regulatory factor in tumour growth. *Nature Reviews. Cancer, 2*, 38–47.

Harrison, C. A., Wiater, E., Gray, P. C., Greenwald, J., Choe, S., & Vale, W. (2004). Modulation of activin and BMP signaling. *Molecular and Cellular Endocrinology, 225*, 19–24.

Haupt, A., Joberty, G., Bantscheff, M., Frohlich, H., Stehr, H., Schweiger, M. R., et al. (2012). Hsp90 inhibition differentially destabilises MAP kinase and TGF-β signalling components in cancer cells revealed by kinase-targeted chemoproteomics. *BMC Cancer, 12*, 38.

Hayes, G. M., Carrigan, P. E., & Miller, L. J. (2007). Serine-arginine protein kinase 1 overexpression is associated with tumorigenic imbalance in mitogen-activated protein kinase pathways in breast, colonic, and pancreatic carcinomas. *Cancer Research, 67*, 2072–2080.

He, X. C., Zhang, J., Tong, W. G., Tawfik, O., Ross, J., Scoville, D. H., et al. (2004). BMP signaling inhibits intestinal stem cell self-renewal through suppression of Wnt-β-catenin signaling. *Nature Genetics, 36*, 1117–1121.

Herath, N. I., Doecke, J., Spanevello, M. D., Leggett, B. A., & Boyd, A. W. (2009). Epigenetic silencing of EphA1 expression in colorectal cancer is correlated with poor survival. *British Journal of Cancer, 100*, 1095–1102.

Hoesel, B., & Schmid, J. A. (2013). The complexity of NF-κB signaling in inflammation and cancer. *Molecular Cancer, 12*, 86.

Howe, J. R., Bair, J. L., Sayed, M. G., Anderson, M. E., Mitros, F. A., Petersen, G. M., et al. (2001). Germline mutations of the gene encoding bone morphogenetic protein receptor 1A in juvenile polyposis. *Nature Genetics*, *28*, 184–187.

Huang, R., Huang, D., Dai, W., & Yang, F. (2015). Overexpression of HMGA1 correlates with the malignant status and prognosis of breast cancer. *Molecular and Cellular Biochemistry*, *404*, 251–257.

Huntoon, C. J., Nye, M. D., Geng, L., Peterson, K. L., Flatten, K. S., Haluska, P., et al. (2010). Heat shock protein 90 inhibition depletes LATS1 and LATS2, two regulators of the mammalian hippo tumor suppressor pathway. *Cancer Research*, *70*, 8642–8650.

Hurst, D. R., Mehta, A., Moore, B. P., Phadke, P. A., Meehan, W. J., Accavitti, M. A., et al. (2006). Breast cancer metastasis suppressor 1 (BRMS1) is stabilized by the Hsp90 chaperone. *Biochemical and Biophysical Research Communications*, *348*, 1429–1435.

Hurst, D. R., & Welch, D. R. (2011a). Metastasis suppressor genes at the interface between the environment and tumor cell growth. *International Review of Cell and Molecular Biology*, *286*, 107–180.

Hurst, D. R., & Welch, D. R. (2011b). Unraveling the enigmatic complexities of BRMS1-mediated metastasis suppression. *FEBS Letters*, *585*, 3185–3190.

Isaacs, J. S., Jung, Y. J., Mimnaugh, E. G., Martinez, A., Cuttitta, F., & Neckers, L. M. (2002). Hsp90 regulates a von Hippel Lindau-independent hypoxia-inducible factor-1 α-degradative pathway. *Journal of Biological Chemistry*, *277*, 29936–29944.

Isaacs, J. S., Jung, Y. J., & Neckers, L. (2004). Aryl hydrocarbon nuclear translocator (ARNT) promotes oxygen-independent stabilization of hypoxia-inducible factor-1α by modulating an Hsp90-dependent regulatory pathway. *Journal of Biological Chemistry*, *279*, 16128–16135.

Jiang, H. B., Yang, T. J., Lu, P., & Ma, Y. J. (2014). Gene expression profiling of gastric cancer. *European Review for Medical and Pharmacological Sciences*, *18*, 2109–2115.

Khurana, N., & Bhattacharyya, S. (2015). Hsp90, the concertmaster: Tuning transcription. *Frontiers in Oncology*, *5*, 100.

Komori, A., Sueoka, E., Fujiki, H., Ishii, M., & Kozu, T. (1999). Association of MTG8 (ETO/CDR), a leukemia-related protein, with serine/threonine protein kinases and heat shock protein HSP90 in human hematopoietic cell lines. *Japanese Journal of Cancer Research*, *90*, 60–68.

Lachmann, S., Jevons, A., De Rycker, M., Casamassima, A., Radtke, S., Collazos, A., et al. (2011). Regulatory domain selectivity in the cell-type specific PKN-dependence of cell migration. *PLoS One*, *6*, e21732.

Lancet, J. E., Gojo, I., Burton, M., Quinn, M., Tighe, S. M., Kersey, K., et al. (2010). Phase I study of the heat shock protein 90 inhibitor alvespimycin (KOS-1022, 17-DMAG) administered intravenously twice weekly to patients with acute myeloid leukemia. *Leukemia*, *24*, 699–705.

Landen, C. N., Kinch, M. S., & Sood, A. K. (2005). EphA2 as a target for ovarian cancer therapy. *Expert Opinion on Therapeutic Targets*, *9*, 1179–1187.

Le, A., Cooper, C. R., Gouw, A. M., Dinavahi, R., Maitra, A., Deck, L. M., et al. (2010). Inhibition of lactate dehydrogenase A induces oxidative stress and inhibits tumor progression. *Proceedings of the National Academy of Sciences of the United States of America*, *107*, 2037–2042.

Ledgerwood, E. C., & Morison, I. M. (2009). Targeting the apoptosome for cancer therapy. *Clinical Cancer Research*, *15*, 420–424.

Lewis, J., Devin, A., Miller, A., Lin, Y., Rodriguez, Y., Neckers, L., et al. (2000). Disruption of hsp90 function results in degradation of the death domain kinase, receptor-interacting protein (RIP), and blockage of tumor necrosis factor-induced nuclear factor-κB activation. *Journal of Biological Chemistry*, *275*, 10519–10526.

Light, J. E., Koyama, H., Minturn, J. E., Ho, R., Simpson, A. M., Iyer, R., et al. (2012). Clinical significance of NTRK family gene expression in neuroblastomas. *Pediatric Blood & Cancer*, *59*, 226–232.
Lin, R. K., & Wang, Y. C. (2014). Dysregulated transcriptional and post-translational control of DNA methyltransferases in cancer. *Cell & Bioscience*, *4*, 46.
Lisabeth, E. M., Falivelli, G., & Pasquale, E. B. (2013). Eph receptor signaling and ephrins. *Cold Spring Harbor Perspectives in Biology*, *5*, a009159.
Liu, W. M., Guerra-Vladusic, F. K., Kurakata, S., Lupu, R., & Kohwi-Shigematsu, T. (1999). HMG-I(Y) recognizes base-unpairing regions of matrix attachment sequences and its increased expression is directly linked to metastatic breast cancer phenotype. *Cancer Research*, *59*, 5695–5703.
Luo, D., Wang, Z., Wu, J., Jiang, C., & Wu, J. (2014). The role of hypoxia inducible factor-1 in hepatocellular carcinoma. *BioMed Research International*, *2014*, 409272.
Mahajan, K., & Mahajan, N. P. (2015). ACK1/TNK2 tyrosine kinase: Molecular signaling and evolving role in cancers. *Oncogene*, *34*, 4162–4167.
Mahajan, N. P., Whang, Y. E., Mohler, J. L., & Earp, H. S. (2005). Activated tyrosine kinase Ack1 promotes prostate tumorigenesis: Role of Ack1 in polyubiquitination of tumor suppressor Wwox. *Cancer Research*, *65*, 10514–10523.
Maloney, A., Clarke, P. A., Naaby-Hansen, S., Stein, R., Koopman, J. O., Akpan, A., et al. (2007). Gene and protein expression profiling of human ovarian cancer cells treated with the heat shock protein 90 inhibitor 17-allylamino-17-demethoxygeldanamycin. *Cancer Research*, *67*, 3239–3253.
Mao, W., Luis, E., Ross, S., Silva, J., Tan, C., Crowley, C., et al. (2004). EphB2 as a therapeutic antibody drug target for the treatment of colorectal cancer. *Cancer Research*, *64*, 781–788.
May, W. S., Hoare, K., Hoare, S., Reinhard, M. K., Lee, Y. J., & Oh, S. P. (2010). Tnk1/Kos1: A novel tumor suppressor. *Transactions of the American Clinical and Climatological Association*, *121*, 281–292, discussion 292–293.
Melotte, V., Qu, X., Ongenaert, M., van Criekinge, W., de Bruine, A. P., Baldwin, H. S., et al. (2010). The N-myc downstream regulated gene (NDRG) family: Diverse functions, multiple applications. *FASEB Journal*, *24*, 4153–4166.
Meyer, T., Xu, L., Chang, J., Liu, E. T., Craven, R. J., & Cance, W. G. (2003). Breast cancer cell line proliferation blocked by the Src-related Rak tyrosine kinase. *International Journal of Cancer*, *104*, 139–146.
Michie, A. M., McCaig, A. M., Nakagawa, R., & Vukovic, M. (2010). Death-associated protein kinase (DAPK) and signal transduction: Regulation in cancer. *The FEBS Journal*, *277*, 74–80.
Mihai, C., Chotani, M., Elton, T. S., & Agarwal, G. (2009). Mapping of DDR1 distribution and oligomerization on the cell surface by FRET microscopy. *Journal of Molecular Biology*, *385*, 432–445.
Moniz, L., Dutt, P., Haider, N., & Stambolic, V. (2011). Nek family of kinases in cell cycle, checkpoint control and cancer. *Cell Division*, *6*, 18.
Mujcic, H., Hill, R. P., Koritzinsky, M., & Wouters, B. G. (2014). Hypoxia signaling and the metastatic phenotype. *Current Molecular Medicine*, *14*, 565–579.
Namiki, T., Tanemura, A., Valencia, J. C., Coelho, S. G., Passeron, T., Kawaguchi, M., et al. (2011). AMP kinase-related kinase NUAK2 affects tumor growth, migration, and clinical outcome of human melanoma. *Proceedings of the National Academy of Sciences of the United States of America*, *108*, 6597–6602.
Namiki, T., Yaguchi, T., Nakamura, K., Valencia, J. C., Coelho, S. G., Yin, L., et al. (2015). NUAK2 amplification coupled with PTEN deficiency promote melanoma development via CDK activation. *Cancer Research*, *75*, 2708–2715.

Nathan, D. F., Vos, M. H., & Lindquist, S. (1997). In vivo functions of the *Saccharomyces cerevisiae* Hsp90 chaperone. *Proceedings of the National Academy of Sciences of the United States of America*, *94*, 12949–12956.

Neckers, L., & Workman, P. (2012). Hsp90 molecular chaperone inhibitors: Are we there yet? *Clinical Cancer Research*, *18*, 64–76.

Pandey, P., Saleh, A., Nakazawa, A., Kumar, S., Srinivasula, S. M., Kumar, V., et al. (2000). Negative regulation of cytochrome c-mediated oligomerization of Apaf-1 and activation of procaspase-9 by heat shock protein 90. *The EMBO Journal*, *19*, 4310–4322.

Pasquale, E. B. (2005). Eph receptor signalling casts a wide net on cell behaviour. *Nature Reviews. Molecular Cell Biology*, *6*, 462–475.

Pawlus, M. R., Wang, L., & Hu, C. J. (2014). STAT3 and HIF1α cooperatively activate HIF1 target genes in MDA-MB-231 and RCC4 cells. *Oncogene*, *33*, 1670–1679.

Peterson, L. F., & Zhang, D. E. (2004). The 8;21 translocation in leukemogenesis. *Oncogene*, *23*, 4255–4262.

Pixley, F. J., & Stanley, E. R. (2004). CSF-1 regulation of the wandering macrophage: Complexity in action. *Trends in Cell Biology*, *14*, 628–638.

Purkayastha, B. P., & Roy, J. K. (2015). Cancer cell metabolism and developmental homeodomain/POU domain transcription factors: A connecting link. *Cancer Letters*, *356*, 315–319.

Qian, B. Z., & Pollard, J. W. (2010). Macrophage diversity enhances tumor progression and metastasis. *Cell*, *141*, 39–51.

Raftopoulou, M., & Hall, A. (2004). Cell migration: Rho GTPases lead the way. *Developmental Biology*, *265*, 23–32.

Renoir, J. M. (2012). Estradiol receptors in breast cancer cells: Associated co-factors as targets for new therapeutic approaches. *Steroids*, 77, 1249–1261.

Rhyasen, G. W., & Starczynowski, D. T. (2015). IRAK signalling in cancer. *British Journal of Cancer*, *112*, 232–237.

Saleh, A., Srinivasula, S. M., Balkir, L., Robbins, P. D., & Alnemri, E. S. (2000). Negative regulation of the Apaf-1 apoptosome by Hsp70. *Nature Cell Biology*, *2*, 476–483.

Senapedis, W. T., Baloglu, E., & Landesman, Y. (2014). Clinical translation of nuclear export inhibitors in cancer. *Seminars in Cancer Biology*, *27*, 74–86.

Shannon, P., Markiel, A., Ozier, O., Baliga, N. S., Wang, J. T., Ramage, D., et al. (2003). Cytoscape: A software environment for integrated models of biomolecular interaction networks. *Genome Research*, *13*, 2498–2504.

Shen, R. R., & Hahn, W. C. (2011). Emerging roles for the non-canonical IKKs in cancer. *Oncogene*, *30*, 631–641.

Shi, Q., Song, X., Wang, J., Gu, J., Zhang, W., Hu, J., et al. (2015). FRK inhibits migration and invasion of human glioma cells by promoting N-cadherin/β-catenin complex formation. *Journal of Molecular Neuroscience*, *55*, 32–41.

Shrivastava, A., Radziejewski, C., Campbell, E., Kovac, L., McGlynn, M., Ryan, T. E., et al. (1997). An orphan receptor tyrosine kinase family whose members serve as nonintegrin collagen receptors. *Molecular Cell*, *1*, 25–34.

Singel, S. M., Batten, K., Cornelius, C., Jia, G., Fasciani, G., Barron, S. L., et al. (2014). Receptor-interacting protein kinase 2 promotes triple-negative breast cancer cell migration and invasion via activation of nuclear factor-κB and c-Jun N-terminal kinase pathways. *Breast Cancer Research*, *16*, R28.

Slipicevic, A., Holm, R., Emilsen, E., Ree Rosnes, A. K., Welch, D. R., Malandsmo, G. M., et al. (2012). Cytoplasmic BRMS1 expression in malignant melanoma is associated with increased disease-free survival. *BMC Cancer*, *12*, 73.

Sobolesky, P. M., & Moussa, O. (2013). The role of β-arrestins in cancer. *Progress in Molecular Biology and Translational Science*, *118*, 395–411.

Soldevilla, B., Millan, C. S., Bonilla, F., & Dominguez, G. (2013). The TP73 complex network: Ready for clinical translation in cancer? *Genes, Chromosomes & Cancer, 52*, 989–1006.
Stephens, P., Edkins, S., Davies, H., Greenman, C., Cox, C., Hunter, C., et al. (2005). A screen of the complete protein kinase gene family identifies diverse patterns of somatic mutations in human breast cancer. *Nature Genetics, 37*, 590–592.
Stopa, N., Krebs, J. E., & Shechter, D. (2015). The PRMT5 arginine methyltransferase: Many roles in development, cancer and beyond. *Cellular and Molecular Life Sciences, 72*, 2041–2059.
Stransky, N., Cerami, E., Schalm, S., Kim, J. L., & Lengauer, C. (2014). The landscape of kinase fusions in cancer. *Nature Communications, 5*, 4846.
Su, G. H., Bansal, R., Murphy, K. M., Montgomery, E., Yeo, C. J., Hruban, R. H., et al. (2001). ACVR1B (ALK4, activin receptor type 1B) gene mutations in pancreatic carcinoma. *Proceedings of the National Academy of Sciences of the United States of America, 98*, 3254–3257.
Subramaniam, D., Thombre, R., Dhar, A., & Anant, S. (2014). DNA methyltransferases: A novel target for prevention and therapy. *Frontiers in Oncology, 4*, 80.
Sullivan, A. R., & Pixley, F. J. (2014). CSF-1R signaling in health and disease: A focus on the mammary gland. *Journal of Mammary Gland Biology and Neoplasia, 19*, 149–159.
Sun, X., Gao, L., Chien, H. Y., Li, W. C., & Zhao, J. (2013). The regulation and function of the NUAK family. *Journal of Molecular Endocrinology, 51*, R15–R22.
Sun, L., Hartson, S. D., & Matts, R. L. (2015). Identification of proteins associated with Aha1 in HeLa cells by quantitative proteomics. *Biochimica et Biophysica Acta, 1854*, 365–380.
Taipale, M., Krykbaeva, I., Koeva, M., Kayatekin, C., Westover, K. D., Karras, G. I., et al. (2012). Quantitative analysis of HSP90-client interactions reveals principles of substrate recognition. *Cell, 150*, 987–1001.
Tandon, M., Vemula, S. V., & Mittal, S. K. (2011). Emerging strategies for EphA2 receptor targeting for cancer therapeutics. *Expert Opinion on Therapeutic Targets, 15*, 31–51.
Tiedemann, R. E., Zhu, Y. X., Schmidt, J., Yin, H., Shi, C.-X., Que, Q., et al. (2010). Kinome-wide RNAi studies in human multiple myeloma identify vulnerable kinase targets, including a lymphoid-restricted kinase, GRK6. *Blood, 115*, 1594–1604.
Toma, M. I., Erdmann, K., Diezel, M., Meinhardt, M., Zastrow, S., Fuessel, S., et al. (2014). Lack of ephrin receptor A1 is a favorable independent prognostic factor in clear cell renal cell carcinoma. *PLoS One, 9*, e102262.
Tu, L. C., Yan, X., Hood, L., & Lin, B. (2007). Proteomics analysis of the interactome of N-myc downstream regulated gene 1 and its interactions with the androgen response program in prostate cancer cells. *Molecular and Cellular Proteomics, 6*, 575–588.
Turley, H., Wu, L., Canamero, M., Gatter, K. C., & Hickson, I. D. (2001). The distribution and expression of the Bloom's syndrome gene product in normal and neoplastic human cells. *British Journal of Cancer, 85*, 261–265.
Turner, J. G., Dawson, J., & Sullivan, D. M. (2012). Nuclear export of proteins and drug resistance in cancer. *Biochemical Pharmacology, 83*, 1021–1032.
Ueda, K., Xu, J., Morimoto, H., Kawabe, A., & Imaoka, S. (2008). MafG controls the hypoxic response of cells by accumulating HIF-1α in the nuclei. *FEBS Letters, 582*, 2357–2364.
Valiathan, R. R., Marco, M., Leitinger, B., Kleer, C. G., & Fridman, R. (2012). Discoidin domain receptor tyrosine kinases: New players in cancer progression. *Cancer Metastasis Reviews, 31*, 295–321.
Verhelst, K., Verstrepen, L., Carpentier, I., & Beyaert, R. (2013). IkB kinase e (IKKe): A therapeutic target in inflammation and cancer. *Biochemical Pharmacology, 85*, 873–880.

Vogel, W., Gish, G. D., Alves, F., & Pawson, T. (1997). The discoidin domain receptor tyrosine kinases are activated by collagen. *Molecular Cell, 1*, 13–23.
Vogelstein, B., Papadopoulos, N., Velculescu, V. E., Zhou, S., Diaz, L. A., Jr., & Kinzler, K. W. (2013). Cancer genome landscapes. *Science, 339*, 1546–1558.
Vrzalikova, K., Woodman, C. B., & Murray, P. G. (2012). BLIMP1α, the master regulator of plasma cell differentiation is a tumor supressor gene in B cell lymphomas. *Biomedical Papers of the Medical Faculty of the University Palacky, Olomouc, Czechoslovakia, 156*, 1–6.
Williams, M. D., Zhang, X., Belton, A. S., Xian, L., Huso, T., Park, J. J., et al. (2015). HMGA1 drives metabolic reprogramming of intestinal epithelium during hyperproliferation, polyposis, and colorectal carcinogenesis. *Journal of Proteome Research, 14*, 1420–1431.
Wu, Z., Moghaddas Gholami, A., & Kuster, B. (2012). Systematic identification of the HSP90 candidate regulated proteome. *Molecular and Cellular Proteomics, 11*. M111.016675.
Xia, L. M., Huang, W. J., Wang, B., Liu, M., Zhang, Q., Yan, W., et al. (2009). Transcriptional up-regulation of FoxM1 in response to hypoxia is mediated by HIF-1. *Journal of Cellular Biochemistry, 106*, 247–256.
Xiao, K., McClatchy, D. B., Shukla, A. K., Zhao, Y., Chen, M., Shenoy, S. K., et al. (2007). Functional specialization of β-arrestin interactions revealed by proteomic analysis. *Proceedings of the National Academy of Sciences of the United States of America, 104*, 12011–12016.
Yang, Y., & Bedford, M. T. (2013). Protein arginine methyltransferases and cancer. *Nature Reviews. Cancer, 13*, 37–50.
Yang, G., Thompson, M. A., Brandt, S. J., & Hiebert, S. W. (2007). Histone deacetylase inhibitors induce the degradation of the t(8;21) fusion oncoprotein. *Oncogene, 26*, 91–101.
Yuan, S., & Akey, C. W. (2013). Apoptosome structure, assembly, and procaspase activation. *Structure, 21*, 501–515.
Zhang, D., Lin, J., & Han, J. (2010). Receptor-interacting protein (RIP) kinase family. *Cellular & Molecular Immunology, 7*, 243–249.
Zhang, L., Nephew, K. P., & Gallagher, P. J. (2007). Regulation of death-associated protein kinase. Stabilization by HSP90 heterocomplexes. *Journal of Biological Chemistry, 282*, 11795–11804.
Zhang, J., Xu, E., & Chen, X. (2013). TAp73 protein stability is controlled by histone deacetylase 1 via regulation of Hsp90 chaperone function. *Journal of Biological Chemistry, 288*, 7727–7737.
Zhong, X. Y., Ding, J. H., Adams, J. A., Ghosh, G., & Fu, X. D. (2009). Regulation of SR protein phosphorylation and alternative splicing by modulating kinetic interactions of SRPK1 with molecular chaperones. *Genes & Development, 23*, 482–495.
Zhou, Q., Agoston, A. T., Atadja, P., Nelson, W. G., & Davidson, N. E. (2008). Inhibition of histone deacetylases promotes ubiquitin-dependent proteasomal degradation of DNA methyltransferase 1 in human breast cancer cells. *Molecular Cancer Research, 6*, 873–883.
Zhou, B., Li, Y., Deng, Q., Wang, H., Wang, Y., Cai, B., et al. (2013). SRPK1 contributes to malignancy of hepatocellular carcinoma through a possible mechanism involving PI3K/Akt. *Molecular and Cellular Biochemistry, 379*, 191–199.
Zhou, Y., Sun, H., Danila, D. C., Johnson, S. R., Sigai, D. P., Zhang, X., et al. (2000). Truncated activin type I receptor Alk4 isoforms are dominant negative receptors inhibiting activin signaling. *Molecular Endocrinology, 14*, 2066–2075.

CHAPTER TWO

Impact of Posttranslational Modifications on the Anticancer Activity of Hsp90 Inhibitors

Mark R. Woodford*[,‡], Diana Dunn*[,†,‡], Jonelle B. Miller*[,‡], Sami Jamal*[,‡], Len Neckers[§], Mehdi Mollapour*[,†,‡,1]

*Department of Urology, SUNY Upstate Medical University, Syracuse, New York, USA
[†]Department of Biochemistry and Molecular Biology, SUNY Upstate Medical University, Syracuse, New York, USA
[‡]Cancer Research Institute, SUNY Upstate Medical University, Syracuse, New York, USA
[§]Urologic Oncology Branch, Center for Cancer Research, National Cancer Institute, Bethesda, Maryland, USA
[1]Corresponding author: e-mail address: mollapom@upstate.edu

Contents

Abstract

Molecular chaperones are essential for guarding proteins that are indispensable for normal cellular functions. Heat shock protein 90 (Hsp90) is a vital molecular chaperone in eukaryotes that participates in stabilizing and activating approximately 200 target proteins, called "clients," many of which are involved in signal transduction pathways. Cancer cells however utilize Hsp90 to chaperone an array of mutated and overexpressed oncoproteins to protect them from misfolding and degradation. Therefore, Hsp90 is an attractive target in cancer therapy. Hsp90 chaperone function relies on ATP binding and hydrolysis, which in turn guides its carefully orchestrated conformational changes. This chaperone cycle is fine-tuned by another group of proteins called co-chaperones. They

Advances in Cancer Research, Volume 129
ISSN 0065-230X
http://dx.doi.org/10.1016/bs.acr.2015.09.002

are able to accelerate or decelerate the cycle, allowing Hsp90 to chaperone different clients. Posttranslational modifications (PTMs) can also regulate the chaperone cycle at an epigenetic level thereby tailoring Hsp90 function to suit a specific cell type or environmental condition. Recent evidence suggests that inhibition of the enzymes that catalyze the PTM of Hsp90 can act synergistically with Hsp90 inhibitors, providing a novel therapeutic strategy to enhance the efficacy of Hsp90 inhibitors in cancer cells.

1. INTRODUCTION

Heat shock protein 90 (Hsp90) is a molecular chaperone and as the name suggests, it protects proteins that are involved in normal cellular function (Schwartz et al., 2015; Trepel, Mollapour, Giaccone, & Neckers, 2010; Fig. 1). However, Hsp90 is also important in folding, stability, and activity of many proteins responsible for tumor initiation, progression, and metastasis. Therefore, Hsp90 is recognized as a crucial facilitator of oncogenic addiction and cancer cell survival (Fig. 1).

The molecular chaperone Hsp90 is a homodimer containing three structural domains: an N-terminal domain for ATP- and drug-binding and a co-chaperone interacting motif; a middle domain for client and co-chaperone binding, and a C-terminal domain containing sites for constitutive dimerization as well as co-chaperone and client interaction (Prodromou, 2012; Rohl, Rohrberg, & Buchner, 2013). The two isoforms

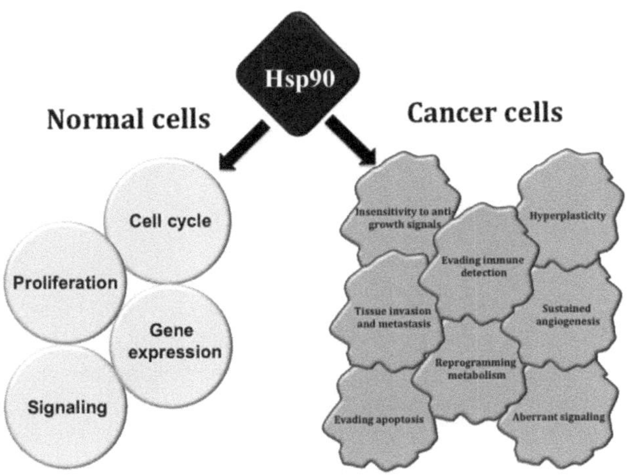

Figure 1 Role of Hsp90 function in cells. Hsp90 chaperones proteins involved in normal cellular functions. It also stabilizes clients essential for the maintenance of each of the proposed hallmarks of cancer. (See the color plate.)

of human Hsp90 (α and β) are encoded by distinct genes, Hsp90α being more stress inducible than the constitutive Hsp90β isoform (Prodromou, 2012). Hsp90 chaperone function is critically dependent on ordered dynamic conformational changes that are coupled to binding and hydrolysis of ATP. Hsp90 inhibitors, such as radicicol (RD) or geldanamycin (GA), disrupt the chaperone cycle by replacing ATP in Hsp90's nucleotide binding pocket (Neckers & Workman, 2012). In eukaryotes, the Hsp90 chaperone cycle is regulated by a group of proteins termed co-chaperones that interact with distinct Hsp90 conformations and serve discrete functions (Li, Soroka, & Buchner, 2012; Fig. 2). Growing evidence suggests that Hsp90 interacts asymmetrically with clients and some co-chaperones and that each Hsp90 protomer can hydrolyze ATP independently (Cunningham, Krukenberg, & Agard, 2008; Mayer & Le Breton, 2015).

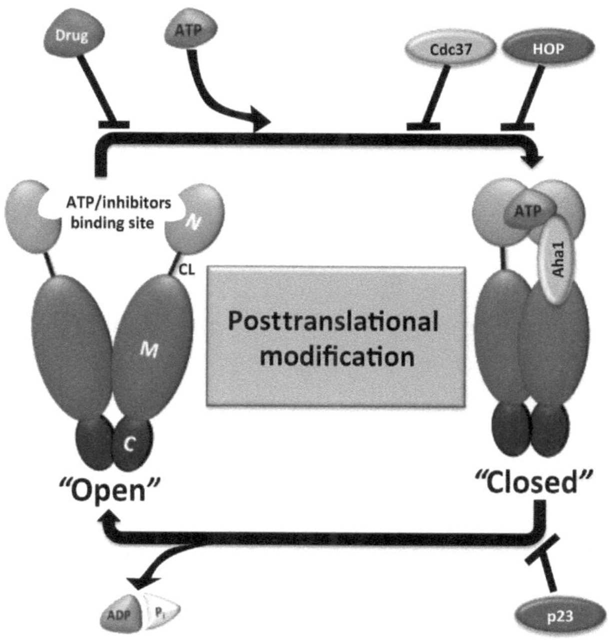

Figure 2 Hsp90 chaperone cycle. ATP binding to the N-terminal domain of Hsp90 (orange) in an "open" conformation promotes transient dimerization of the N-domains to a "closed" conformation capable of ATP hydrolysis. The co-chaperone Aha1 increases the ATPase activity while Cdc37, HOP, and Hsp90 inhibitors exert the opposite effect by preventing N-domain dimerization. p23 slows ATPase activity at a late stage of the chaperone cycle. Domain labeling is as follows: N, N-domain (orange); CL, charged linker (black); M, M-domain (brown); C, C-domain (black). (See the color plate.)

A number of co-chaperones (e.g., p50/Cdc37, HOP/Sti1, p23, Aha1) modulate the rate of Hsp90-mediated ATP hydrolysis, thereby regulating the rate of the chaperone cycle, tailoring and synchronizing Hsp90 chaperone activity to the requirements of specific clients (Cox & Johnson, 2011). For example, the co-chaperone Aha1 facilitates energy-intensive conformational changes necessary to establish Hsp90 ATPase competence, markedly increasing the weak endogenous ATPase activity of Hsp90, and is thus considered to be a crucial component of active Hsp90 chaperone complexes (Abisambra et al., 2013).

Hsp90 is subject to several posttranslational modifications (PTMs) that regulate its chaperone function (Fig. 2; Walton-Diaz et al., 2013). Recent studies suggest that inhibition of the enzymes that modify Hsp90 can act synergistically with Hsp90 inhibitors to impair its function (Iwai et al., 2012). This suggests a novel therapeutic strategy to enhance the efficacy of Hsp90 inhibitors against cancer cells.

2. INHIBITORS OF HSP90 N-DOMAIN

Detailed coverage of the Hsp90 N-domain inhibitors and their clinical development are reviewed elsewhere (Neckers & Workman, 2012). This topic will only be discussed briefly here.

GA (a benzaquinoid ansamycin antibiotic) and RD (a macrocyclic antifungal antibiotic) were the first identified Hsp90 inhibitors (Neckers & Workman, 2012). They mimic the unusual structure that ATP adopts when bound to the N-terminal nucleotide binding pocket, thereby blocking ATP binding and hydrolysis. This inhibits the chaperone cycle, and the degradation of most client proteins ensues. 17-Allylamino-demethoxygeldanamycin (17-AAG, also known as Tanespimycin) is a GA derivative with reduced liver toxicity that in combination with trastuzumab has shown significant clinical activity against $HER2^+$ breast cancer. 17-AAG is also effective in combination with bortezomib in relapsed/refractory multiple myeloma (Modi et al., 2011). 17-Dimethylaminoethylamino-17-demethoxygeldanamycin (17-DMAG, alvespimycin) and the stabilized hydroquinone form of 17-AAG, IPI-504 (retaspimycin) are both water-soluble GA derivatives with reduced hepatotoxicity (Fig. 3; Neckers & Workman, 2012). Unfortunately, none of these compounds were felt to show sufficiently promising activity in phase II clinical testing and their further development has been abandoned. Compounds based on a resorcinol scaffold such as ganetespib (formerly STA-9090 developed by Synta Pharmaceuticals), NVP-AUY922 (developed by Vernalis plc. and

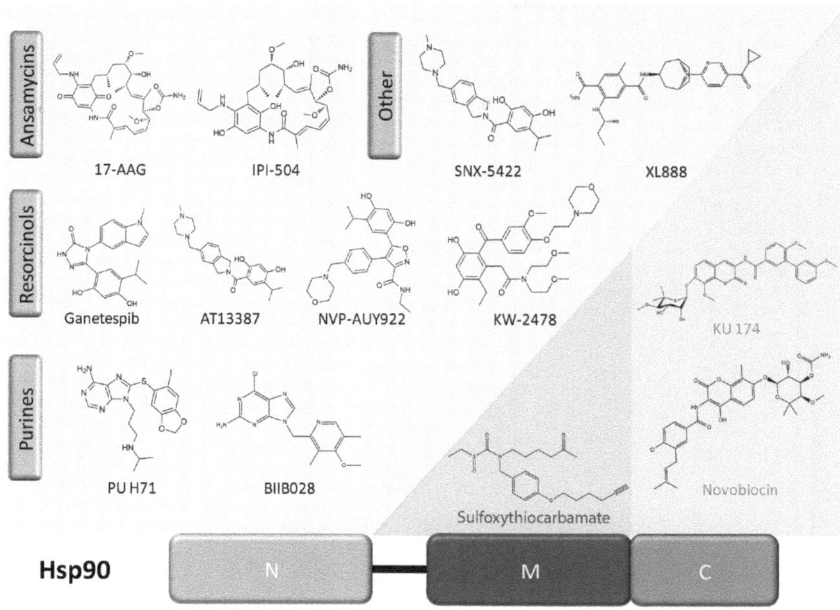

Figure 3 Small molecule inhibitors of Hsp90. (See the color plate.)

Novartis Pharmaceuticals), KW-2478, and AT13387 have been found to be more active in multiple clinical trials. As well, PF-04929113 (SNX-5422) (Esanex Inc.), PF-04928473 (SNX-2112) (Esanex Inc.), and XL888 (Exelixis Inc.) are all orally bioavailable Hsp90 inhibitors demonstrating good antitumor activity in preclinical models (Fig. 3 and Table 1; Neckers & Workman, 2012; Scaltriti, Dawood, & Cortes, 2012).

Based on Hsp90's crystal structure and the specific fold adopted by ATP upon binding to Hsp90, purine-scaffold class derivatives (PU-class) were designed and synthesized. BIIB021 and BIIB028, as well as PU-H71 are all purine-based small molecules with Hsp90 inhibitory activities (Fig. 3; Neckers & Workman, 2012). PU-H71 is currently being evaluated in phase I clinical trials (Neckers & Workman, 2012; Scaltriti et al., 2012).

In addition to these compounds, TAS-116 and FW-04-806 are newly developed drugs that have been found active in preclinical studies. TAS-116 (Taiho Pharmaceutical) is a small molecule ATP-competitive inhibitor of cytosolic Hsp90α and β that has been shown to compete with geldanamycin-FITC binding in a fluorescence polarization assay at nanomolar concentrations. Like ganetespib, TAS-116 has less retinal toxicity compared to some second-generation Hsp90 inhibitors in clinical trials,

Table 1 Current Hsp90 Inhibitors in Clinical Trials

Domain	Inhibitor	Structure	Phase	Mono/Combo	Disease Target	Class
N	AT13387	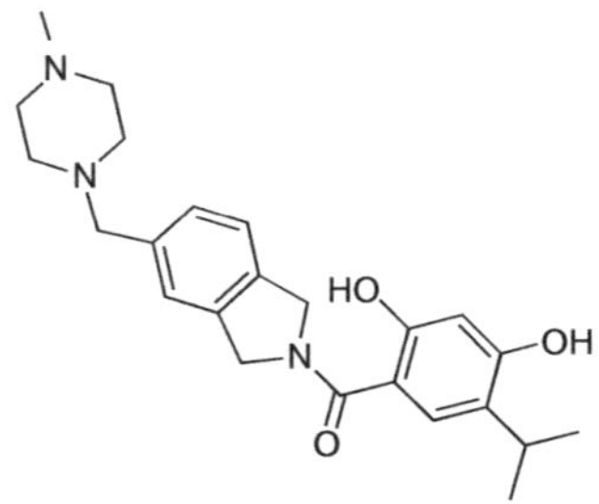	I, II	Both	CRPC, NSCLC, solid tumors	Resorcinol analog
N	Ganetespib	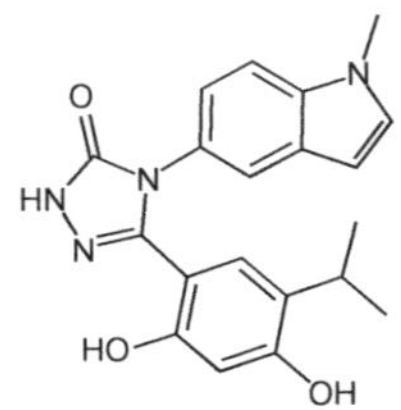	I, II, III	Both	MDS, EOC, HCC, MPNST, ocular melanoma, breast, ovarian and fallopian tube cancer, NSCLC	Resorcinol analog
N	PU-H71	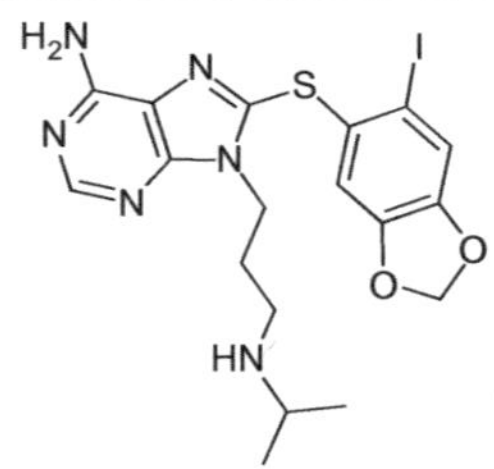	I	Mono	Low-grade non-Hodgkins lymphoma, solid tumors	Purine analog

N	SNX-5422	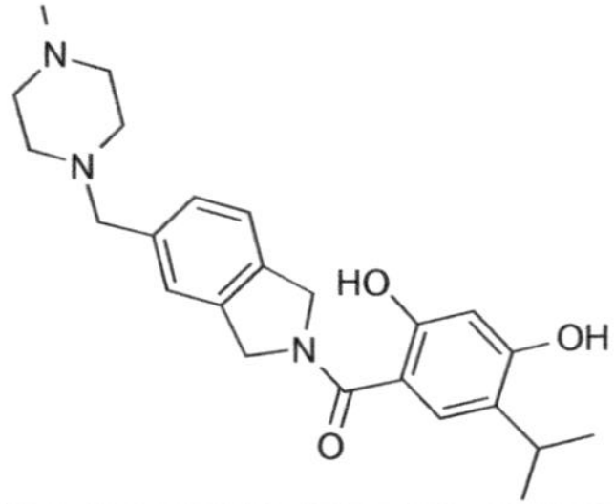	I	Both	Lung adenocarcinoma, solid tumors, HER2^{+} cancers	Benzamide
N	XL888	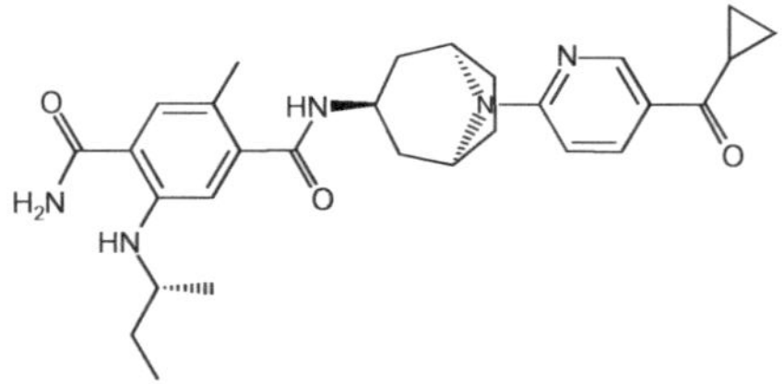	I	Combo	BRAF-mutant stage III/IV melanoma	Other

CRPC, castration-resistant prostate cancer; NSCLC, nonsmall cell lung cancer; MDS, myelodysplastic syndrome; EOC, epithelial ovarian cancer; HCC, hepatocellular carcinoma; MPNST, malignant peripheral nerve sheath tumor.

notably NVP-AUY922 (Seggewiss-Bernhardt et al., 2015). TAS-116 also displays good antitumor activity at well-tolerated drug exposures in human tumor xenograft mouse models (Ohkubo et al., 2015). It is currently under evaluation in a phase I clinical trial (JapicCTI-142444) in Japan (www.clinicaltrials.jp).

FW-04-806 is a natural product extracted from *Streptomyces* FIM-04-806 with an identical structure to the known macrolide antibiotic Conglobatin. This compound does not disrupt Hsp90 ATPase activity, but rather binds to the N-domain of Hsp90 and prevents formation of the Hsp90–Cdc37 complex. This form of Hsp90 inhibition arrests the cell cycle in G_2/M phase and halts tumor cell proliferation. It also promotes degradation of client proteins and subsequent induction of apoptosis (Huang et al., 2014). FW-04-806 is reported to have antitumor activity as monotherapy against SK-BR-3 and MCF-7 xenografts in mice; however, it exhibits increased efficacy in combination with the HER2 kinase inhibitor lapatinib in an *ex vivo* HER2$^+$ tumor xenograft model (Huang et al., 2015).

In summary, targeting the Hsp90 N-domain is a well-researched area of investigation; however, there remain opportunities for further exploration, optimization of existing compounds, and refinement of therapeutic strategies. This refinement includes increasing the efficacy of Hsp90 inhibitors by combining them with other targeted therapies specific to the molecular genetic abnormalities driving a particular tumor type.

3. TARGETING HSP90 M-DOMAIN

A recently identified class of compounds, sulfoxythiocarbamate alkynes (STCAs), selectively target cysteine residues in the Hsp90 middle domain and form stable thiocarbamate adducts in a process called thiocarbamylation. STCAs cause conformational changes in Hsp90 that inhibit chaperoning of clients but do not interfere with ATP hydrolysis (Zhang et al., 2014).

S-Nitrosylation of Cys597-Hsp90α, or its substitution with *S*-nitrosylation-mimicking residues such as asparagine and aspartic acid, decreases chaperone activity (Martinez-Ruiz et al., 2005; Retzlaff et al., 2009). Some of the antitumor activity of nitric oxide has been attributed to inhibition of Hsp90 by this mechanism. It is therefore likely that STCAs are similarly responsible for inhibition of Hsp90 chaperone function by thiocarbamylation of this residue. Targeting the Hsp90 middle domain provides a novel strategy to inhibit its chaperone function, but the poor specificity of compounds reported to date limits their utility as chemical biological probes and their potential as therapeutics (Fig. 3).

4. INHIBITORS OF HSP90 C-DOMAIN

Blagg and Donnelly have reviewed in detail the Hsp90 C-domain inhibitors (Blagg & Donnelly, 2008). A brief summary of this topic is provided below.

Previous reports by the Neckers' group have identified a second drug-binding site in the C-domain of Hsp90 (Marcu, Chadli, Bouhouche, Catelli, & Neckers, 2000; Fig. 3). Coumarin antibiotics such as novobiocin target this site and Blagg's group has made advances in improving the affinity of these compounds for Hsp90. In a recent study, Blagg and colleagues performed computational studies to design and synthesize several compounds using a biphenyl scaffold system (similar to novobiocin), replacing the coumarin ring with various functional groups. Biaryl side chain analogs demonstrate enhanced antitumor properties against breast cancer cell lines (SK-BR-3 and MCF-7) and are promising lead compounds for future drug studies (Zhao et al., 2015).

One of the potential benefits associated with some of the Hsp90 C-domain inhibitors is decreased activation of heat shock factor 1 (*HSF1*). This property avoids the feedback effect caused by Hsp90 N-domain inhibitors whereby HSF1-dependent transcriptional induction of Hsp70, Hsp27, and to some degree Hsp90 itself acts to protect cancer cells from apoptosis (Zuehlke, Beebe, Neckers, & Prince, 2015).

Existing data strongly support further medicinal chemistry optimization and preclinical evaluation of C-terminal Hsp90 inhibitors (Matts et al., 2011).

5. PTMs OF HSP90

Phosphorylation, acetylation, *S*-nitrosylation, methylation, oxidation, ubiquitination, and SUMOylation are the major PTMs of Hsp90 that are known to impact its chaperone function (Walton-Diaz et al., 2013). Here, we review those modifications that are known to impact the binding of inhibitors to the chaperone.

5.1 Hsp90 Phosphorylation

Hsp90 is a well-characterized phosphoprotein. An early study showed that the serine/threonine phosphatase inhibitor, okadaic acid, causes hyperphosphorylation of Hsp90 in NIH3T3 cells transformed with p60$^{v\text{-}src}$. Hsp90 is also disassociated from its client kinase p60$^{v\text{-}src}$ by this compound, therefore linking Hsp90 phosphorylation with its ability to chaperone client proteins (Mimnaugh, Worland, Whitesell, & Neckers, 1995). Supporting these data, the co-chaperone PP5/Ppt1 has been shown to dephosphorylate Hsp90 *in vitro* and to positively regulate its chaperone activity (Wandinger, Suhre, Wegele, & Buchner, 2006). Recent work by Soroka et al. has shown Ppt1-mediated dephosphorylation of S379, S485, and S604 in yeast Hsp90. Hyperphosphorylation of these residues is an important determinant of yeast Hsp90 sensitivity to RD (Soroka et al., 2012).

During the past 4 years, a series of proteomic studies have identified a large number of phosphorylated sites on Hsp90 (www.phosphosite.org/). Wee1, double-stranded DNA-dependent protein kinase, Akt, B-Raf, protein kinase A, protein kinase C, casein kinase 1 (CK1), casein kinase 2 (CK2), GSK3-β, c-Src, and c-Yes are some of the kinases known to phosphorylate Hsp90 (Mollapour & Neckers, 2012).

The serine/threonine kinase CK2 phosphorylates a conserved residue T36 in human Hsp90α (yeast T22). Nonphosphorylatable (T22A, T36A) and phosphomimetic (T22E, T36E) mutants can show decreased Hsp90 chaperone function. But in yeast, only the phosphomimetic mutant demonstrates reduced ATPase activity. Further, the phosphorylation state of this residue affected sensitivity to GA, RD, SNX-2112, and ganetespib (Table 2; Mollapour & Neckers, 2012).

It is noteworthy that CK2 also phosphorylates S13 in the human co-chaperone Cdc37. PP5/Ppt1-mediated dephosphorylation of S13 directly affects Cdc37 interaction with Hsp90 and negatively impacts the chaperoning of numerous kinase clients by the Hsp90–Cdc37 complex, mainly protein kinases. Since CK2-mediated serine phosphorylation of Cdc37 is important for its interaction with Hsp90 (Mollapour & Neckers, 2012), Cdc37 phosphorylation status is likely to affect cellular sensitivity to Hsp90 inhibition. Indeed, the cytotoxicity of GA markedly increased in cells expressing nonphosphorylatable Cdc37-S13A (Mollapour & Neckers, 2012). Further, overexpression of PP5/Ppt1 is synthetically lethal with GA in yeast. PP5 overexpression in cancer cells correlated with reduced Cdc37 phosphorylation and reduced activity of clients such as Raf-1 and those involved in the MAP kinase pathway (Mollapour & Neckers, 2012).

In addition to modification of co-chaperones themselves, phosphorylation of Hsp90 and Hsp70 has been shown to affect chaperone complex formation. Hsp90α can be phosphorylated by CK1, CK2, and GSK3-β at

Table 2 Posttranslational Modifications of Human Hsp90α

Residue	Modification	Enzyme	Drug
T36	Phosphorylation	CK2	GA, RD, GB, SNX-2112
Y38	Phosphorylation	Wee1	GA
K69	Acetylation	p300	GA
K100	Acetylation	p300	GA
T115	Phosphorylation	PKCγ	17-AAG
K191	SUMOylation	SUMO-1	GA, RD
K292	Acetylation	p300	GA
K327	Acetylation	p300	GA
S399	Phosphorylation		RD
C420	*S*-Nitrosylation		STCA
K478	Acetylation	p300	GA
T425	Phosphorylation	PKCγ	17-AAG
S505	Phosphorylation		RD
K546	Acetylation	p300	GA
K558	Acetylation	p300	GA
C572	*S*-Nitrosylation		STCA
C598	*S*-Nitrosylation		STCA
T603	Phosphorylation	PKCγ	17-AAG
T725	Phosphorylation	CK1, CK2, GSK3-β	17-AAG
S726	Phosphorylation	CK1, CK2, GSK3-β	17-AAG

Human Hsp90α posttranslational modification sites that affect drug sensitivity.
RD, radicicol; GA, geldanamycin; GB, ganetespib; STCA, sulfoxythiocarbamate alkyne.

C-terminal residues Thr725 and Ser726 (and Ser718 in Hsp90β). These residues are immediately upstream of the tetratricopeptide (TPR) motif EEVD, thus their phosphorylation state could potentially regulate the binding of TPR-containing co-chaperones. This possibility was explored for two such co-chaperones, the E3-ubiquitin ligase CHIP and HOP, which is critical for protein folding. Indeed CHIP was shown to bind more readily to nonphosphorylated C-terminus of Hsp90, while HOP prefers C-terminal phosphorylation of Hsp90 (Muller et al., 2013).

Importantly, hyperphosphorylation of the Hsp90 C-terminus was noted in primary human breast tumors when compared to normal tissue. Perhaps, this imbalance in favor of protein folding is one mechanism that helps enable the hyperproliferative state of tumor tissues. The phosphorylation status of these residues is regulated in an as-yet unknown way by p38MAPK, though a synergistic decrease in client protein levels is observed upon combined inhibition of Hsp90 by 17-AAG and p38MAPK by SB202190 in T47D breast cancer cells. Taken together, this phosphorylation and dephosphorylation may ultimately determine the balance between stabilization and degradation for relevant client proteins (Muller et al., 2013).

The tyrosine kinase Wee1 is an Hsp90 client and also a key regulator of cell cycle progression. Wee1 phosphorylates a conserved tyrosine residue in the N-domain of yeast (Y24) and human (Y38) Hsp90. This phosphorylation occurs in the nucleus and leads to ubiquitination and degradation of Hsp90 by the proteasome. This phosphorylation also regulates the chaperoning of kinase clients. Lack of phosphorylation as a result of knockdown of *WEE1* in prostate or cervical cancer cell lines led to hypersensitization of these cells to Hsp90 inhibitors (GA, RD, SNX2112, and ganetespib) (Table 2). Pharmacological inhibition of Wee1 also sensitizes prostate cancer cells to Hsp90 inhibitors both *in vivo* and *in vitro* and enhances the binding of these inhibitors to Hsp90. The downstream effect of this combination treatment of prostate cancer cells is the transcriptional downregulation of Survivin and Wee1 with subsequent activation of the intrinsic apoptotic pathway (Iwai et al., 2012).

Recently, Lu et al. demonstrated that PKCγ (protein kinase C γ) is involved in the regulation of Hsp90α (Lu et al., 2014). Interaction with Hsp90 allows for phosphorylation of PKCγ at Thr514, thus activating its kinase activity. In turn, PKCγ phosphorylates Hsp90α at three key threonine residues, Thr115, Thr425, and Thr603, which regulate Hsp90 ATPase activity and the release of PKCγ from the chaperone complex. Overexpression of PKCγ increases the migration of HCT116 colon carcinoma cells, and inhibition of Hsp90 by 17-AAG negates this phenotype. PKCγ is a known suppressor of apoptosis (Parsons & Adams, 2008), and indeed overexpression of PKCγ leads to decreased apoptosis in HCT116 cells. Inhibition of PKCγ with chelerythrine chloride causes a marked increase in apoptosis as measured by cleaved caspase 3. A synergistic increase in apoptosis has been demonstrated with combined use of Hsp90 and PKCγ inhibitors (Lu et al., 2014).

These findings suggest novel therapeutic strategies based on combining Wee1/Hsp90, p38MAPK/Hsp90 or PKCγ/Hsp90 inhibitors and also suggest a mechanism by which the pro-apoptotic activity of Hsp90 inhibitors might be enhanced.

5.2 Hsp90 Phosphorylation Differentially Affects Drug Binding

Duval et al. has shown previously that the tyrosine kinase c-Src is responsible for phosphorylating the highly conserved residue Y301 in HSP90β (Y309 in HSP90α). This event increases the ability of HSP90 to activate VEGF-induced eNOS and thus stimulate NO release from endothelial cells during pro-angiogenic signaling (Duval, Le Boeuf, Huot, & Gratton, 2007). Recent work by Beebe et al. has shown that the binding of the Hsp90 N-domain inhibitor PU-H71 was not affected by mutation of this residue in either Hsp90 isoform. Phosphomimetic mutation of this tyrosine in Hsp90β (Y301E), however, reduced GA binding. Interestingly, neither phosphomimetic nor nonphosphorylatable mutation (Y309E and Y309F, respectively) in Hsp90α affected GA binding. These observations suggest that the phosphorylation status of this residue may provide isoform specificity to GA but not PU-H71 (Beebe et al., 2013).

Beebe et al. have also provided evidence that nonphosphorylated Hsp90 prefers to bind to GA, whereas PU-H71 does not discriminate between phospho- and nonphosphorylated Hsp90. First, it was observed that, in the presence of the phosphotyrosine phosphatase inhibitor bpV(phen), PU-H71-beads captured tyrosine-phosphorylated Hsp90 with greater efficiency than did GA-beads. Second, using an anti-phospho-Y197-Hsp90α antibody, it was shown that phospho-Y197-Hsp90α bound to PU-H71-beads but not to GA-beads. Likewise, when purified Hsp90α was phosphorylated *in vitro* with v-Src kinase prior to capture by either PU-H71-beads or GA-beads, it was shown that GA-beads preferentially bound non-phosphorylated Hsp90, whereas PU-H71 did not show any preference (Beebe et al., 2013).

These findings were further confirmed using phosphomimetic (Y197E) and nonphosphorylatable (Y197F) Hsp90α mutants as well as the equivalent Hsp90β mutants (Y192E and Y192F) (Xu et al., 2012). Beebe et al. has shown that interaction of either phosphomimetic mutant (Hsp90α-Y197E or Hsp90β-Y192E) with GA-beads was undetectable, while GA-beads bound to nonphosphorylatable mutants (Hsp90α-Y197F and Hsp90β-Y192F) with similar affinity to wild-type Hsp90. In contrast, binding of PU-beads to both Hsp90α and Hsp90β was minimally affected by these mutations. These findings provide further evidence that phospho-Hsp90 has reduced affinity for GA. One can envisage a scenario where inhibition of protein kinases that target Hsp90 could make cancer cells hypersensitive to GA-like drugs (Beebe et al., 2013).

5.3 Hsp90 Acetylation and Deacetylation

Acetylation is a PTM involving the addition of an acetyl group to a lysine residue. This is a reversible process catalyzed by acetyltransferases and deacetylases also known as HATs (histone acetyltransferases) and HDACs (histone deacetylases), because they were first discovered in association with histones (Mollapour & Neckers, 2012).

An HDAC inhibitory depsipeptide (romidepsin) has been shown to cause hyperacetylation of Hsp90, reduce ATP binding and abrogate the chaperoning of several clients including ErbB2, Raf-1, and mutant p53. HAT p300 catalyzes acetylation, and HDAC1, 6, and 10 deacetylate Hsp90, thereby affecting its chaperone function. Rao et al. have specifically shown that siRNA-mediated silencing of HDAC6 caused hyperacetylation of Hsp90 and also increased binding to 17-AAG (Rao et al., 2008). It was later shown that treating HEK293 cells with the pan-HDAC inhibitor panobinostat (LBH589) leads to hyperacetylation of seven lysine residues in Hsp90α. Acetylation of all these lysine residues enhanced the binding of Hsp90α to 17-AAG (Rao et al., 2008). These findings support a model in which acetylation–deacetylation of Hsp90 impacts its chaperone function and also influences its ability to bind N-domain inhibitors (Table 2).

5.4 SUMOylation

SUMOylation is a reversible PTM in which small ubiquitin-like modifier (SUMO) proteins are covalently attached to lysine residues, in a process similar to ubiquitylation. However, unlike ubiquitylation, which most commonly targets proteins for degradation by proteasomes, SUMOylation is more likely to affect the function or subcellular location of the modified protein. We have recently identified SUMOylation of a conserved lysine residue in the N-domains of both yeast (lysine 178) and human (lysine 191) Hsp90 (Mollapour et al., 2014). Importantly, N-domain SUMOylation of the Hsp90 dimer occurs on only one protomer, thereby facilitating the recruitment of Aha1 to Hsp90. Surprisingly, it was also shown that Hsp90 inhibitors (GA, RD, ganetespib, and SNX-2112) preferred binding

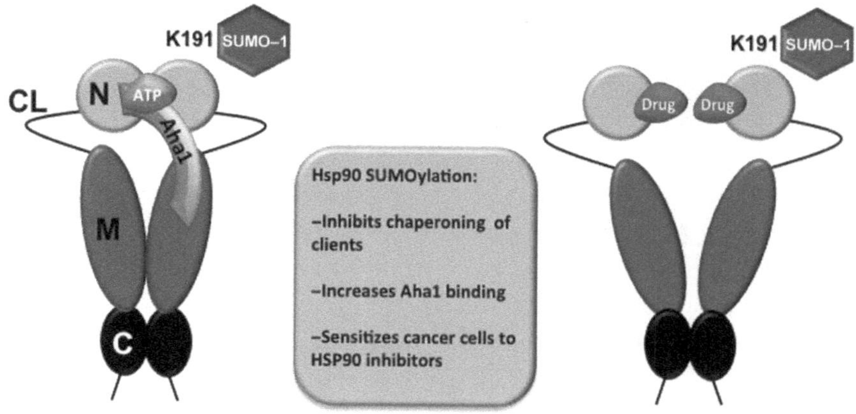

Figure 4 Hsp90 SUMOylation. Asymmetric N-domain SUMOylation of K191 in human Hsp90α not only facilitates Aha1 association but also enhances binding of ATP-competitive inhibitors. Domain labeling is as follows: N, N-domain (orange); CL, charged linker (black); M, M-domain (brown); C, C-domain (black). (See the color plate.)

to SUMOylated Hsp90. Overexpression of SUMO-1 in yeast increases Hsp90 affinity for biotinylated ganetespib by 10-fold compared to Hsp90 from wild-type cells. These findings support a model in which N-terminal inhibitors compete with Aha1 for binding to SUMOylated Hsp90 by displacing previously bound ATP at a point in the chaperoning cycle prior to Hsp90's commitment to ATP hydrolysis (and N-domain dimerization) (Mollapour et al., 2014).

Consistent with this model, overexpression of SUMO-1 sensitizes yeast and NIH 3T3 cells to a panel of chemically distinct Hsp90 inhibitors. Importantly, both steady-state N-domain SUMOylation and Hsp90 inhibitor sensitivity increased upon tumorigenic transformation of NIH3T3 cells with v-Src or mutated (constitutively active) MET (MET-Y1428H) (Table 2). Our recent report of these data provides mechanistic insights that may explain why Hsp90 isolated from tumor cell lysate simultaneously demonstrates greater ATPase activity and greater affinity for Hsp90 inhibitors as compared to Hsp90 isolated from nontransformed cells (Fig. 4; Mollapour et al., 2014).

6. CONCLUSION

Hsp90 inhibitors are currently being evaluated in multiple clinical trials (Table 1), and single-agent activity has been seen in certain cases in which the tumor is driven by an Hsp90 client protein. These include

HER2-positive breast cancer and EML4-ALK-positive nonsmall cell lung cancer. In other cancers, such as multiple myeloma, it is still unclear why Hsp90 inhibitors are effective. Better understanding of the underlying pathways in this disease which make it unusually susceptible to Hsp90 inhibition is needed (Usmani & Chiosis, 2011). Unfortunately, in most cases, Hsp90 inhibitor monotherapy has proven less effective than expected. Particularly disappointing, AUY-922, which recently failed in a large lung cancer trial, largely due to poor therapeutic index has been dropped from further development (Johnson et al., 2015). This is despite the fact that Hsp90 is involved in numerous signaling pathways whose activity is essential for cancer cell survival and proliferation.

A recurring theme in the disappointing activity of Hsp90 inhibitors during their development has been their primarily cytostatic rather than cytotoxic effects in both preclinical models and in patients. Therefore, strategies to enhance tumor cell death in response to Hsp90 inhibitors are being actively sought. Hsp90 clients, mainly protein kinases, have been shown to target and posttranslationally modify both Hsp90 and its co-chaperones, thereby fine-tuning their chaperone function. Pharmacologic inhibition of these kinase clients has been shown to synergize with Hsp90 inhibitors (Schwartz et al., 2015), therefore providing a novel strategy to enhance the anticancer activity of Hsp90 inhibitors. For example, recent work has shown that c-Abl kinase phosphorylates Y223 in the Aha1 co-chaperone therefore promoting its interaction with Hsp90. Pharmacologic inhibition of c-Abl disrupts this interaction thereby, hypersensitizing cancer cells to Hsp90 inhibitors both *in vitro* and *ex vivo* (Dunn et al., 2015). Conversely, increased SUMOylation of Hsp90 has been recently shown to sensitize mammalian cells to Hsp90 inhibitors. Further dissection of the cross talk between various Hsp90 PTMs will greatly add to our understanding of the interconnected, multilayered control mechanisms that normal and cancer cells adopt to regulate Hsp90 function (Fig. 5). Such improved understanding will be invaluable in realizing the goal of targeting HSP90 to cure cancers.

7. SUMMARY BOX

- Hsp90 is an essential molecular chaperone in eukaryotes.
- Cancers cells are addicted to Hsp90, as they have an increased dependence on its chaperoning activity.
- Hsp90 inhibitors have shown promising activity in clinical trials in patients with nonsmall cell lung cancer and $HER2^{+}$ breast cancer.

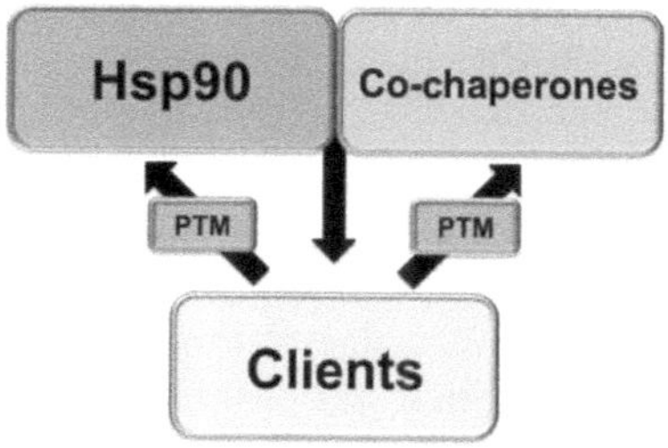

Figure 5 Reciprocal regulatory mechanisms possible between Hsp90 and its "clients" by PTMs. (See the color plate.)

- PTMs and co-chaperones regulate Hsp90 chaperone function.
- Targeting the kinases that regulate Hsp90's function through PTM may provide a strategy to increase the anticancer activity of Hsp90 inhibitors without increasing their toxicity to normal tissues.

ACKNOWLEDGMENTS

We thank all our collaborators Jane Trepel, Chris Prodromou, Laurence H. Pearl, Cara Vaughan, Johannes Buchner, Matthias Mayer, Brian Blagg, William G. Stetler-Stevenson, and Andrew Truman for their scientific contributions. This work was supported with funds from SUNY Upstate Medical University, Carol M. Baldwin Breast Cancer Fund of Central New York and One Square Mile of Hope Foundation (M.M.).

REFERENCES

Abisambra, J., Jinwal, U. K., Miyata, Y., Rogers, J., Blair, L., Li, X., et al. (2013). Allosteric heat shock protein 70 inhibitors rapidly rescue synaptic plasticity deficits by reducing aberrant tau. *Biological Psychiatry*, *74*(5), 367–374. http://dx.doi.org/10.1016/j.biopsych.2013.02.027.

Beebe, K., et al. (2013). Post-translational modification and conformational state of heat shock protein 90 differentially affect binding of chemically diverse small molecule inhibitors. *Oncotarget*, *4*(7), 1065–1074.

Blagg, B. S., & Donnelly, A. (2008). Novobiocin and additional inhibitors of the Hsp90 C-terminal nucleotide binding pocket. *Current Medicinal Chemistry*, *15*(26), 2702–2717.

Cox, M. B., & Johnson, J. L. (2011). The role of p23, Hop, immunophilins, and other co-chaperones in regulating Hsp90 function. *Methods in Molecular Biology*, *787*, 45–66. http://dx.doi.org/10.1007/978-1-61779-295-3_4.

Cunningham, C. N., Krukenberg, K. A., & Agard, D. A. (2008). Intra- and intermonomer interactions are required to synergistically facilitate ATP hydrolysis in Hsp90. *The Journal of Biological Chemistry*, *283*(30), 21170–21178. http://dx.doi.org/10.1074/jbc.M800046200. M800046200 [pii].

Dunn, D. M., Woodford, M. R., Truman, A. W., Jensen, S. M., Schulman, J., Caza, T., et al. (2015). c-Abl mediated tyrosine phosphorylation of Aha1 activates its co-chaperone function in cancer cells. *Cell Reports*, *12*(6), 1006–1018. http://dx.doi.org/10.1016/j.celrep.2015.07.004.

Duval, M., Le Boeuf, F., Huot, J., & Gratton, J. P. (2007). Src-mediated phosphorylation of Hsp90 in response to vascular endothelial growth factor (VEGF) is required for VEGF receptor-2 signaling to endothelial NO synthase. *Molecular Biology of the Cell*, *18*, 4659–4668. http://dx.doi.org/10.1091/mbc.E07-05-0467.

Huang, W., Wu, Q. D., Zhang, M., Kong, Y. L., Cao, P. R., Zheng, W., et al. (2015). Novel Hsp90 inhibitor FW-04-806 displays potent antitumor effects in HER2-positive breast cancer cells as a single agent or in combination with lapatinib. *Cancer Letters*, *356*(2 Pt. B), 862–871. http://dx.doi.org/10.1016/j.canlet.2014.10.040.

Huang, W., Ye, M., Zhang, L. R., Wu, Q. D., Zhang, M., Xu, J. H., et al. (2014). FW-04-806 inhibits proliferation and induces apoptosis in human breast cancer cells by binding to N-terminus of Hsp90 and disrupting Hsp90-Cdc37 complex formation. *Molecular Cancer*, *13*, 150. http://dx.doi.org/10.1186/1476-4598-13-150.

Iwai, A., Bourboulia, D., Mollapour, M., Jensen-Taubman, S., Lee, S., Donnelly, A. C., et al. (2012). Combined inhibition of Wee1 and Hsp90 activates intrinsic apoptosis in cancer cells. *Cell Cycle*, *11*(19), 3649–3655. http://dx.doi.org/10.4161/cc.21926.

Johnson, M. L., Yu, H. A., Hart, E. M., Weitner, B. B., Rademaker, A. W., Patel, J. D., et al. (2015). Phase I/II study of HSP90 inhibitor AUY922 and erlotinib for EGFR-mutant lung cancer with acquired resistance to epidermal growth factor receptor tyrosine kinase inhibitors. *Journal of Clinical Oncology*, *33*(15), 1666–1673. http://dx.doi.org/10.1200/jco.2014.59.7328.

Li, J., Soroka, J., & Buchner, J. (2012). The Hsp90 chaperone machinery: Conformational dynamics and regulation by co-chaperones. *Biochimica et Biophysica Acta*, *1823*(3), 624–635. http://dx.doi.org/10.1016/j.bbamcr.2011.09.003. S0167-4889(11)00261-8 [pii].

Lu, X. A., Wang, X., Zhuo, W., Jia, L., Jiang, Y., Fu, Y., et al. (2014). The regulatory mechanism of a client kinase controlling its own release from Hsp90 chaperone machinery through phosphorylation. *The Biochemical Journal*, *457*(1), 171–183. http://dx.doi.org/10.1042/BJ20130963.

Marcu, M. G., Chadli, A., Bouhouche, I., Catelli, M., & Neckers, L. M. (2000). The heat shock protein 90 antagonist novobiocin interacts with a previously unrecognized ATP-binding domain in the carboxyl terminus of the chaperone. *The Journal of Biological Chemistry*, *275*(47), 37181–37186. http://dx.doi.org/10.1074/jbc.M003701200. M003701200 [pii].

Martinez-Ruiz, A., Villanueva, L., Gonzalez de Orduna, C., Lopez-Ferrer, D., Higueras, M. A., Tarin, C., et al. (2005). S-Nitrosylation of Hsp90 promotes the inhibition of its ATPase and endothelial nitric oxide synthase regulatory activities. *Proceedings of the National Academy of Sciences of the United States of America*, *102*(24), 8525–8530. http://dx.doi.org/10.1073/pnas.0407294102. 0407294102 [pii].

Matts, R. L., Brandt, G. E., Lu, Y., Dixit, A., Mollapour, M., Wang, S., et al. (2011). A systematic protocol for the characterization of Hsp90 modulators. *Bioorganic & Medicinal Chemistry*, *19*(1), 684–692. http://dx.doi.org/10.1016/j.bmc.2010.10.029. S0968-0896(10)00952-1 [pii].

Mayer, M. P., & Le Breton, L. (2015). Hsp90: Breaking the symmetry. *Molecular Cell*, *58*(1), 8–20. http://dx.doi.org/10.1016/j.molcel.2015.02.022.

Mimnaugh, E. G., Worland, P. J., Whitesell, L., & Neckers, L. M. (1995). Possible role for serine/threonine phosphorylation in the regulation of the heteroprotein complex between the hsp90 stress protein and the pp60v-src tyrosine kinase. *The Journal of Biological Chemistry*, *270*(48), 28654–28659.

Modi, S., Stopeck, A., Linden, H., Solit, D., Chandarlapaty, S., Rosen, N., et al. (2011). HSP90 inhibition is effective in breast cancer: A phase II trial of tanespimycin (17-AAG) plus trastuzumab in patients with HER2-positive metastatic breast cancer progressing on trastuzumab. *Clinical Cancer Research*, *17*(15), 5132–5139. http://dx.doi.org/10.1158/1078-0432.CCR-11-0072.

Mollapour, M., Bourboulia, D., Beebe, K., Woodford, M. R., Polier, S., Hoang, A., et al. (2014). Asymmetric Hsp90 N domain SUMOylation recruits Aha1 and ATP-competitive inhibitors. *Molecular Cell*, *53*(2), 317–329. http://dx.doi.org/10.1016/j.molcel.2013.12.007.

Mollapour, M., & Neckers, L. (2012). Post-translational modifications of Hsp90 and their contributions to chaperone regulation. *Biochimica et Biophysica Acta*, *1823*(3), 648–655. http://dx.doi.org/10.1016/j.bbamcr.2011.07.018. S0167-4889(11)00217-5 [pii].

Muller, P., Ruckova, E., Halada, P., Coates, P. J., Hrstka, R., Lane, D. P., et al. (2013). C-terminal phosphorylation of Hsp70 and Hsp90 regulates alternate binding to co-chaperones CHIP and HOP to determine cellular protein folding/degradation balances. *Oncogene*, *32*(25), 3101–3110. http://dx.doi.org/10.1038/onc.2012.314.

Neckers, L., & Workman, P. (2012). Hsp90 molecular chaperone inhibitors: Are we there yet? *Clinical Cancer Research*, *18*(1), 64–76. http://dx.doi.org/10.1158/1078-0432.CCR-11-1000. 18/1/64 [pii].

Ohkubo, S., Kodama, Y., Muraoka, H., Hitotsumachi, H., Yoshimura, C., Kitade, M., et al. (2015). TAS-116, a highly selective inhibitor of heat shock protein 90alpha and beta, demonstrates potent antitumor activity and minimal ocular toxicity in preclinical models. *Molecular Cancer Therapeutics*, *14*(1), 14–22. http://dx.doi.org/10.1158/1535-7163.MCT-14-0219.

Parsons, M., & Adams, J. C. (2008). Rac regulates the interaction of fascin with protein kinase C in cell migration. *Journal of Cell Science*, *121*(Pt. 17), 2805–2813. http://dx.doi.org/10.1242/jcs.022509.

Prodromou, C. (2012). The 'active life' of Hsp90 complexes. *Biochimica et Biophysica Acta*, *1823*(3), 614–623. http://dx.doi.org/10.1016/j.bbamcr.2011.07.020. S0167-4889(11)00219-9 [pii].

Rao, R., Fiskus, W., Yang, Y., Lee, P., Joshi, R., Fernandez, P., et al. (2008). HDAC6 inhibition enhances 17-AAG—Mediated abrogation of hsp90 chaperone function in human leukemia cells. *Blood*, *112*(5), 1886–1893. http://dx.doi.org/10.1182/blood-2008-03-143644. blood-2008-03-143644 [pii].

Retzlaff, M., Stahl, M., Eberl, H. C., Lagleder, S., Beck, J., Kessler, H., et al. (2009). Hsp90 is regulated by a switch point in the C-terminal domain. *EMBO Reports*, *10*(10), 1147–1153. http://dx.doi.org/10.1038/embor.2009.153. embor2009153 [pii].

Rohl, A., Rohrberg, J., & Buchner, J. (2013). The chaperone Hsp90: Changing partners for demanding clients. *Trends in Biochemical Sciences*, *38*(5), 253–262. http://dx.doi.org/10.1016/j.tibs.2013.02.003.

Scaltriti, M., Dawood, S., & Cortes, J. (2012). Molecular pathways: Targeting hsp90—Who benefits and who does not. *Clinical Cancer Research*, *18*(17), 4508–4513. http://dx.doi.org/10.1158/1078-0432.CCR-11-2138.

Schwartz, H., Scroggins, B., Zuehkle, A., Kijima, T., Beebe, K., Mishra, A., et al. (2015). Combined HSP90 and kinase inhibitor therapy: Insights from The Cancer Genome Atlas. *Cell Stress & Chaperones*, *20*, 729–741. http://dx.doi.org/10.1007/s12192-015-0604-1.

Seggewiss-Bernhardt, R., Bargou, R. C., Goh, Y. T., Stewart, A. K., Spencer, A., Alegre, A., et al. (2015). Phase 1/1B trial of the heat shock protein 90 inhibitor NVP-AUY922 as monotherapy or in combination with bortezomib in patients with relapsed or refractory multiple myeloma. *Cancer*. http://dx.doi.org/10.1002/cncr.29339.

Soroka, J., Wandinger, S. K., Mausbacher, N., Schreiber, T., Richter, K., Daub, H., et al. (2012). Conformational switching of the molecular chaperone Hsp90 via regulated phosphorylation. *Molecular Cell*, *45*(4), 517–528. http://dx.doi.org/10.1016/j.molcel.2011.12.031. S1097-2765(12)00040-8 [pii].

Trepel, J., Mollapour, M., Giaccone, G., & Neckers, L. (2010). Targeting the dynamic HSP90 complex in cancer. *Nature Reviews Cancer, 10*(8), 537–549. http://dx.doi.org/10.1038/nrc2887. nrc2887 [pii].

Usmani, S. Z., & Chiosis, G. (2011). HSP90 inhibitors as therapy for multiple myeloma. *Clinical Lymphoma, Myeloma & Leukemia, 11*(Suppl. 1), S77–S81. http://dx.doi.org/10.1016/j.clml.2011.03.027.

Walton-Diaz, A., Khan, S., Bourboulia, D., Trepel, J. B., Neckers, L., & Mollapour, M. (2013). Contributions of co-chaperones and post-translational modifications towards Hsp90 drug sensitivity. *Future Medicinal Chemistry, 5*(9), 1059–1071. http://dx.doi.org/10.4155/fmc.13.88.

Wandinger, S. K., Suhre, M. H., Wegele, H., & Buchner, J. (2006). The phosphatase Ppt1 is a dedicated regulator of the molecular chaperone Hsp90. *The EMBO Journal, 25*(2), 367–376. http://dx.doi.org/10.1038/sj.emboj.7600930. 7600930 [pii].

Xu, W., Mollapour, M., Prodromou, C., Wang, S., Scroggins, B. T., Palchick, Z., et al. (2012). Dynamic tyrosine phosphorylation modulates cycling of the HSP90-P50 (CDC37)-AHA1 chaperone machine. *Molecular Cell, 47*(3), 434–443. http://dx.doi.org/10.1016/j.molcel.2012.05.015.

Zhang, Y., Dayalan Naidu, S., Samarasinghe, K., Van Hecke, G. C., Pheely, A., Boronina, T. N., et al. (2014). Sulphoxythiocarbamates modify cysteine residues in HSP90 causing degradation of client proteins and inhibition of cancer cell proliferation. *British Journal of Cancer, 110*(1), 71–82. http://dx.doi.org/10.1038/bjc.2013.710.

Zhao, H., Garg, G., Zhao, J., Moroni, E., Girgis, A., Franco, L. S., et al. (2015). Design, synthesis and biological evaluation of biphenylamide derivatives as Hsp90 C-terminal inhibitors. *European Journal of Medicinal Chemistry, 89*, 442–466. http://dx.doi.org/10.1016/j.ejmech.2014.10.034.

Zuehlke, A. D., Beebe, K., Neckers, L., & Prince, T. (2015). Regulation and function of the human HSP90AA1 gene. *Gene, 570*, 8–16. http://dx.doi.org/10.1016/j.gene.2015.06.018.

CHAPTER THREE

Anticancer Inhibitors of Hsp90 Function: Beyond the Usual Suspects

Gaurav Garg, Anuj Khandelwal, Brian S.J. Blagg[1]
Department of Medicinal Chemistry, The University of Kansas, Lawrence, Kansas, USA
[1]Corresponding author: e-mail address: bblagg@ku.edu

Contents

Abstract

The 90-kDa heat-shock protein (Hsp90) is a molecular chaperone responsible for the stability and function of a wide variety of client proteins that are critical for cell growth and survival. Many of these client proteins are frequently mutated and/or overexpressed in cancer cells and are therefore being actively pursued as individual therapeutic targets. Consequently, Hsp90 inhibition offers a promising strategy for simultaneous degradation of several anticancer targets. Currently, most Hsp90 inhibitors under clinical evaluation act by blocking the binding of ATP to the Hsp90 N-terminal domain and thereby, induce the degradation of many Hsp90-dependent oncoproteins. Although, they have shown some promising initial results, clinical challenges such as induction of the heat-shock response, retinopathy, and gastrointestinal tract toxicity are emerging from human trials, which constantly raise concerns about the future development of these inhibitors. Novobiocin derivatives, which do not bind the chaperone's N-terminal ATPase pocket, have emerged over the past decade as an alternative strategy to inhibit Hsp90, but to date, no derivative has been investigated in the clinical setting. In recent years, a number of natural or synthetic compounds have been identified that modulate Hsp90 function via various mechanisms. These compounds not only offer new chemotypes for the development of future Hsp90 inhibitors but can also serve as chemical probes to unravel the biology of Hsp90. This chapter presents a synopsis of inhibitors that directly, allosterically, or even indirectly alters Hsp90 function, and highlights their proposed mechanisms of action.

Advances in Cancer Research, Volume 129
ISSN 0065-230X
http://dx.doi.org/10.1016/bs.acr.2015.12.001

1. INTRODUCTION

An efficient protein quality control system is fundamental to all cellular processes and is critical for protein homeostasis within the crowded cellular environment (Taipale, Jarosz, & Lindquist, 2010). Since the cellular environment undergoes rapid change, numerous adaptive mechanisms have evolved to manage protein folding and quality control. Upon exposure to environmental stresses, such as high temperature, oxidative stress, hypoxia, acidosis, or malignant transformation, cells induce the expression of a diverse set of proteins, including molecular chaperones, which maintain the dynamic equilibrium between protein folding and degradation (Caplan, Mandal, & Theodoraki, 2007). Molecular chaperones are a highly conserved class of proteins that modulate the folding, intracellular disposition, and degradation of client protein substrates (Whitesell & Lindquist, 2005). The heat-shock proteins (Hsp's) represent a class of molecular chaperones that are constitutively expressed under normal physiological conditions, but are upregulated in response to cellular stress to sustain cell viability by maintaining the structural and functional integrity of key regulators of cell growth, differentiation, and survival (Jolly & Morimoto, 2000).

The 90-kDa heat-shock protein (Hsp90) is a highly abundant molecular chaperone that is responsible for the maintenance of protein homeostasis under basal conditions and during the stress response (Young, Agashe, Siegers, & Hartl, 2004). Hsp90 comprises ~1–2% of total cell protein in unstressed cells, but is overexpressed (~4–6%) under hostile conditions to buffer proteotoxic stresses (Donnelly & Blagg, 2008). In humans, Hsp90 exists as four isoforms: Hsp90α (inducible form) and Hsp90β (constitutive form) are mainly found in the cytosol, while the 94-kDa glucose-regulated protein (GRP94) and Hsp75/tumor necrosis factor receptor associated protein 1 (TRAP-1) are localized in the endoplasmic reticulum and mitochondria, respectively (Blagg & Kerr, 2006). In addition, a fraction of Hsp90 is found on the cell surface of cancer cells as well as in the extracellular milieu (Sidera & Patsavoudi, 2008; Trepel, Mollapour, Giaccone, & Neckers, 2010). Hsp90 plays a central role in the conformational maturation, activation, cellular trafficking, and proteolytic turnover of a wide range of substrates, referred to as client proteins (Neckers & Workman, 2012; Taipale et al., 2010). In fact, recent studies indicate that there are ~400 client proteins that depend upon the Hsp90 protein folding machinery to achieve and maintain their active conformations (Taipale et al., 2012). Hsp90 client proteins regulate a vast array of cellular functions, including signal transduction, protein trafficking, chromatin remodeling, autophagy, cell proliferation, and survival (Zuehlke & Johnson, 2010). However, many Hsp90 client proteins are frequently mutated and/or overexpressed in cancer cells and are consequently pursued as individual therapeutic targets for cancer treatment (Whitesell & Lindquist, 2005). As a result, Hsp90 inhibition can provide a unique opportunity to simultaneous deplete multiple anticancer targets (Koga, Kihara, & Neckers, 2009). Therefore, current Hsp90 research has focused on its therapeutic potential as a target for the development of cancer chemotherapeutics. In contrast to its role in driving oncoprotein degradation, Hsp90 inhibition has been shown to induce the prosurvival heat-shock response, which increases molecular chaperone levels (Luo, Sun, Taldone, Rodina, & Chiosis, 2010; Whitesell, Bagatell, & Falsey, 2003). The upregulation of molecular chaperones appears beneficial for neurodegenerative disorders, such as Alzheimer's and Parkinson disease, where they protect cells from the accumulation of neurotoxic proteins (Paul & Mahanta, 2014). As a result of its broad participation in cell biology, Hsp90 has emerged as a promising therapeutic target for the treatment of multiple disease states, including cancer. To date, however, there is no FDA-approved Hsp90 inhibitor. Given the essential role played by Hsp90 in multiple cellular processes, unanticipated adverse effects resulting from Hsp90 inhibition cannot be ruled out in future.

2. HSP90 STRUCTURE, FUNCTION, CHAPERONE CYCLE, AND POINTS OF DISRUPTION BY INHIBITORS

Hsp90 belongs to the GHKL (*G*yrase, *H*sp90, Histidine *K*inase, and Mut*L*) superfamily of ATPases that contain a Bergerat ATP-binding fold. GHKL family members feature an ATP-binding pocket in which ATP is bound in a unique, bent conformation that is distinct from the typical extended conformation exhibited by protein kinases (Dutta & Inouye, 2000). In humans, Hsp90 exists as a homodimer with each monomer consisting of a highly conserved N-terminal ATP-binding domain (NTD) connected to a middle domain (MD) and a C-terminal dimerization domain (CTD) (Donnelly & Blagg, 2008). The NTD of Hsp90 contains an ATP-binding site that is responsible for its ATPase activity and provides the requisite energy for the chaperone cycle (Dutta & Inouye, 2000; Panaretou et al., 1998). Historically, the NTD has been the major binding site for the development of Hsp90 inhibitors; e.g., natural products geldanamycin (GDA) and radicicol (RDC) compete with ATP for N-terminal ATP-binding and block Hsp90 function. The NTD is connected to the MD of Hsp90 by a flexible, highly charged linker. This domain plays a key role in modulating Hsp90 ATPase activity by binding the γ-phosphate of ATP when bound to the N-terminus (Meyer et al., 2003). Structural and mutagenesis studies indicate that this site serves for the recognition and binding of client proteins and cochaperones (e.g., Aha1) (Huai et al., 2005). The CTD is important for the homodimerization of Hsp90 into its biologically active conformation (Pearl & Prodromou, 2006). The CTD contains a second nucleotide-binding site that allosterically regulates N-terminal ATPase activity (Prodromou et al., 1999; Sőti, Vermes, Haystead, & Csermely, 2003). This domain also features a conserved MEEVD sequence that is responsible for recruiting TPR-domain (tetratricopeptide-containing repeats) containing cochaperones, such as Hsp70–Hsp90 organizing protein (HOP) and immunophilins. Natural products such as novobiocin (NB) and epigallocatechin-3-gallate (EGCG) bind the CTD and modulate Hsp90 function (Marcu, Chadli, Bouhouche, Catelli, & Neckers, 2000; Yin, Henry, & Gasiewicz, 2009).

The Hsp90-mediated protein folding process is complex and has been reviewed extensively (Blagg & Kerr, 2006; Donnelly & Blagg, 2008; Hall, Forsberg, & Blagg, 2014; Li, Soroka, & Buchner, 2012; Wandinger, Richter, & Buchner, 2008). Although the complete mechanism of the Hsp90-mediated protein folding cycle is still not fully understood, accumulating evidence indicates that this cycle requires the interaction of Hsp90 with a number of cochaperones, partner proteins, and immunophilins to form the multiprotein complexes that enable proper function of the machinery (Fig. 1; Peterson & Blagg, 2009).

The chaperone cycle begins with the binding of nascent polypeptides to the Hsp70/Hsp40/ADP complex (**I**) to prevent aggregation (Walter & Buchner, 2002). This complex can be stabilized by the Hsp70-interacting protein (HIP) or, alternatively, Bcl2-associated athanogene homologs that bind and stimulate the exchange of ATP for ADP (Chaudhury, Welch, & Blagg, 2006). The Hsp70/Hsp40/client complex (**II**) then associates with Hsp90 (**III**) to deliver the unfolded protein. The association between Hsp70 and Hsp90 is mediated by HOP/Sti1 (Hsp90–Hsp70 organizing protein), which serves as an adaptor protein (Murphy, Kanelakis, Galigniana, Morishima, & Pratt, 2001). In the case of protein kinases, the cochaperone Cdc37 (cell-division-cycle 37 homologue, also known as p50) is often recruited to the Hsp70/Hsp40/client complex, which promotes the loading of client kinases onto Hsp90 with the aid of HOP

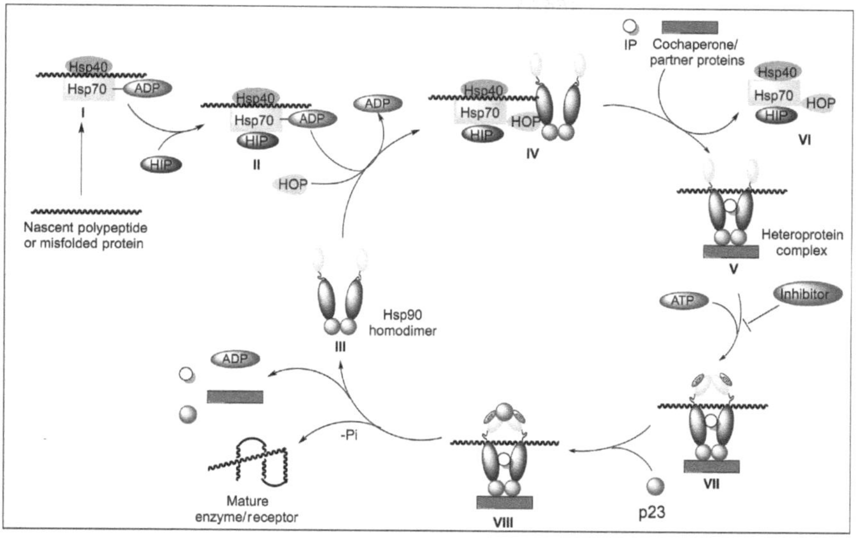

Figure 1 The Hsp90-mediated protein folding process. (See the color plate.)

(Caplan et al., 2007). Following client substrate loading, various immunophilins (FKBP51, FKBP52), cochaperones, and partner proteins bind the Hsp90 homodimer (**IV**) to form an activated heteroprotein complex (**V**) with concomitant release of Hsp70, HIP, and HOP (Kosano, Stensgard, Charlesworth, McMahon, & Toft, 1998). ATP binds to the "open" heteroprotein complex (**V**) at the N-terminus of Hsp90 and promotes a structural reorganization of Hsp90 that results in a "closed" conformation (**VIII**) (Prodromou et al., 2000). At this stage, Hsp90 inhibitors can compete with ATP for the N-terminal binding site, which prevents formation of the closed conformation, and ultimately, leads to the degradation of many clients via the ubiquitin–proteasome pathway (Donnelly & Blagg, 2008). In the absence of an inhibitor, this multiprotein assembly is stabilized by the association of cochaperone p23, followed by the recruitment of Aha1 (activator of Hsp90 ATPase homologue 1) to the MD of each Hsp90 monomer. Binding of Aha1 stimulates the hydrolysis of ATP and promotes folding of the bound client, followed by the dissociation of immunophilins and cochaperones (Ali et al., 2006).

As depicted in Fig. 1, the chaperone cycle is a multistage process that requires the participation of various cochaperones and coactivators that work in conjunction with Hsp90 to modulate the activity of the machinery (Table 1; Peterson & Blagg, 2009; Zuehlke & Johnson, 2010). These cochaperones and partner proteins enter the chaperone cycle at different stages and assist in the conformational maturation of specific client protein classes.

Table 1 Cochaperones and Partner Proteins That Participate in the Hsp90 Protein Folding Cycle (Peterson & Blagg, 2009; Zuehlke & Johnson, 2010)

Cochaperone or Partner Proteins	Description
Aha1	Stimulates ATPase activity
Cdc37	Mediates activation of protein kinase substrates
CHIP	Involved in degradation of unfolded client proteins
Cyclophilin-40	Peptidyl propyl isomerase
FKBP51 and 52	Peptidyl propyl isomerase
HOP	Mediates interaction between Hsp90 and Hsp70
Hsp40	Stabilizes and delivers client proteins to Hsp90 complex
Hsp70	Stabilizes and delivers client proteins to Hsp90 complex
p23	Stabilizes closed, clamped substrate bound conformation
HIP	Inhibits ATPase activity of Hsp70
PP5	Protein phosphatase 5
Sgt1	Client adaptor, involved in client recruitment
Tom70	Facilitates translocation of pre-proteins into mitochondrial matrix
WISp39	Regulates p21 stability

For example, Cdc37 is required for the recruitment of kinase clients to the Hsp90 machinery and is overexpressed in some kinase-driven cancers. In addition, numerous posttranslational modifications, including phosphorylation, S-nitrosylation, and SUMOylation, regulate Hsp90 function by modulating its affinity for cochaperones and/or client proteins (see Impact of Posttranslational Modifications on the Anticancer Activity of Hsp90 Inhibitors by Woodford et al.) (Hall, Forsberg, et al., 2014; Trepel et al., 2010).

3. THE ROLES OF HSP90 IN CANCER

As the field of cancer research has progressed, new approaches to cancer chemotherapy have emerged. Molecularly targeted therapeutic strategies initially focused on the inhibition of specific enzymes and/or receptors associated with cell signaling, but off-target effects and/or resistance have limited their efficacy against most advanced solid tumors. Through better understanding of cancer biology, it has become increasingly evident that clinical cancers result from dysregulation of multiple interconnected pathways (Logue & Morrison, 2012). In 2000 and 2011, Hanahan and Weinberg proposed 10 hallmarks of cancer that result from genetic and epigenetic alterations of key regulatory proteins, enzymes, and receptors (Hanahan & Weinberg, 2000, 2011). These hallmarks include (1) sustaining proliferative signaling, (2) evading growth suppressors, (3) resisting cell death, (4) enabling replicative immortality, (5) inducing angiogenesis, (6) activating invasion and metastasis, (7) deregulated cellular energetics, (8) avoiding immune destruction, (9) tumor-promoting inflammation, and (10) genome instability and mutation. In light of this understanding, Hsp90 inhibition is particularly appealing because it has the potential to simultaneously disrupt multiple pathways by acting on a single target and thereby exerting a multipronged attack on malignant cells (Xu & Neckers, 2007). Hsp90 is essential for the stability and function of a wide range of oncogenic proteins, such as signaling kinases, steroid hormone receptors, telomerase, and many others that contribute directly to the hallmarks of cancer (Table 2) (and chapter by Vartholomaiou et al., 2016) (Blagg & Kerr, 2006; Koga et al., 2009). Therefore, inhibition of Hsp90 by a small molecule represents an exciting strategy for development of new cancer chemotherapeutics (Fig. 2).

The ubiquitous and essential nature of Hsp90 function raised concerns about the potential for selective use of Hsp90 inhibitors. However, studies have indicated that Hsp90 inhibitors are more toxic to malignant cells than to normal tissues, and therefore the potential does exist for selective cytotoxicity (Chiosis & Neckers, 2006; Chiosis et al., 2003; Kamal et al., 2003; Workman, 2004). Moreover, at least some Hsp90 inhibitors accumulate in malignant cells to a greater extent than surrounding tissue (Chiosis et al., 2003). In support of this observation, Kamal and coworkers demonstrated that the apparent accumulation of Hsp90 inhibitors in tumor cells following systemic administration could result from the increased affinity of Hsp90 present in cancer cells as compared to normal cells (Kamal et al., 2003). Although the concept remains debated, it has been suggested that Hsp90 in cancer cells is engaged in an activated heteroprotein complex that exhibits both enhanced ATPase activity and higher affinity for Hsp90 inhibitors, compared to the inactivated, homodimeric complex found in normal cells. Moreover, a number of Hsp90 inhibitors have shown promising results

Table 2 Hsp90 Client Proteins Associated with the Hallmarks of Cancer (Blagg & Kerr, 2006; Vartholomaiou et al., 2016)

Hallmarks of Cancer	Hsp90 Client Protein(s)
1. Sustaining proliferative signaling	Raf-1, AKT, Her2, MEK, Bcr-Abl
2. Evading growth suppressors	Plk, Wee1, Myc1, CDK4, CDK6, Myt1
3. Resisting cell death	NF-κ, AKT, p53, c-MET, Apaf-1, Survivin
4. Enabling replicative immortality	Telomerase (h-Tert)
5. Inducing angiogenesis	HIF-1α, VEGFR, PI3K/AKT, RTKs, flt-3
6. Activating invasion and metastasis	c-MET, SSDF-1, MMP-2
7. Deregulated cellular energetics	ARNT, ARRB1, HIF-1α, HMG1, SREBF1
8. Avoiding immune destruction	IRAK3
9. Tumor-promoting inflammation	IL-6, IL-8, IRAK1, IRAK2, IRAK3
10. Genome instability and mutation	FANCA, MAFG, NEK8, NEK9, NEK11

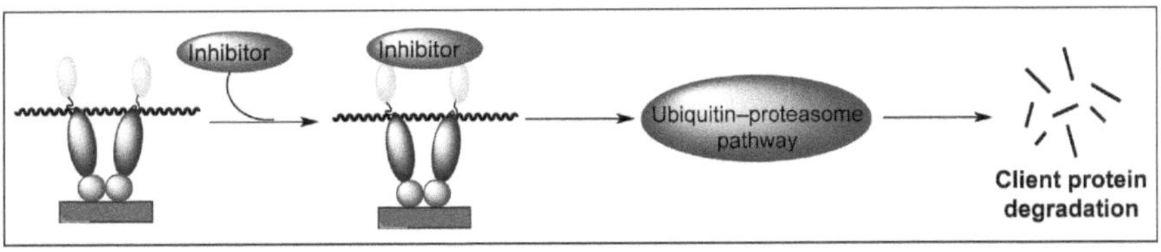

Figure 2 Rationale for Hsp90 inhibition as an anticancer strategy. Classical Hsp90 inhibitors compete with ATP for the nucleotide-binding domain of Hsp90 and halt the progression of the chaperone cycle. Consequently, the client protein is often directed to the ubiquitin-mediated degradation pathway. (See the color plate.)

in clinical trials and have been relatively well tolerated at drug exposures that clearly impair Hsp90 function as judged by induction of a systemic heat-shock response and depletion of client proteins in both tumor and normal tissues (Jhaveri, Taldone, Modi, & Chiosis, 2012; Neckers & Workman, 2012). Taken together, preclinical studies and a decade of clinical experience with various Hsp90 inhibitor chemotypes indicate that Hsp90 can be safely targeted for the development of cancer chemotherapeutics.

Classical Hsp90 inhibitors act by competitive binding to the ATP-binding site at the N-terminal domain of Hsp90, and consequently halt progression of the protein folding machinery, which leads to the degradation of most client proteins (Khandelwal, Crowley, & Blagg, 2016). The natural products, GDA and RDC, were the first Hsp90 inhibitors identified in the early 1990s. Upon their identification, both GDA and RDC served as starting points for various drug discovery programs (Bagatell & Whitesell, 2004), which ultimately led to the investigation of 17 distinct chemical entities in clinical trials (Neckers & Workman, 2012). Some of these investigational new drugs that inhibit the Hsp90 N-terminus are shown in Fig. 3. Clinical experience to date has provided a proof of concept for the use of Hsp90 inhibitors in cancer patients as a novel approach to inhibit multiple cancer pathways via Hsp90 modulation. Although there have been some encouraging clinical responses, concerns regarding concomitant induction of the prosurvival response, disruption of apoptotic mechanisms, impairment of antitumor immune mechanisms, cardiac arrhythmia, and hepatotoxicity have emerged from clinical trials (Whitesell & Lindquist, 2005). While isoform-selective N-terminal inhibitors may address some of the toxic liabilities, induction of the prosurvival

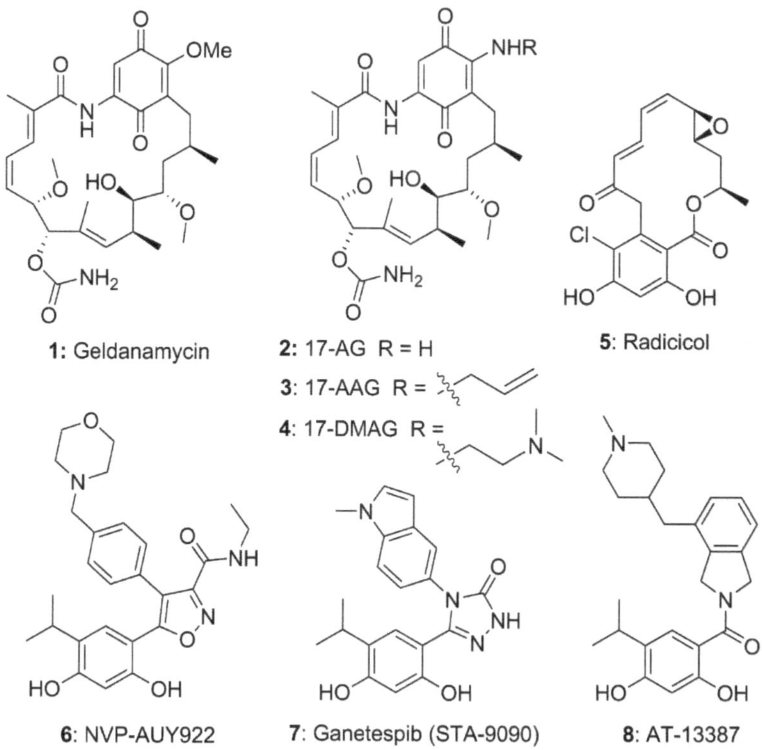

Figure 3 Structure of representative examples of Hsp90 N-terminal inhibitors.

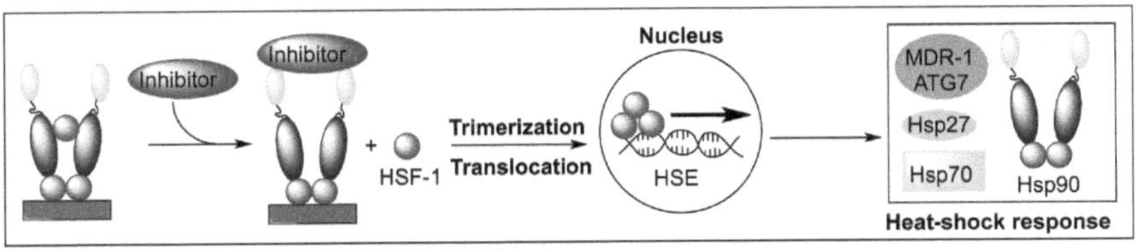

Figure 4 A proposed mechanism for induction of heat-shock protein expression by Hsp90 N-terminal inhibitors. Hsp90 inhibitors bind to the Hsp90 N-terminus, which result in the release of a transcription factor, HSF-1. Upon release, HSF-1 becomes trimerized, phosphorylated, and translocated to nucleus, wherein it binds to consensus sequences and upregulates many prosurvival mechanisms, including overexpression of prosurvival chaperones such as Hsp27, Hsp40, Hsp70, and Hsp90. (See the color plate.)

heat-shock response (Fig. 4) by inhibitors when given at or near their maximally tolerated dose may represent a fundamental impediment to their clinical efficacy (Whitesell & Lindquist, 2005). Therefore, Hsp90 inhibitors that do not induce the heat-shock response represent a promising new direction for the Hsp90 field of research.

In 2000, Neckers and coworkers discovered that NB, a coumarin antibiotic, bound to a previously unrecognized Hsp90 C-terminal nucleotide-binding site ($IC_{50} \sim 700$ μM in SKBr3 cells) and induced the degradation of Hsp90-dependent client proteins, v-src, Raf-1, and Erb2 (Marcu, Chadli, et al., 2000; Marcu, Schulte, & Neckers, 2000). Interestingly, NB did not induce a prosurvival heat-shock response, one of the major drawbacks associated with N-terminal inhibition. Moreover, it was observed that Hsp90 C-terminal inhibitors allosterically modulate the N-terminal ATPase activity. Encouraged by the initial findings, attempts were made to improve the selectivity and potency of NB. In 2005, initial structure–activity relationship studies were performed to reveal several key structural features of NB that are required for Hsp90 inhibitory activity (Burlison, Neckers, Smith, Maxwell, & Blagg, 2006; Yu et al., 2005). It was observed that both the coumarin 4-hydroxy and the 3′-carbamoyl group on NB are detrimental to Hsp90 inhibition and upon their removal, the first Hsp90 C-terminal inhibitors (DHN1 and DHN2) with improved potency and selectivity were reported by our group.

Subsequent SAR studies explored the coumarin core and benzamide side chain of NB and revealed optimal appendages for these moieties (Burlison et al., 2008; Donnelly et al., 2008; Garg, Zhao, & Blagg, 2015; Zhao et al., 2011, 2015). It was observed that the coumarin core of NB serves as a backbone for orientation of the sugar and benzamide side chains within the binding pocket and could be replaced with other aromatic/heteroaromatic cores (Burlison et al., 2008; Donnelly et al., 2008). In fact, introduction of a biphenyl or quinolinone ring system in lieu of the coumarin core not only improved efficacy but also provided insights into the nature of the C-terminal binding pocket (Zhao, Moroni, Colombo, & Blagg, 2013; Zhao et al., 2015). In addition, it was discovered that replacement of benzamide side chain with biaryl and triazole moieties led to analogs that manifested improved antiproliferative activities (Burlison et al., 2008; Zhao et al., 2014). Furthermore, recent studies indicate that the sugar moiety, although important for enhancing solubility (and hence activity), could be replaced with other sugars or sugar surrogates without compromising inhibitory activity (Donnelly, Zhao, Reddy Kusuma, & Blagg, 2010; Shelton et al., 2009; Zhao, Reddy Kusuma, & Blagg, 2010; Zhao et al., 2011). SAR studies on the NB scaffold have led to the development of several promising lead molecules such as KU135, KU174, and KU675, which manifest potent antiproliferative activity against multiple cancer cell lines (Donnelly et al., 2008; Eskew et al., 2011; Ghosh et al., 2015; Liu et al., 2015; Samadi et al., 2011; Zhao et al., 2010, 2011). A summary of SAR for NB and its derivatives based on their cytotoxicity is presented in Fig. 5.

4. NOVEL HSP90 INHIBITORS: BEYOND THE USUAL SUSPECTS

Current Hsp90 inhibitors derived from the natural products, GDA, RDC, or a purine scaffold have been reviewed extensively in literature (Hong et al., 2013; Khandelwal et al., 2016; Trepel et al., 2010). Although significant progress has been made toward the development of highly active N-terminal inhibitors, their clinical application has been hampered by undesired side effects in many cases. Several NB analogs have been developed during the past decade (Burlison et al., 2006; Donnelly & Blagg, 2008; Zhao et al., 2011), and these have shown promising results in preclinical studies. Their evaluation in clinical trials, however, has not yet been undertaken. More recently, new compounds have been identified that disrupt Hsp90 chaperone activity via yet other mechanisms (Brandt & Blagg, 2009; Piaz, Terracciano, De Tommasi, & Braca, 2015). These compounds can be broadly divided into two main categories: (1) direct Hsp90 inhibitors or (2) disruptors of Hsp90/cochaperone interactions, both of which will be discussed in detail below.

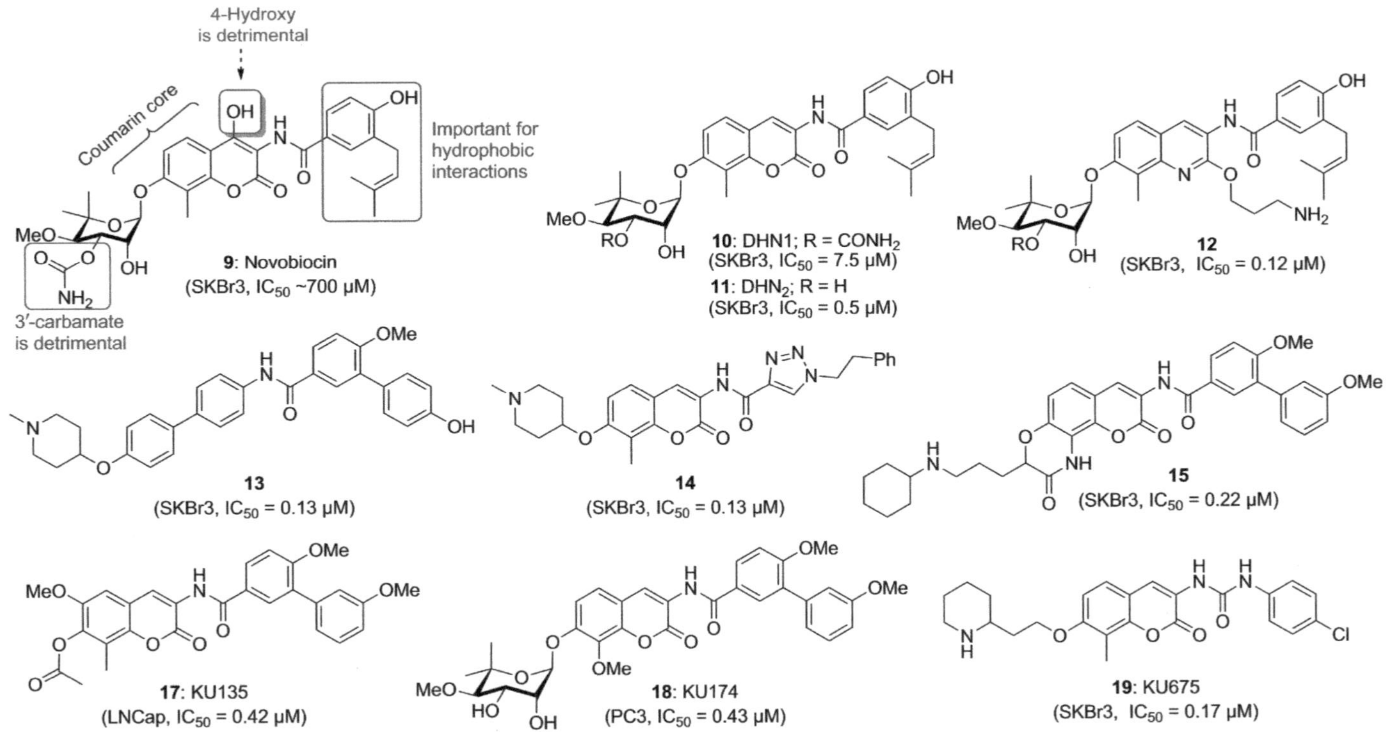

Figure 5 Structure–activity relationships for novobiocin and its derivatives. (See the color plate.)

4.1 Compounds that Bind Directly to Hsp90

4.1.1 Epigallocatechin-3-Gallate

EGCG is a polyphenolic compound found in green tea and is well known for its antioxidant, antimicrobial, and anticancer activities (Zaveri, 2006). Previous studies have shown that EGCG inhibits the activity of a wide range of proteins at 70 μM, including telomerase, the aryl hydrocarbon receptor (AhR), several kinases, and transcription factors, all of which are well-known Hsp90 client proteins (Yin, Henry, & Gasiewicz, 2008). In 2005, using affinity purification experiments, Palermo and coworkers revealed that EGCG exhibits its inhibitory activity against AhR, at least in part via Hsp90 inhibition (Palermo, Westlake, & Gasiewicz, 2005). Subsequent studies by Yin and coworkers demonstrated that EGCG binds near the C-terminal ATP-binding site (residues 538–738) of Hsp90 and unlike NB and other inhibitors, stabilizes the association of cochaperones Hsp70, Cyp40, and XAP-2 to Hsp90 (Yin et al., 2008). Furthermore, it was found that EGCG induces concentration-dependent degradation of the Hsp90-dependent oncoproteins ErbB2, Raf-1, and pAkt along with a slight increase in Hsp70 levels.

Recently, Khandelwal and coworkers published the first structure–activity relationships studies on EGCG using cytotoxicity and the depletion of several Hsp90 client proteins as endpoints (Khandelwal, Hall, & Blagg, 2013). Results are summarized in Fig. 6. In these studies, they observed that the phenols on the B- and the D-rings are detrimental to inhibitory activity, while syn-stereochemistry of the linker that connects the B- and D-rings with the benzopyran core is beneficial. The prenylated benzamide present in NB was shown to represent an ideal replacement for the gallic acid moiety of EGCG and resulted in ~15-fold improvement in antiproliferative activity and ultimately led to the development of compound **21** (MCF-7, $IC_{50}=4$ μM). Further studies by Bhat and coworkers confirmed the non-essential nature of the B- and D-ring phenols and established the ester linker connecting the C- and D-rings could be replaced with an amide or sulfonamide without compromising anticancer activity (Bhat et al., 2014). How these modifications affect interactions with Hsp90 remain unknown.

4.1.2 Silybin

Silybin is the major component of the flavonolignan extract isolated from the seed of milk thistle plants (*Silybum marianum*) and has demonstrated hepato-protective effects and growth inhibitory activity against various

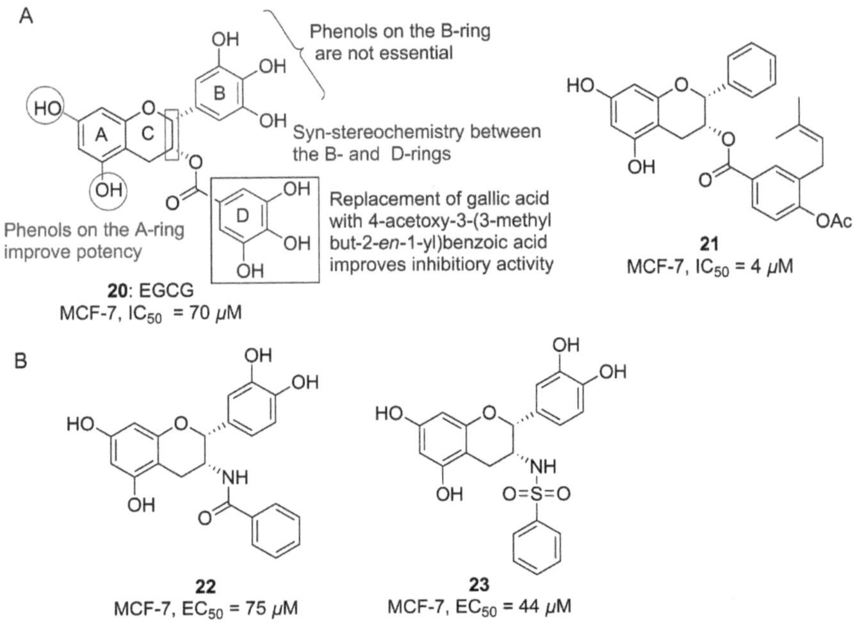

Figure 6 Summary of cytotoxicity structure–activity relationships for EGCG and its analogs. (See the color plate.)

cancer cells (Gazak, Walterova, & Kren, 2007). Early studies demonstrated that silybin induced cell cycle arrest and caused the depletion of CDK2, CDK4, cyclin E, and cyclin D1 proteins in colon cancer cells (Agarwal et al., 2003). CDK2 and CDK4 are well-known Hsp90-dependent clients, which led to speculation that Hsp90 could be the primary target of silybin. In an effort to determine whether silybin can bind Hsp90, Zhao and coworkers performed a luciferase-refolding assay with silybin and demonstrated that silybin inhibited the renaturation of heat-denatured luciferase, suggesting that Hsp90 could be a biochemical target for silybin (Zhao, Brandt, Galam, Matts, & Blagg, 2011). Subsequent studies demonstrated that silybin induced a concentration-dependent degradation of the Hsp90-dependent client proteins Her2, Raf-1, and Akt, without affecting Hsp70 or Hsp90 levels (Zhao, Brandt, et al., 2011). SAR studies by the same research team identified key structural features required for the scaffold's cytotoxic activity in which Hsp90 inhibition could play a part (Fig. 7; Zhao, Brandt, et al., 2011). Their studies showed that the C-3 and C-23 hydroxyl groups were not required for activity; however, at least one substitution (preferably 4-hydoxyl) on the E-ring was important for activity. Furthermore, SAR studies suggested that

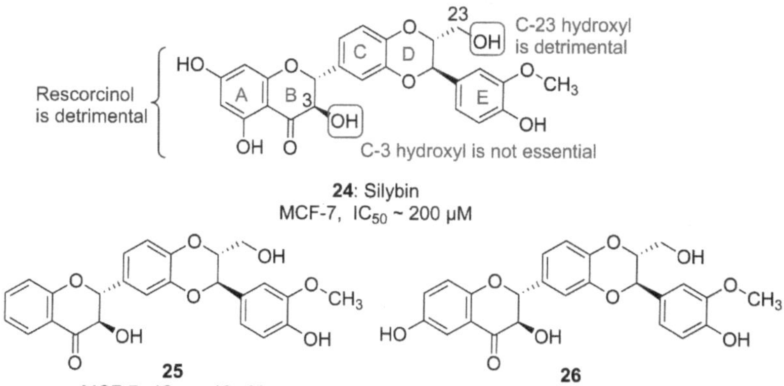

Figure 7 Summary of SAR for silybin and its analogs. (See the color plate.)

the A-ring phenol was not required and its removal led to the development of compounds **25** and **26**, which manifest IC_{50}s of 13 and 16 μM against MCF-7 cell line, respectively, versus an IC_{50} of ~200 μM for silybin. Recently, Riebold and coworkers demonstrated that silybin binds to the C-terminal domain of Hsp90 and releases mature glucocorticoid receptors from the Hsp90 complex as demonstrated by NMR analysis (Riebold et al., 2015).

4.1.3 Cisplatin and LA-12

Cisplatin (Fig. 8) is a platinum-containing chemotherapeutic agent that is widely used for the treatment of ovarian, testicular, bladder, cervical, and other solid tumors (Galanski, 2006). The anticancer activity of cisplatin has been ascribed to its ability to form intrastrand and/or interstrand DNA adducts which are particularly lethal in several cancer cell types (Jordan & Carmo-Fonseca, 2000). However, it has also been shown that due to its chemical reactivity, cisplatin interacts with various proteins, phospholipids, and RNA (Sreedhar, Soti, & Csermely, 2004). In 1999, Itoh and coworkers reported that cisplatin inhibits Hsp90 chaperone activity (Itoh et al., 1999). Affinity purification and protein fingerprinting studies were used to demonstrate that cisplatin binds to the Hsp90 C-terminal domain. Subsequently, Csermely and coworkers demonstrated that cisplatin is a C-terminal inhibitor that binds near the previously identified C-terminal nucleotide-binding site (Söt, Rácz, & Csermely, 2002). Studies by Rosenhagen and colleagues indicated that the administration of cisplatin to neuroblastoma cells resulted in the degradation of steroid hormone

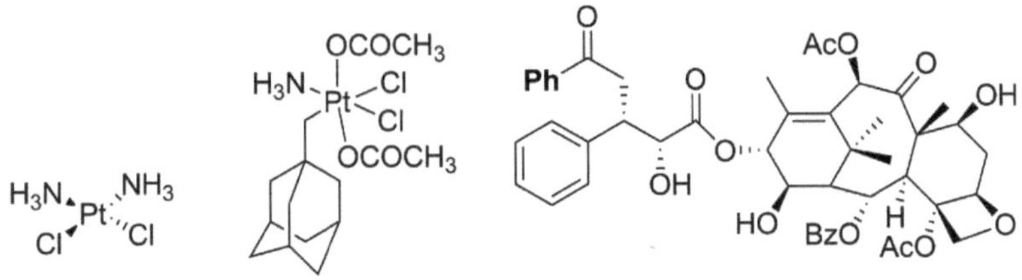

27: Cisplatin **28**: LA-12 **29**: Taxol

Figure 8 Structures of cisplatin, LA-12, and taxol.

receptors (androgen and glucocorticoid receptors), but no other Hsp90-dependent clients, such as Raf-1, lck, and c-rac (Rosenhagen et al., 2003). Moreover, by use of a heat-shock factor (HSF)-dependent luciferase reporter assay, they showed that cisplatin does not induce the heat-shock response. These results suggest that unlike compounds that bind Hsp90, cisplatin selectively inhibits some Hsp90 functions and thus, could provide insights into novel ways to modulate its chaperone activity. Recently, it was shown that LA-12 (Fig. 8), an optimized derivative of cisplatin, exhibits higher affinity for Hsp90 than cisplatin and moreover, induces the degradation of additional Hsp90 client proteins, such as mutant p53, Cyclin D1, and estrogen receptors (Kvardova et al., 2010). In addition, LA-12 exhibits a more favorable pharmacokinetic profile as compared to cisplatin and demonstrates enhanced cytotoxicity against multiple cancer cell lines, including those that are cisplatin resistant (Kvardova et al., 2010; Zak et al., 2004).

4.1.4 Taxol

Taxol (Fig. 8) is a frequently used chemotherapeutic agent for the treatment of various cancers and its anticancer activity has been attributed to the inhibition of mitosis via stabilization of microtubules (Wani, Taylor, Wall, Coggon, & McPhail, 1971). In addition, taxol produces many lipopolysaccharide (LPS)-like cellular responses, such as induction of cytokines, activation of kinases, and transcription factors, suggesting that it exhibits a multifaceted effect on cancer cells (Ding, Porteu, Sanchez, & Nathan, 1990; Ding, Sanchez, & Nathan, 1993). Byrd and coworkers performed affinity purification experiments with biotinylated taxol and identified Hsp90 and Hsp70, as potential mediators of its LPS-mimicking activity (Byrd et al., 1999). Surprisingly, unlike classical Hsp90 inhibitors (e.g., GDA), taxol appears to stimulate Hsp90 function and induces macrophage activation (Byrd et al., 1999). No study describing the region in which taxol binds Hsp90 has been reported. Further work is needed to investigate the effect of taxol on Hsp90-dependent client proteins and to determine whether its binding to Hsp90 causes disruption of cochaperone interactions. Interestingly, a follow-up study showed that taxol and 17-AAG act synergistically in breast cancer xenografts and that 17-AAG sensitizes cancer cells to taxol-induced apoptosis through suppression of the Hsp90 client, Akt kinase (Solit, Basso, Olshen, Scher, & Rosen, 2003). Combination therapies with taxol and other Hsp90 inhibitors may represent new avenues for cancer chemotherapy, but benefit has yet to be demonstrated in the clinic. A recent phase-3 study of ganetespib in combination with the taxane, docetaxel, in recurrent lung cancer (GALAXY 2) failed to show any significant clinical benefit and was terminated.

4.1.5 Sansalvamide A-Amide

Sansalvamide A (San A) is a depsipeptide isolated from a marine fungus of the genus *Fusarium* and exhibits moderate antitumor activity (IC_{50} 45 μg/mL against HCT-29 colon cancer cells) (Belofsky, Jensen, & Fenical, 1999). San A is a cyclic pentapeptide containing a lactone moiety (Fig. 9), which is susceptible to ring opening by the esterases present in plasma and in cells. In an effort to improve stability, Silvermann and coworkers synthesized a peptide derivative of San A, sansalvamide A-amide (San A-amide), which was found to be 10-fold more potent (IC_{50} 4.5 μg/mL) than the natural product (Gu, Liu, & Silverman, 2002; Sellers et al., 2010). Biochemical studies by Vasko and coworkers revealed that San A-amide binds the N-MD of Hsp90 and disrupts Hsp90 chaperone activity (Vasko et al., 2010). Interestingly, like 17-AAG, San A-amide induces Hsp70 levels (Ardi, Alexander, Johnson, & McAlpine, 2011); however, it shows no effect on the binding of the client protein Her2, suggesting a unique mechanism of action for this compound (Vasko et al., 2010). In addition, it was found that San A-amide disrupts interactions between Hsp90 and various C-terminal domain-binding cochaperones, including IP6K2, FKBP52, and HOP, suggesting an allosteric mechanism for its modulation of Hsp90 function (Kunicki et al., 2011; Vasko et al., 2010). Moreover, San A-amide shows no effect on Hsp90 ATPase activity and preferentially binds to the closed conformation of Hsp90, further supporting the notion that San A-amide, among many other biological activities might act as an allosteric modulator of Hsp90 function (Alexander, Partridge, Agard, & McAlpine, 2011).

Over the past decade, McAlpine and coworkers have conducted several structure–activity relationship studies on San A-amide and have developed a

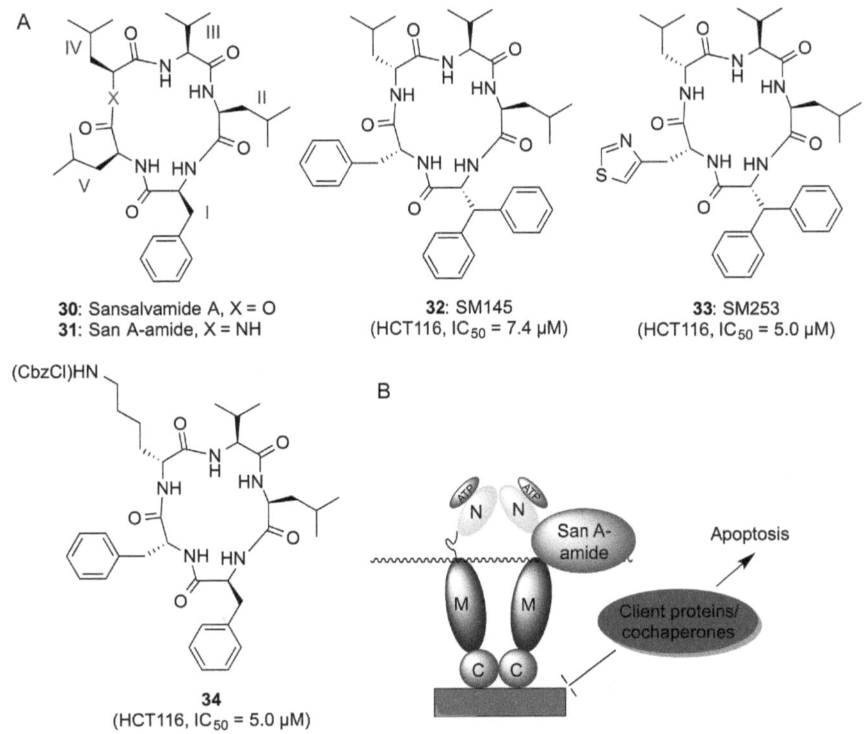

Figure 9 (A) Structures of sansalvamide A and its analogs. (B) Proposed mechanism of action of sansalvamide A derivatives. (See the color plate.)

number of analogs that manifest potent cytotoxicity against several cancer cell lines, including pancreatic, breast, prostate, and colon (Ardi et al., 2011; Carroll et al., 2005; Davis et al., 2012; Rodriguez et al., 2007; Sellers et al., 2010). Like San A-amide, these compounds have been reported to allosterically inhibit interaction between Hsp90 and multiple TPR-containing proteins, and also, induce caspase-dependent apoptosis in cancer cells. Interestingly, some of these analogs (**32** and **33**) manifest anti-proliferative activities without inducing the heat-shock response, a major drawback associated with the parent compounds (Koay et al., 2014; McConnell, Alexander, & McAlpine, 2014). Recently, Ramsey and coworkers reported a novel San A-amide derivative, **34**, which induces apoptosis and interacts with Hsp90 in biochemical pull-down assays, but has no effect on interaction between Hsp90 and C-terminal domain proteins, suggesting a novel mechanism by which it might modulate Hsp90 function (Ramsey et al., 2012).

4.1.6 Deguelin and L80

Deguelin is a naturally occurring flavonoid isolated from *Derris trifoliata* Lour. or *Mundulea sericea* (Leguminosae). It has demonstrated potent antiproliferative, antimetastatic, and apoptotic activity against several cancers both *in vitro* and *in vivo* (Chang et al., 2012; Hastings, Hadden, & Blagg, 2008; Lee et al., 2015). A number of reports indicate deguelin induces cell death by inhibiting several cell signaling pathways, such as PI3K-Akt, IKK-IκBα-NF-κB, AMPK-mTOR-survivin, and HIF-1α-VEGF. In 2007, Oh and coworkers reported that deguelin disrupts interactions between Hsp90 and its client protein, HIF-α (Oh et al., 2007). Subsequent biochemical analysis and molecular docking studies suggested that deguelin binds the C-terminal ATP-binding pocket of Hsp90 and suppresses Hsp90 function, which leads to proteasome-mediated degradation of Hsp90 client proteins, without inducing Hsp90 expression (Lee et al., 2015). In preclinical studies, administration of deguelin significantly reduced tumor growth by inducing apoptosis. However, it was observed that deguelin produces Parkinson's disease-like syndrome in rats at high doses, which may limit its therapeutic application (Caboni et al., 2004). In an attempt to circumvent this detrimental feature and to develop simpler analogs, Chang and coworkers reported structure–activity relationships studies for deguelin using HIF-1α reduction and cytotoxicity as endpoints (Fig. 10; Chang et al., 2012). Their studies revealed that the 2,2-dimethyl-2*H*-chromene moiety and both methoxy groups at the C9 and C10 positions of deguelin are critical for these activities. The SAR insights reported led to the development of compounds **36** and **37**, which have cytotoxicity IC_{50}s of 0.14 and 0.49 μM, against H1299 cell line, respectively. The extent to which these effects result from Hsp90 inhibition remain unclear. Recently, Lee and coworkers reported a novel deguelin derivative, L80 that manifests antiproliferative and apoptotic activities both *in vitro* and *in vivo* without systemic toxicity (Lee et al., 2015). Consistent with earlier observations for this scaffold, L80 was found to bind

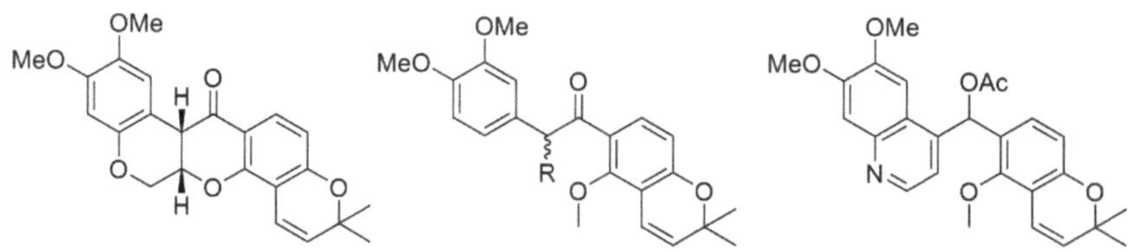

35: Deguelin (H1299, IC_{50} = 0.11 μM)

36: R = H, (H1299, IC_{50} = 0.14 μM)
37: R = (S)-Me, (H1299, IC_{50} = 0.49 μM)

38: L80

Figure 10 Structures of deguelin, L80, and its derivatives.

directly to Hsp90 in biochemical assays and disrupt the Hsp90 chaperone cycle. Computational studies suggest L80 might form key interactions with Ser677 and Lys615 within the Hsp90 C-terminal domain. However, the exact mechanism remains unclear in cells.

4.2 Disruptors of Hsp90 Interaction with Cochaperones and Client Proteins

Currently, Hsp90 inhibitors undergoing clinical evaluation are pan-inhibitors and induce the degradation of many Hsp90-dependent client proteins (Jhaveri et al., 2012; Patel et al., 2013). Although pan-Hsp90 inhibition could be beneficial by providing a multifaceted attack on cancer cells, it may also produce detrimental side effects. For example, it has been found that inhibition of the Hsp90-dependent trafficking of cardiac potassium channel hERG could be responsible for the cardiac arrhythmias seen in clinical trials of some Hsp90 inhibitors (Peterson, Eskew, Vielhauer, & Blagg, 2012). Consequently, there is growing interest in identifying new Hsp90 modulators that could provide alternative mechanisms of action without the negative effects of pan-inhibition. Over the past decade, new compounds have emerged that disrupt interactions between Hsp90 and its cochaperones (Brandt & Blagg, 2009; Hall, Forsberg, et al., 2014; Piaz et al., 2015). These compounds not only represent new chemotypes for the development of future Hsp90 inhibitors but also appear to block the maturation of more restricted subsets of Hsp90 client proteins, which could limit target-related toxicities induced by pan-inhibitors. Several natural products, such as celastrol, gedunin, and cruentaren A, have been shown, among a plethora of other biochemical and biological effects to disrupt interactions between Hsp90 and its partner proteins.

4.2.1 Celastrol, a Disruptor of Cdc37–Hsp90 Interaction

Celastrol (Fig. 11) is a pentacyclic triterpenoid isolated from the root extract of *Tripterygium wilfordii* Hook F. (as known as Thunder of God Vine) and has been used for centuries in oriental traditional medicine to treat inflammatory and autoimmune disorders, such as rheumatoid arthritis (Allison, Cacabelos, Lombardi, Álvarez, & Vigo, 2001; Duan et al., 2000; Salminen, Lehtonen, Paimela, & Kaarniranta, 2010; Tao & Lipsky, 2000). In recent years, a renewed interest in the therapeutic application of celastrol has increased due to its diverse biological activities, especially for the treatment of various inflammatory diseases and cancer (Allison et al., 2001; Duan et al., 2000; Kannaiyan et al., 2011). Studies have indicated that celastrol manifests

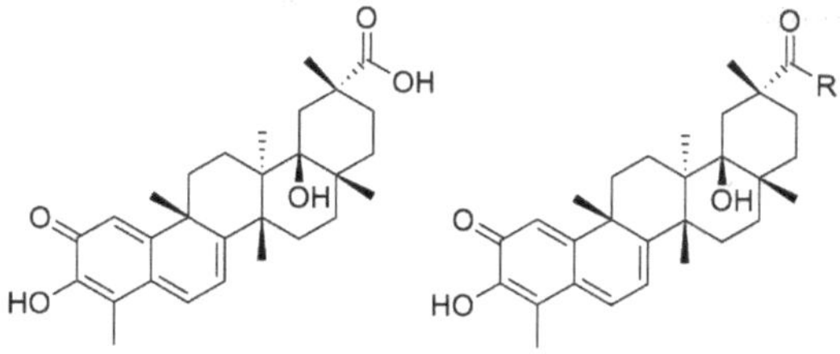

39: Celastrol
(Hep3B HCC, IC_{50} = 0.30 μM)

40: R =Me
(Hep3B HCC, IC_{50} = 0.85 μM)

41: R = -$NHCH_2CH_2Ph$
(Hep3B HCC, IC_{50} = 1.06 μM)

42: R = -OCH_2CH_2Ph
(Hep3B HCC, IC_{50} = 2.77μM)

Figure 11 Structures of celastrol and its derivatives.

cytotoxicity against different cancer cell lines, including prostate, multiple myeloma, lung, gliomas, pancreatic, and cervical cancers (Liu, Ma, & Zhou, 2011; Liu et al., 2010; Yang, Chen, Cui, Yuan, & Dou, 2006; Zhou & Huang, 2009). In addition, it suppresses metastatic invasion, induces apoptosis (Kannaiyan et al., 2011; Sethi, Ahn, Pandey, & Aggarwal, 2007), and sensitizes drug-resistant cancer cells to combination therapy (Chen, Rose, Doudican, Osman, & Orlow, 2009). Celastrol is a quinone methide triterpene and exhibits a high propensity to form covalent Michael adducts with cysteine residues (Salminen et al., 2010). In fact, studies with celastrol have identified numerous intracellular targets with relevance to cancer, such as NF-κB/IKKβ (Sethi et al., 2007), the proteasome (Yang et al., 2006), topoisomerase II (Nagase et al., 2003), and a variety of signaling pathways that are essential to the survival of cancer cells (Liu et al., 2011).

In 2006, Lamb and coworkers developed a novel chemical genomic approach called connectivity maps to discover and predict the biological pathways targeted by various anticancer agents (Lamb et al., 2006). Briefly, these researchers generated a database of gene expression changes resulting from the exposure of tumor cells to drugs with well-characterized modes of action. Using this database to identify compounds with similar modes of action, Hieronymus and coworkers reported that celastrol exerts its antiproliferative activity, at least in part via disruption of Hsp90-related pathways (Hieronymus et al., 2006). Since this initial work, extensive research has been conducted to investigate the effect of celastrol on Hsp90. Using molecular docking studies and coimmunoprecipitation assays, Zhao and coworkers revealed that celastrol disrupts the association between Hsp90 and Cdc37, which leads to the degradation of Hsp90-dependent client kinases, such as Akt and Cdk4 (Zhang et al., 2008). It was also observed that celastrol induces the heat-shock response by activation of HSF-1

(Westerheide et al., 2004). In another study, Chadli and coworkers showed that celastrol can inactivate the cochaperone p23 and cause amyloid-like fibril formation, which in turn halts the chaperoning of steroid hormone receptors (Chadli et al., 2010). Early studies indicated that celastrol binds the Hsp90 C-terminal domain and allosterically modulates its chaperoning activity (Zhang et al., 2009). However, HSQC NMR-based studies by Sreeramulu and coworkers suggested that celastrol disrupts Hsp90/Cdc37 interactions by covalently binding to cysteine residues on Cdc37, without direct interactions with Hsp90 (Sreeramulu, Gande, Göbel, & Schwalbe, 2009). Recently, Zanphorlin and coworkers conducted detailed studies on celastrol and proposed a new model for celastrol/Hsp90 binding, suggesting that celastrol binds the C-terminal domain of Hsp90 and interferes with Hsp90 function by induction of its oligomerization (Zanphorlin, Alves, & Ramos, 2014). Taken together, these studies suggest that celastrol may modulate Hsp90 function through multiple mechanisms. Additional studies are needed to further clarify the exact nature of its effects on Hsp90, but promiscuity limits its utility as a probe for studying Hsp90 function specifically in whole cells.

In 2014, Wei and coworkers performed a limited structure–activity relationship study on the celastrol scaffold using cytotoxicity and depletion of kinase levels as endpoints (summarized in Fig. 11; Wei et al., 2014). Seven celastrol derivatives were prepared and their antitumor activity evaluated against human hepatocellular carcinoma (HCC) cell line *in vitro* and in HCC patient-derived xenografts. These celastrol derivatives were shown to deplete cellular levels of protein kinases involved in the Raf/MEK/ERK and PI3K/AKT/mTOR signaling pathways and induce apoptosis. Although no derivative was found more active than the natural product (cytotoxicity IC_{50}, 0.30 μM against Hep3B HCC cell line), it was revealed that modifications to the carboxylic acid moiety of celastrol were tolerated.

4.2.2 Gedunin, a Disruptor of Hsp90–p23 Interaction

Gedunin (Fig. 12) is a tetranortriterpenoid isolated from the Indian neem tree (*Azadirachta indica*, Meliacae) and has been used for the treatment of malaria and other infectious diseases in traditional Indian medicine (Patwardhan et al., 2013). In addition, gedunin has demonstrated antiproliferative activity against various cancer cell line including prostate, colon, and ovarian (Hieronymus et al., 2006; Kamath et al., 2009; Uddin et al., 2007). Like celastrol, gedunin is a strong, thiol-reactive electrophile that

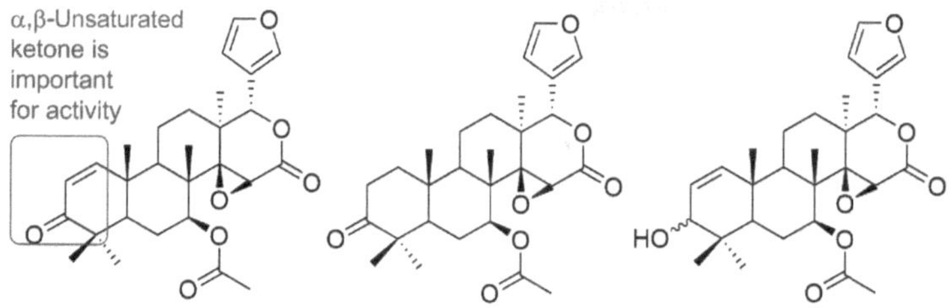

43: Gedunin (MCF-7, IC_{50} = 8.84 μM) **44** (MCF-7, IC_{50} >100 μM) **45** (MCF-7, IC_{50} = 54.3 μM)

Figure 12 Summary of SAR for gedunin and derivatives. (See the color plate.)

activates the heat-shock response. In 2006, Hieronymus and coworkers used connectivity map analysis to report that gedunin exerts its antiproliferative activity at least in part via modulation of Hsp90-dependent pathways, which results in the depletion of cellular levels of Hsp90-dependent client proteins (Hieronymus et al., 2006; Lamb et al., 2006). Subsequent biochemical studies showed that gedunin inhibits Hsp90 ATPase activity and disrupts the Hsp90 chaperone cycle. However, unlike most Hsp90 inhibitors, gedunin was unable to compete with GDA for binding to the N-terminal ATP-binding pocket in florescence polarization assays, suggesting a novel mechanism for Hsp90 modulation (Hieronymus et al., 2006). Recent studies by Patwardhan and coworkers revealed that gedunin binds to the cochaperone p23 and blocks its interaction with Hsp90, which leads to deactivation of the Hsp90 folding machinery (Patwardhan et al., 2013). Interestingly, unlike GDA, gedunin induced relatively modest overexpression of Hsp70. Furthermore, it was observed that gedunin selectively destabilizes steroid receptors such as GR and induces apoptotic cell death through the activation of caspase 7. In 2008, Brandt and coworkers synthesized a series of compounds with chemical modifications to the gedunin scaffold and revealed key structural features that are required for cytotoxic activity (Brandt, Schmidt, Prisinzano, & Blagg, 2008). Nineteen semisynthetic derivatives of gedunin were prepared and their antiproliferative activity evaluated against MCF-7 and SKBr3 breast cancer cells. No analog was found to be more active than the natural product (MCF-7, IC_{50}=8.84 μM). Furthermore, it was shown that the α,β-unsaturated ketone within the A-ring, although important for antiproliferative activity, did not necessarily serve as a Michael acceptor and therefore, may avoid potential toxicities associated with such motifs.

4.2.3 Withaferin A, a Disruptor of Hsp90–Cdc37 Interaction

Withaferin A (WA, Fig. 13) is a withanolide isolated from the Indian medicinal plant of *Withania somnifera* (commonly known as "Ashwagandha" or "Indian Winter Cherry" in Ayurvedic medicine) and possesses diverse biological activities, such as anti-inflammatory (Kaileh et al., 2007), antistress, antioxidant, immunomodulatory (Mishra, Singh, & Dagenais, 2000), anti-angiogenesis (Mohan et al., 2004), and anticancer activities (Yang, Shi, & Dou, 2007). Since its discovery in the late 1960s, withaferin A has been extensively studied for its anticancer activity, and numerous mechanisms and molecular targets proposed (Falsey et al., 2006; Kaileh et al., 2007; Shohat, Gitter, Abraham, & Lavie, 1967; Srinivasan, Ranga, Burikhanov, Han, & Chendil, 2007; Yang et al., 2007; Yokota, Bargagna-Mohan, Ravindranath, Kim, & Mohan, 2006). It has been reported that withaferin A inhibits nuclear factor-κB (NF-κB) activation of IκB kinase via a thioalkylation-sensitive redox mechanism (Kaileh et al., 2007) induces apoptosis in prostate cancer cells through Par-4 induction (Srinivasan et al., 2007), targets β5 subunit of tumor proteasome (Yang et al., 2007), and covalently binds to Annexin II to alter cytoskeletal architecture (Falsey et al., 2006). In 2010, Yu and coworkers demonstrated that withaferin A exhibits antiproliferative activity and inhibits Hsp90 in pancreatic cells where it was reported to deplete cellular levels of Hsp90-dependent client proteins (Akt, Cdk4, and GR) (Yu et al., 2010). In addition, it was observed that withaferin A induces Hsp70 expression, without affecting Hsp90 levels. Moreover, these researchers found that withaferin A binds Hsp90 and halts the Hsp90 chaperone cycle through a novel ATP-independent mechanism. To identify the domain to which withaferin A binds in Hsp90, a pull-down assay using WA-biotin was used, which suggested interaction with the chaperone's C-terminal domain (Yu et al., 2010). Coimmunoprecipitation

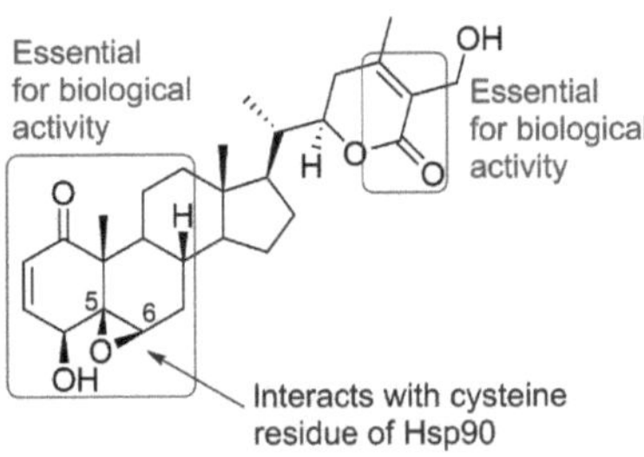

46: Withaferin A
(Panc1, IC_{50} = 1.24 μM)

Figure 13 Summary of SAR for withaferin A. (See the color plate.)

studies showed that withaferin A disrupts formation of the Hsp90/Cdc37 complex in pancreatic cancer cells (Yu et al., 2010). Structure–activity relationship studies have identified a pharmacophore of WA that involves the 4-hydroxy-5,6-epoxy-22-*en*-1-one moiety and its unsaturated lactone as critical for cytotoxic activity (Mohan et al., 2004; Yousuf et al., 2011). Recent studies with the withanolides indicate that the 5,6-epoxide may react with reactive cysteine residues in Hsp90 and induce aggregation, leading to disruption of Hsp90 function (Gu et al., 2014).

4.2.4 Derrubone, a Disruptor of Hsp90–Cdc37 Interaction

Derrubone is a prenylated isoflavone that was originally isolated from the Indian tree *Debrris robusta* in 1969 (East, Ollis, & Wheeler, 1969). However, its biological activities remained uncharacterized until recently. In 2007, high-throughput screening of a library of natural products identified derrubone as a potential Hsp90 inhibitor (Hadden, Galam, Gestwicki, Matts, & Blagg, 2007). The screening was based on the ability of natural products to inhibit the Hsp90-dependent refolding of thermally denatured firefly luciferase (Galam et al., 2007). In this study, derrubone potently inhibited refolding with an IC_{50} value of 0.23 ± 0.04 μM. Subsequent cellular studies revealed that it exhibits antiproliferative activity against various cancer cell lines (MCF-7 $IC_{50} = 11.9$ μM; HCT116 $IC_{50} = 13.7$ μM), and depletes cellular levels of Hsp90-dependent client proteins, including Her2, Raf-1, Akt, and ERα, without altering Hsp90 levels (Hadden et al., 2007; Mays, Hill, Moyers, & Blagg, 2010). Using purified recombinant Hsp70, Hadden and coworkers demonstrated that derrubone has no effect on Hsp70 ATPase activity, suggesting Hsp90 inhibition as a plausible mechanism for its inhibitory activity on refolding (Hadden et al., 2007). Moreover, it was observed that derrubone inhibits Hsp90 function by stabilizing the Hsp90 hetero-complex (Hsp90/cochaperone/client complex) formed between Hsp90, Cdc37, and heme-regulated eIF2α kinase (HRI), and consequently, halts progression of the chaperone cycle. In follow-up, a small library of derrubone analogs was prepared and evaluated to elucidate the critical structural features of derrubone for cytotoxicity and Hsp90 client protein depletion (Hastings et al., 2008; Mays et al., 2010). These structure–activity relationship studies revealed the importance of the 6-prenyl and 3-aryl side chains for activity (Fig. 14). Recent molecular docking studies by Khalid and coworkers suggest derrubone binds to the Hsp90 C-terminal domain and interacts with Leu665, Leu666, and Leu694 (Khalid & Paul, 2014).

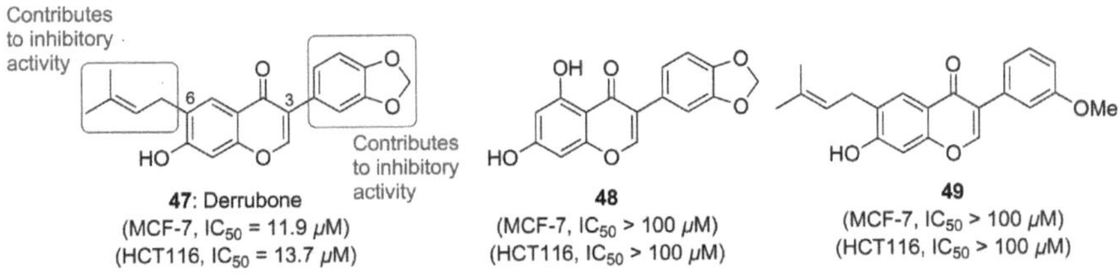

Figure 14 Summary of cytotoxicity SAR for derrubone and its derivatives. (See the color plate.)

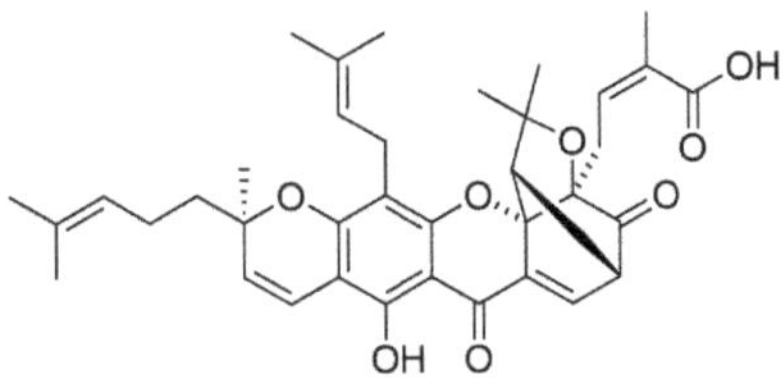

50: Gambogic acid
(MCF-7, IC_{50} = 2.0 μM)
(SKBr3, IC_{50} = 0.8 μM)

Figure 15 Structure of gambogic acid.

4.2.5 Gambogic Acid, a Disruptor of Hsp90–Cdc37 Interaction

Gambogic acid (Fig. 15) is a xanthonoid isolated from the exudate of *Garcinia hanburyi* Hook f. (Clusiaceae) and has been used for centuries to treat infections and tumors (Ren et al., 2011). In recent decades, interest in gambogic acid as a potential anticancer agent has increased because it demonstrates antitumor, antiangiogenic, and antimetastatic activities against multiple cancer cell lines (Ren et al., 2011). Recently, it entered phase II clinical trials in China for metastatic cancers (Chi et al., 2013). Several studies have shown that gambogic acid exerts its anticancer activity via numerous targets and signaling pathways, such as apoptosis induction (Pandey et al., 2007), antiangiogenesis (Yi et al., 2008), inhibition of human topoisomerase-Iiα (Qin et al., 2007), and telomerase (Zhao et al., 2008). In 2010, Zhang and coworkers ascribed the antiproliferative activity of gambogic acid to Hsp90 inhibition in HeLa cells (Zhang et al., 2010). Using fluorescence-quenching assays and spectroscopic tools, they demonstrated that gambogic acid binds the Hsp90 N-terminal domain and inhibits its ATPase activity. In addition, it was found that gambogic acid causes down-regulation of the TNF-α/NF-κB signaling pathway, which in turn induces apoptosis in HeLa cells. Their findings were further supported by a contemporaneous study by Davenport and coworkers, which demonstrated that gambogic acid inhibits Hsp90-dependent refolding of thermally denatured luciferase in a high-throughput screening assay previously developed by the same group (Davenport et al., 2011; Galam et al., 2007). In these studies, it was found that gambogic acid inhibits the proliferation and survival of two breast cancer cell lines (MCF-7, IC_{50} 2.0 μM and SKBr3, IC_{50} 0.8 μM), and depletes cellular levels of the Hsp90-dependent client proteins Her2, Akt, and Raf-1 in a concentration-dependent manner. Importantly, it was observed that gambogic acid induces Hsp90 and Hsp70 expression, a hallmark of Hsp90 N-terminal inhibition, but also a general feature of many thiol-reactive compounds. Like celastrol, gambogic acid was found to block the association of Hsp90, Hsp70, and Cdc37 with HRI. Recent surface plasmon resonance (SPR) analysis and virtual docking studies suggest gambogic acid binds the Hsp90 N-terminal domain; however, it does not compete with GDA for binding, suggesting a site of interaction distinct from the ATP-binding pocket (Davenport et al., 2011).

4.2.6 Cruentaren A, a Hsp90/F1F0 ATP Synthase Disruptor

Cruentaren A (Fig. 16) is a macrolide isolated from the myxobacterium *Byssovorax cruenta* which is highly cytotoxic to various cancer cell lines (Kunze, Sasse, Wieczorek, & Huss, 2007). It has been reported that cruentaren A exerts its cytotoxicity through selective inhibition of F_1F_0-ATP synthase, the enzyme responsible for the mitochondrial production of ATP (Kunze et al., 2007, 2006). In 2006, Papathanassiu and coworkers discovered that F_1F_0-ATP synthase can act as an Hsp90 cochaperone that provides the energy required for the maturation of client proteins (Papathanassiu, MacDonald, Bencsura, & Vu, 2006). They also demonstrated that fungal peptides, known as efrapeptins, disrupt interaction between Hsp90 and F_1F_0-ATPase synthase, which leads to the depletion

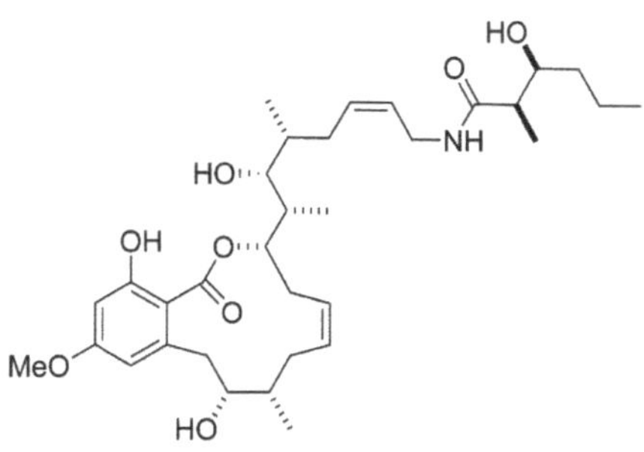

51: Cruentaren A

EC_{50} values of cruentaren A

Cancer cell lines

MCF-7 = 8 nm
A549 = 0.19 nm

Normal human cells

HEK293 >500 nm
MRC5 >500 nm

Figure 16 Structure and cytotoxic activity of cruentaren A.

of Hsp90-dependent client proteins, including ERα, mutated p53, and caspase-3, along with downregulation of Hsp27, Hsp70, and even Hsp90 levels. However, the complex peptide structure and promiscuous nature of efrapeptins render them unsuitable for further development. Recently, Hall and coworkers demonstrated that incubation of cruentaren A, a more selective F_1F_0-ATP synthase inhibitor, disrupts interactions between Hsp90α and F_1F_0-ATPase synthase (Hall, Kusuma, Brandt, & Blagg, 2014). Interestingly, it was observed that inhibition of F_1F_0-ATPase synthase via cruentaren A reduces cellular levels of select Hsp90 client proteins without induction of the heat-shock response and thus, could provide a novel approach to modulating Hsp90 function. However, limited synthetic accessibility to cruentaren A represents an obstacle that has yet to be overcome.

5. CONCLUSIONS AND FUTURE PROSPECTIVE

Since recognition of Hsp90 as a critical mediator of oncogenic survival and proliferation decades ago, significant progress toward the development of N-terminal Hsp90 inhibitors has been made and numerous compounds have undergone clinical evaluation. Although these inhibitors have shown promising results in a limited number of disease settings, problems, such as induction of prosurvival responses, and a range of dose-limiting systemic toxicities have become apparent. In contrast, C-terminal Hsp90 inhibitors derived from NB have shown promising results in preclinical studies, but their therapeutic potential has yet to be tested in humans. In recent years, many new compounds that modulate Hsp90 function by a variety of mechanisms have been reported. Natural products, such as EGCG, taxol, and silybin, have been identified as exhibiting the potential to inhibit Hsp90, but the extent to which such activity contributes to their broad-spectrum of antitumor activities remains largely unknown. In addition, small molecules have been reported that disrupt the protein–protein interactions that occur between Hsp90 and its cochaperones and client proteins. These may provide useful insights for the development of compounds that can alter Hsp90 function in more subtle ways than direct inhibition of its N-terminal ATP-binding pocket, such as selective disruption of only certain client proteins which might lead to fewer undesirable effects. Interestingly, many of these natural products, including celastrol and withaferin, feature thiol-reactive motifs, which react with cysteine residues in Hsp90 or its cochaperones. Such compounds represent "soft spots" for Hsp90 manipulation, and suggest a potential role for Hsp90 in global protein homeostasis as a sensor of redox stress. This chapter has highlighted the broad range of compounds emerging in recent times that impact Hsp90 function and are now being used to probe more deeply into the biology of Hsp90. The insights gained from such studies should enable the development of new Hsp90 inhibitors with improved properties for clinical applications in the near future.

REFERENCES

Agarwal, C., et al. (2003). Silibinin upregulates the expression of cyclin-dependent kinase inhibitors and causes cell cycle arrest and apoptosis in human colon carcinoma HT-29 cells. *Oncogene*, *22*, 8271–8282. http://dx.doi.org/10.1038/sj.onc.1207158.

Alexander, L. D., Partridge, J. R., Agard, D. A., & McAlpine, S. R. (2011). A small molecule that preferentially binds the closed conformation of Hsp90. *Bioorganic & Medicinal Chemistry Letters*, *21*, 7068–7071. http://dx.doi.org/10.1016/j.bmcl.2011.09.096.

Ali, M. M. U., et al. (2006). Crystal structure of an Hsp90–nucleotide–p23/Sba1 closed chaperone complex. *Nature*, *440*, 1013–1017. http://www.nature.com/nature/journal/v440/n7087/suppinfo/nature04716_S1.html.

Allison, A. C., Cacabelos, R., Lombardi, V. R. M., Álvarez, X. A., & Vigo, C. (2001). Celastrol, a potent antioxidant and anti-inflammatory drug, as a possible treatment for Alzheimer's disease. *Progress in Neuro-Psychopharmacology and Biological Psychiatry*, *25*, 1341–1357. http://dx.doi.org/10.1016/S0278-5846(01)00192-0.

Ardi, V. C., Alexander, L. D., Johnson, V. A., & McAlpine, S. R. (2011). Macrocycles that inhibit the binding between heat shock protein 90 and TPR-containing proteins. *ACS Chemical Biology*, *6*, 1357–1366. http://dx.doi.org/10.1021/cb200203m.

Bagatell, R., & Whitesell, L. (2004). Altered Hsp90 function in cancer: A unique therapeutic opportunity. *Molecular Cancer Therapeutics*, *3*, 1021–1030.

Belofsky, G. N., Jensen, P. R., & Fenical, W. (1999). Sansalvamide: A new cytotoxic cyclic depsipeptide produced by a marine fungus of the genus Fusarium. *Tetrahedron Letters*, *40*, 2913–2916. http://dx.doi.org/10.1016/S0040-4039(99)00393-7.

Bhat, R., et al. (2014). Structure–activity studies of (−)-epigallocatechin gallate derivatives as HCV entry inhibitors. *Bioorganic & Medicinal Chemistry Letters*, *24*, 4162–4165. http://dx.doi.org/10.1016/j.bmcl.2014.07.051.

Blagg, B. S. J., & Kerr, T. D. (2006). Hsp90 inhibitors: Small molecules that transform the Hsp90 protein folding machinery into a catalyst for protein degradation. *Medicinal Research Reviews*, *26*, 310–338. http://dx.doi.org/10.1002/med.20052.

Brandt, G. E., & Blagg, B. S. (2009). Alternate strategies of Hsp90 modulation for the treatment of cancer and other diseases. *Current Topics in Medicinal Chemistry*, *9*, 1447–1461.

Brandt, G. E. L., Schmidt, M. D., Prisinzano, T. E., & Blagg, B. S. J. (2008). Gedunin, a novel Hsp90 inhibitor: Semisynthesis of derivatives and preliminary structure–activity relationships. *Journal of Medicinal Chemistry*, *51*, 6495–6502. http://dx.doi.org/10.1021/jm8007486.

Burlison, J. A., Neckers, L., Smith, A. B., Maxwell, A., & Blagg, B. S. J. (2006). Novobiocin: Redesigning a DNA gyrase inhibitor for selective inhibition of Hsp90. *Journal of the American Chemical Society*, *128*, 15529–15536. http://dx.doi.org/10.1021/ja065793p.

Burlison, J. A., et al. (2008). Development of novobiocin analogues that manifest anti-proliferative activity against several cancer cell lines. *The Journal of Organic Chemistry*, *73*, 2130–2137. http://dx.doi.org/10.1021/jo702191a.

Byrd, C. A., et al. (1999). Heat shock protein 90 mediates macrophage activation by Taxol and bacterial lipopolysaccharide. *Proceedings of the National Academy of Sciences of the United States of America*, *96*, 5645–5650. http://dx.doi.org/10.1073/pnas.96.10.5645.

Caboni, P., et al. (2004). Rotenone, deguelin, their metabolites, and the rat model of Parkinson's disease. *Chemical Research in Toxicology*, *17*, 1540–1548. http://dx.doi.org/10.1021/tx049867r.

Caplan, A. J., Mandal, A. K., & Theodoraki, M. A. (2007). Molecular chaperones and protein kinase quality control. *Trends in Cell Biology*, *17*, 87–92. http://dx.doi.org/10.1016/j.tcb.2006.12.002.

Carroll, C. L., et al. (2005). Synthesis and cytotoxicity of novel sansalvamide A derivatives. *Organic Letters*, *7*, 3481–3484. http://dx.doi.org/10.1021/ol051161g.

Chadli, A., et al. (2010). Celastrol inhibits Hsp90 chaperoning of steroid receptors by inducing fibrillization of the co-chaperone p23. *Journal of Biological Chemistry*, *285*, 4224–4231. http://dx.doi.org/10.1074/jbc.M109.081018.

Chang, D.-J., et al. (2012). Design, synthesis, and biological evaluation of novel deguelin-based heat shock protein 90 (HSP90) inhibitors targeting proliferation and angiogenesis. *Journal of Medicinal Chemistry*, *55*, 10863–10884. http://dx.doi.org/10.1021/jm301488q.

Chaudhury, S., Welch, T. R., & Blagg, B. S. J. (2006). Hsp90 as a target for drug development. *ChemMedChem*, *1*, 1331–1340. http://dx.doi.org/10.1002/cmdc.200600112.

Chen, M., Rose, A. E., Doudican, N., Osman, I., & Orlow, S. J. (2009). Celastrol synergistically enhances temozolomide cytotoxicity in melanoma cells. *Molecular Cancer Research*, *7*, 1946–1953. http://dx.doi.org/10.1158/1541-7786.MCR-09-0243.

Chi, Y., et al. (2013). An open-labeled, randomized, multicenter phase IIa study of gambogic acid injection for advanced malignant tumors. *Chinese Medical Journal*, *126*, 1642–1646.

Chiosis, G., & Neckers, L. (2006). Tumor selectivity of Hsp90 inhibitors: The explanation remains elusive. *ACS Chemical Biology*, *1*, 279–284. http://dx.doi.org/10.1021/cb600224w.

Chiosis, G., et al. (2003). 17AAG: Low target binding affinity and potent cell activity—Finding an explanation. *Molecular Cancer Therapeutics*, *2*, 123–129.

Davenport, J., et al. (2011). Gambogic acid, a natural product inhibitor of Hsp90. *Journal of Natural Products*, *74*, 1085–1092. http://dx.doi.org/10.1021/np200029q.

Davis, M. R., et al. (2012). Synthesis of sansalvamide A peptidomimetics: Triazole, oxazole, thiazole, and pseudoproline containing compounds. *Tetrahedron*, *68*, 1029–1051. http://dx.doi.org/10.1016/j.tet.2011.11.089.

Ding, A. H., Porteu, F., Sanchez, E., & Nathan, C. F. (1990). Shared actions of endotoxin and taxol on TNF receptors and TNF release. *Science*, *248*, 370–372.

Ding, A., Sanchez, E., & Nathan, C. F. (1993). Taxol shares the ability of bacterial lipopolysaccharide to induce tyrosine phosphorylation of microtubule-associated protein kinase. *The Journal of Immunology*, *151*, 5596–5602.

Donnelly, A., & Blagg, B. S. J. (2008). Novobiocin and additional inhibitors of the Hsp90 C-terminal nucleotide-binding pocket. *Current Medicinal Chemistry*, *15*, 2702–2717. http://dx.doi.org/10.2174/092986708786242895.

Donnelly, A. C., Zhao, H., Reddy Kusuma, B., & Blagg, B. S. J. (2010). Cytotoxic sugar analogues of an optimized novobiocin scaffold. *Medicinal Chemistry Communications*, *1*, 165–170. http://dx.doi.org/10.1039/C0MD00063A.

Donnelly, A. C., et al. (2008). The design, synthesis, and evaluation of coumarin ring derivatives of the novobiocin scaffold that exhibit antiproliferative activity. *The Journal of Organic Chemistry*, *73*, 8901–8920. http://dx.doi.org/10.1021/jo801312r.

Duan, H., et al. (2000). Triterpenoids from *Tripterygium wilfordii*. *Phytochemistry*, *53*, 805–810.

Dutta, R., & Inouye, M. (2000). GHKL, an emergent ATPase/kinase superfamily. *Trends in Biochemical Sciences*, *25*, 24–28.

East, A. J., Ollis, W. D., & Wheeler, R. E. (1969). Natural occurrence of 3-aryl-4-hydroxycoumarins. Part I. Phytochemical examination of Derris robusta (roxb.) benth. *Journal of the Chemical Society C: Organic, 365–374.* http://dx.doi.org/10.1039/J39690000365.

Eskew, J. D., et al. (2011). Development and characterization of a novel C-terminal inhibitor of Hsp90 in androgen dependent and independent prostate cancer cells. *BMC Cancer, 11,* 468. http://dx.doi.org/10.1186/1471-2407-11-468.

Falsey, R. R., et al. (2006). Actin microfilament aggregation induced by withaferin A is mediated by annexin II. *Nature Chemical Biology, 2,* 33–38. http://dx.doi.org/10.1038/nchembio755.

Galam, L., et al. (2007). High-throughput assay for the identification of Hsp90 inhibitors based on Hsp90-dependent refolding of firefly luciferase. *Bioorganic & Medicinal Chemistry, 15,* 1939–1946. http://dx.doi.org/10.1016/j.bmc.2007.01.004.

Galanski, M. (2006). Recent developments in the field of anticancer platinum complexes. *Recent Patents on Anti-Cancer Drug Discovery, 1,* 285–295.

Garg, G., Zhao, H., & Blagg, B. S. J. (2015). Design, synthesis, and biological evaluation of ring-constrained novobiocin analogues as Hsp90 C-terminal inhibitors. *ACS Medical Chemistry Letters, 6,* 204–209. http://dx.doi.org/10.1021/ml5004475.

Gazak, R., Walterova, D., & Kren, V. (2007). Silybin and silymarin—New and emerging applications in medicine. *Current Medicinal Chemistry, 14,* 315–338.

Ghosh, S., et al. (2015). Hsp90 C-terminal inhibitors exhibit antimigratory activity by disrupting the Hsp90α/Aha1 complex in PC3-MM2 cells. *ACS Chemical Biology, 10,* 577–590. http://dx.doi.org/10.1021/cb5008713.

Gu, W., Liu, S. S., & Silverman, R. B. (2002). Solid-phase, Pd-catalyzed silicon-aryl carbon bond formation. Synthesis of sansalvamide A peptide. *Organic Letters, 4,* 4171–4174.

Gu, M., et al. (2014). Structure-activity relationship (SAR) of withanolides to inhibit Hsp90 for its activity in pancreatic cancer cells. *Investigational New Drugs, 32,* 68–74. http://dx.doi.org/10.1007/s10637-013-9987-y.

Hadden, M. K., Galam, L., Gestwicki, J. E., Matts, R. L., & Blagg, B. S. J. (2007). Derrubone, an inhibitor of the Hsp90 protein folding machinery. *Journal of Natural Products, 70,* 2014–2018. http://dx.doi.org/10.1021/np070190s.

Hall, J. A., Forsberg, L. K., & Blagg, B. S. (2014). Alternative approaches to Hsp90 modulation for the treatment of cancer. *Future Medicinal Chemistry, 6,* 1587–1605. http://dx.doi.org/10.4155/fmc.14.89.

Hall, J. A., Kusuma, B. R., Brandt, G. E. L., & Blagg, B. S. J. (2014). Cruentaren A binds F1F0 ATP synthase to modulate the Hsp90 protein folding machinery. *ACS Chemical Biology, 9,* 976–985. http://dx.doi.org/10.1021/cb400906e.

Hanahan, D., & Weinberg, R. A. (2000). The hallmarks of cancer. *Cell, 100,* 57–70. http://dx.doi.org/10.1016/S0092-8674(00)81683-9.

Hanahan, D., & Weinberg, Robert A. (2011). Hallmarks of cancer: The next generation. *Cell, 144,* 646–674.

Hastings, J. M., Hadden, M. K., & Blagg, B. S. J. (2008). Synthesis and evaluation of derrubone and select analogues. *The Journal of Organic Chemistry, 73,* 369–373. http://dx.doi.org/10.1021/jo702366g.

Hieronymus, H., et al. (2006). Gene expression signature-based chemical genomic prediction identifies a novel class of HSP90 pathway modulators. *Cancer Cell, 10,* 321–330. http://dx.doi.org/10.1016/j.ccr.2006.09.005.

Hong, D. S., et al. (2013). Targeting the molecular chaperone heat shock protein 90 (HSP90): Lessons learned and future directions. *Cancer Treatment Reviews, 39,* 375–387. http://dx.doi.org/10.1016/j.ctrv.2012.10.001.

Huai, Q., et al. (2005). Structures of the N-terminal and middle domains of *E. coli* Hsp90 and conformation changes upon ADP binding. *Structure*, *13*, 579–590. http://dx.doi.org/10.1016/j.str.2004.12.018.

Itoh, H., et al. (1999). A novel chaperone-activity-reducing mechanism of the 90-kDa molecular chaperone HSP90. *Biochemical Journal*, *343*(Pt. 3), 697–703.

Jhaveri, K., Taldone, T., Modi, S., & Chiosis, G. (2012). Advances in the clinical development of heat shock protein 90 (Hsp90) inhibitors in cancers. *Biochimica et Biophysica Acta*, *1823*, 742–755. http://dx.doi.org/10.1016/j.bbamcr.2011.10.008.

Jolly, C., & Morimoto, R. I. (2000). Role of the heat shock response and molecular chaperones in oncogenesis and cell death. *Journal of the National Cancer Institute*, *92*, 1564–1572. http://dx.doi.org/10.1093/jnci/92.19.1564.

Jordan, P., & Carmo-Fonseca, M. (2000). Molecular mechanisms involved in cisplatin cytotoxicity. *Cellular and Molecular Life Sciences: CMLS*, *57*, 1229–1235. http://dx.doi.org/10.1007/PL00000762.

Kaileh, M., et al. (2007). Withaferin A strongly elicits IκB kinase β hyperphosphorylation concomitant with potent inhibition of its kinase activity. *Journal of Biological Chemistry*, *282*, 4253–4264. http://dx.doi.org/10.1074/jbc.M606728200.

Kamal, A., et al. (2003). A high-affinity conformation of Hsp90 confers tumour selectivity on Hsp90 inhibitors. *Nature*, *425*, 407–410. http://www.nature.com/nature/journal/v425/n6956/suppinfo/nature01913_S1.html.

Kamath, S. G., et al. (2009). Gedunin, a novel natural substance, inhibits ovarian cancer cell proliferation. *International Journal of Gynecological Cancer*, *19*, 1564–1569. http://dx.doi.org/10.1111/IGC.0b013e3181a83135.

Kannaiyan, R., et al. (2011). Celastrol inhibits tumor cell proliferation and promotes apoptosis through the activation of c-Jun N-terminal kinase and suppression of PI3 K/Akt signaling pathways. *Apoptosis*, *16*, 1028–1041. http://dx.doi.org/10.1007/s10495-011-0629-6.

Khalid, S., & Paul, S. (2014). Identifying a C-terminal ATP binding sites-based novel Hsp90-Inhibitor in silico: A plausible therapeutic approach in Alzheimer's disease. *Medical Hypotheses*, *83*, 39–46. http://dx.doi.org/10.1016/j.mehy.2014.04.013.

Khandelwal, A., Crowley, V. M., & Blagg, B. S. (2016). Natural product inspired N-terminal Hsp90 inhibitors: From bench to bedside? *Medicinal Research Reviews*, *36*, 92–118. http://dx.doi.org/10.1002/med.21351.

Khandelwal, A., Hall, J. A., & Blagg, B. S. J. (2013). Synthesis and structure–activity relationships of EGCG analogues, a recently identified Hsp90 inhibitor. *The Journal of Organic Chemistry*, *78*, 7859–7884. http://dx.doi.org/10.1021/jo401027r.

Koay, Y. C., et al. (2014). Chemically accessible hsp90 inhibitor that does not induce a heat shock response. *ACS Medicinal Chemistry Letters*, *5*, 771–776. http://dx.doi.org/10.1021/ml500114p.

Koga, F., Kihara, K., & Neckers, L. (2009). Inhibition of cancer invasion and metastasis by targeting the molecular chaperone heat-shock protein 90. *Anticancer Research*, *29*, 797–807.

Kosano, H., Stensgard, B., Charlesworth, M. C., McMahon, N., & Toft, D. (1998). The assembly of progesterone receptor-hsp90 complexes using purified proteins. *Journal of Biological Chemistry*, *273*, 32973–32979. http://dx.doi.org/10.1074/jbc.273.49.32973.

Kunicki, J. B., et al. (2011). Synthesis and evaluation of biotinylated sansalvamide A analogs and their modulation of Hsp90. *Bioorganic & Medicinal Chemistry Letters*, *21*, 4716–4719. http://dx.doi.org/10.1016/j.bmcl.2011.06.083.

Kunze, B., Sasse, F., Wieczorek, H., & Huss, M. (2007). Cruentaren A, a highly cytotoxic benzolactone from Myxobacteria is a novel selective inhibitor of mitochondrial F1-ATPases. *FEBS Letters*, *581*, 3523–3527. http://dx.doi.org/10.1016/j.febslet.2007.06.069.

Kunze, B., et al. (2006). Cruentaren, a new antifungal salicylate-type macrolide from Byssovorax cruenta (myxobacteria) with inhibitory effect on mitochondrial ATPase activity. Fermentation and biological properties. *The Journal of Antibiotics*, *59*, 664–668. http://dx.doi.org/10.1038/ja.2006.89.

Kvardova, V., et al. (2010). The new platinum(IV) derivative LA-12 shows stronger inhibitory effect on Hsp90 function compared to cisplatin. *Molecular Cancer*, *9*, 147. http://dx.doi.org/10.1186/1476-4598-9-147.

Lamb, J., et al. (2006). The Connectivity Map: Using gene-expression signatures to connect small molecules, genes, and disease. *Science*, *313*, 1929–1935. http://dx.doi.org/10.1126/science.1132939.

Lee, S. C., et al. (2015). Synthesis and evaluation of a novel deguelin derivative, L80, which disrupts ATP binding to the C-terminal domain of heat shock protein 90. *Molecular Pharmacology*, *88*, 245–255. http://dx.doi.org/10.1124/mol.114.096883.

Li, J., Soroka, J., & Buchner, J. (2012). The Hsp90 chaperone machinery: Conformational dynamics and regulation by co-chaperones. *Biochimica et Biophysica Acta (BBA)—Molecular Cell Research*, 624–635. http://dx.doi.org/10.1016/j.bbamcr.2011.09.003.

Liu, Z., Ma, L., & Zhou, G. B. (2011). The main anticancer bullets of the Chinese medicinal herb, thunder god vine. *Molecules*, *16*, 5283–5297. http://dx.doi.org/10.3390/molecules16065283.

Liu, Y., et al. (2010). Triptolide inhibits cell growth and induces G0–G1 arrest by regulating P21wap1/cip1 and P27 kip1 in human multiple myeloma RPMI-8226 cells. *Chinese Journal of Cancer Research*, *22*, 141–147. http://dx.doi.org/10.1007/s11670-010-0141-5.

Liu, W., et al. (2015). KU675, a concomitant heat-shock protein inhibitor of Hsp90 and Hsc70 that manifests isoform selectivity for Hsp90alpha in prostate cancer cells. *Molecular Pharmacology*, *88*, 121–130. http://dx.doi.org/10.1124/mol.114.097303.

Logue, J. S., & Morrison, D. K. (2012). Complexity in the signaling network: Insights from the use of targeted inhibitors in cancer therapy. *Genes and Development*, *26*, 641–650. http://dx.doi.org/10.1101/gad.186965.112.

Luo, W., Sun, W., Taldone, T., Rodina, A., & Chiosis, G. (2010). Heat shock protein 90 in neurodegenerative diseases. *Molecular Neurodegeneration*, *5*, 24.

Marcu, M. G., Chadli, A., Bouhouche, I., Catelli, M., & Neckers, L. M. (2000). The heat shock protein 90 antagonist novobiocin interacts with a previously unrecognized ATP-binding domain in the carboxyl terminus of the chaperone. *Journal of Biological Chemistry*, *275*, 37181–37186. http://dx.doi.org/10.1074/jbc.M003701200.

Marcu, M. G., Schulte, T. W., & Neckers, L. (2000). Novobiocin and related coumarins and depletion of heat shock protein 90-dependent signaling proteins. *Journal of the National Cancer Institute*, *92*, 242–248. http://dx.doi.org/10.1093/jnci/92.3.242.

Mays, J. R., Hill, S. A., Moyers, J. T., & Blagg, B. S. J. (2010). The synthesis and evaluation of flavone and isoflavone chimeras of novobiocin and derrubone. *Bioorganic & Medicinal Chemistry*, *18*, 249–266. http://dx.doi.org/10.1016/j.bmc.2009.10.061.

McConnell, J. R., Alexander, L. A., & McAlpine, S. R. (2014). A heat shock protein 90 inhibitor that modulates the immunophilins and regulates hormone receptors without inducing the heat shock response. *Bioorganic & Medicinal Chemistry Letters*, *24*, 661–666. http://dx.doi.org/10.1016/j.bmcl.2013.11.059.

Meyer, P., et al. (2003). Structural and functional analysis of the middle segment of Hsp90: Implications for ATP hydrolysis and client protein and cochaperone interactions. *Molecular Cell*, *11*, 647–658. http://dx.doi.org/10.1016/S1097-2765(03)00065-0.

Mishra, L. C., Singh, B. B., & Dagenais, S. (2000). Scientific basis for the therapeutic use of Withania somnifera (ashwagandha): A review. *Alternative Medicine Review*, *5*, 334–346.

Mohan, R., et al. (2004). Withaferin A is a potent inhibitor of angiogenesis. *Angiogenesis*, 7, 115–122. http://dx.doi.org/10.1007/s10456-004-1026-3.

Murphy, P. J. M., Kanelakis, K. C., Galigniana, M. D., Morishima, Y., & Pratt, W. B. (2001). Stoichiometry, abundance, and functional significance of the hsp90/hsp70-based multiprotein chaperone machinery in reticulocyte lysate. *Journal of Biological Chemistry*, *276*, 30092–30098. http://dx.doi.org/10.1074/jbc.M103773200.

Nagase, M., et al. (2003). Apoptosis induction in HL-60 cells and inhibition of topoisomerase II by triterpene celastrol. *Bioscience, Biotechnology, and Biochemistry*, *67*, 1883–1887.

Neckers, L., & Workman, P. (2012). Hsp90 molecular chaperone inhibitors: Are we there yet? *Clinical Cancer Research*, *18*, 64–76. http://dx.doi.org/10.1158/1078-0432.ccr-11-1000.

Oh, S. H., et al. (2007). Structural basis for depletion of heat shock protein 90 client proteins by deguelin. *Journal of the National Cancer Institute*, *99*, 949–961. http://dx.doi.org/10.1093/jnci/djm007.

Palermo, C. M., Westlake, C. A., & Gasiewicz, T. A. (2005). Epigallocatechin gallate inhibits aryl hydrocarbon receptor gene transcription through an indirect mechanism involving binding to a 90 kDa heat shock protein. *Biochemistry*, *44*, 5041–5052. http://dx.doi.org/10.1021/bi047433p.

Panaretou, B., et al. (1998). ATP binding and hydrolysis are essential to the function of the Hsp90 molecular chaperone in vivo. *EMBO Journal*, *17*, 4829–4836. http://dx.doi.org/10.1093/emboj/17.16.4829.

Pandey, M. K., et al. (2007). Gambogic acid, a novel ligand for transferrin receptor, potentiates TNF-induced apoptosis through modulation of the nuclear factor-kappaB signaling pathway. *Blood*, *110*, 3517–3525. http://dx.doi.org/10.1182/blood-2007-03-079616.

Papathanassiu, A. E., MacDonald, N. J., Bencsura, A., & Vu, H. A. (2006). F1F0-ATP synthase functions as a co-chaperone of Hsp90-substrate protein complexes. *Biochemical and Biophysical Research Communications*, *345*, 419–429. http://dx.doi.org/10.1016/j.bbrc.2006.04.104.

Patel, P. D., et al. (2013). Paralog-selective Hsp90 inhibitors define tumor-specific regulation of HER2. *Nature Chemical Biology*, *9*, 677–684. http://dx.doi.org/10.1038/nchembio.1335. http://www.nature.com/nchembio/journal/v9/n11/abs/nchembio.1335.html, supplementary-information (2013).

Patwardhan, C. A., et al. (2013). Gedunin inactivates the co-chaperone p23 protein causing cancer cell death by apoptosis. *Journal of Biological Chemistry*, *288*, 7313–7325. http://dx.doi.org/10.1074/jbc.M112.427328.

Paul, S., & Mahanta, S. (2014). Association of heat-shock proteins in various neurodegenerative disorders: Is it a master key to open the therapeutic door? *Molecular and Cellular Biochemistry*, *386*, 45–61. http://dx.doi.org/10.1007/s11010-013-1844-y.

Pearl, L. H., & Prodromou, C. (2006). Structure and mechanism of the Hsp90 molecular chaperone machinery. *Annual Review of Biochemistry*, *75*, 271–294. http://dx.doi.org/10.1146/annurev.biochem.75.103004.142738.

Peterson, L. B., & Blagg, B. S. J. (2009). To fold or not to fold: Modulation and consequences of Hsp90 inhibition. *Future Medicinal Chemistry*, *1*, 267–283. http://dx.doi.org/10.4155/Fmc.09.17.

Peterson, L. B., Eskew, J. D., Vielhauer, G. A., & Blagg, B. S. J. (2012). The hERG channel is dependent upon the Hsp90α isoform for maturation and trafficking. *Molecular Pharmaceutics*, *9*, 1841–1846. http://dx.doi.org/10.1021/mp300138n.

Piaz, F. D., Terracciano, S., De Tommasi, N., & Braca, A. (2015). Hsp90 activity modulation by plant secondary metabolites. *Planta Medica*, *81*, 1223–1239. http://dx.doi.org/10.1055/s-0035-1546251.

Prodromou, C., et al. (1999). Regulation of Hsp90 ATPase activity by tetratricopeptide repeat (TPR)-domain co-chaperones. *EMBO Journal*, *18*, 754–762. http://dx.doi.org/10.1093/emboj/18.3.754.

Prodromou, C., et al. (2000). The ATPase cycle of Hsp90 drives a molecular 'clamp' via transient dimerization of the N-terminal domains. *The EMBO Journal*, *19*, 4383–4392.
Qin, Y., et al. (2007). Gambogic acid inhibits the catalytic activity of human topoisomerase IIalpha by binding to its ATPase domain. *Molecular Cancer Therapeutics*, *6*, 2429–2440. http://dx.doi.org/10.1158/1535-7163.MCT-07-0147.
Ramsey, D. M., et al. (2012). An Hsp90 modulator that exhibits a unique mechanistic profile. *Bioorganic and Medicinal Chemistry Letters*, *22*, 3287–3290. http://dx.doi.org/10.1016/j.bmcl.2012.03.012.
Ren, Y., et al. (2011). Absolute configuration of (−)-gambogic acid, an antitumor agent. *Journal of Natural Products*, *74*, 460–463. http://dx.doi.org/10.1021/np100422z.
Riebold, M., et al. (2015). A C-terminal HSP90 inhibitor restores glucocorticoid sensitivity and relieves a mouse allograft model of Cushing disease. *Nature Medicine*, *21*, 276–280. http://dx.doi.org/10.1038/nm.3776.
Rodriguez, R. A., et al. (2007). Synthesis of second-generation sansalvamide A derivatives: Novel templates as potential antitumor agents. *Journal of Organic Chemistry*, *72*, 1980–2002. http://dx.doi.org/10.1021/jo061830j.
Rosenhagen, M. C., et al. (2003). The heat shock protein 90-targeting drug cisplatin selectively inhibits steroid receptor activation. *Molecular Endocrinology*, *17*, 1991–2001. http://dx.doi.org/10.1210/me.2003-0141.
Salminen, A., Lehtonen, M., Paimela, T., & Kaarniranta, K. (2010). Celastrol: Molecular targets of thunder god vine. *Biochemical and Biophysical Research Communications*, *394*, 439–442. http://dx.doi.org/10.1016/j.bbrc.2010.03.050.
Samadi, A. K., et al. (2011). A novel C-terminal HSP90 inhibitor KU135 induces apoptosis and cell cycle arrest in melanoma cells. *Cancer Letters*, *312*, 158–167. http://dx.doi.org/10.1016/j.canlet.2011.07.031.
Sellers, R. P., et al. (2010). Design and synthesis of Hsp90 inhibitors: Exploring the SAR of Sansalvamide A derivatives. *Bioorganic & Medicinal Chemistry*, *18*, 6822–6856. http://dx.doi.org/10.1016/j.bmc.2010.07.042.
Sethi, G., Ahn, K. S., Pandey, M. K., & Aggarwal, B. B. (2007). Celastrol, a novel triterpene, potentiates TNF-induced apoptosis and suppresses invasion of tumor cells by inhibiting NF-kappaB-regulated gene products and TAK1-mediated NF-kappaB activation. *Blood*, *109*, 2727–2735. http://dx.doi.org/10.1182/blood-2006-10-050807.
Shelton, S. N., et al. (2009). KU135, a novel novobiocin-derived C-terminal inhibitor of the 90-kDa heat shock protein, exerts potent antiproliferative effects in human leukemic cells. *Molecular Pharmacology*, *76*, 1314–1322. http://dx.doi.org/10.1124/mol.109.058545.
Shohat, B., Gitter, S., Abraham, A., & Lavie, D. (1967). Antitumor activity of withaferin A (NSC-101088). *Cancer Chemotherapy Reports*, *51*, 271–276.
Sidera, K., & Patsavoudi, E. (2008). Extracellular HSP90: Conquering the cell surface. *Cell Cycle*, *7*, 1564–1568.
Solit, D. B., Basso, A. D., Olshen, A. B., Scher, H. I., & Rosen, N. (2003). Inhibition of heat shock protein 90 function down-regulates Akt kinase and sensitizes tumors to Taxol. *Cancer Research*, *63*, 2139–2144.
Söti, C., Rácz, A., & Csermely, P. (2002). A nucleotide-dependent molecular switch controls ATP binding at the C-terminal domain of Hsp90: N-terminal nucleotide binding unmasks a C-terminal binding pocket. *Journal of Biological Chemistry*, *277*, 7066–7075. http://dx.doi.org/10.1074/jbc.M105568200.
Sőti, C., Vermes, Á., Haystead, T. A. J., & Csermely, P. (2003). Comparative analysis of the ATP-binding sites of Hsp90 by nucleotide affinity cleavage: A distinct nucleotide specificity of the C-terminal ATP-binding site. *European Journal of Biochemistry*, *270*, 2421–2428. http://dx.doi.org/10.1046/j.1432-1033.2003.03610.x.

Sreedhar, A. S., Soti, C., & Csermely, P. (2004). Inhibition of Hsp90: A new strategy for inhibiting protein kinases. *Biochimica et Biophysica Acta, 1697*, 233–242. http://dx.doi.org/10.1016/j.bbapap.2003.11.027.

Sreeramulu, S., Gande, S. L., Göbel, M., & Schwalbe, H. (2009). Molecular mechanism of inhibition of the human protein complex Hsp90–Cdc37, a kinome chaperone–cochaperone, by triterpene celastrol. *Angewandte Chemie, International Edition, 48*, 5853–5855. http://dx.doi.org/10.1002/anie.200900929.

Srinivasan, S., Ranga, R. S., Burikhanov, R., Han, S. S., & Chendil, D. (2007). Par-4-dependent apoptosis by the dietary compound withaferin A in prostate cancer cells. *Cancer Research, 67*, 246–253. http://dx.doi.org/10.1158/0008-5472.CAN-06-2430.

Taipale, M., Jarosz, D. F., & Lindquist, S. (2010). HSP90 at the hub of protein homeostasis: Emerging mechanistic insights. *Nature Reviews. Molecular Cell Biology, 11*, 515–528.

Taipale, M., et al. (2012). Quantitative analysis of Hsp90-client interactions reveals principles of substrate recognition. *Cell, 150*, 987–1001. http://dx.doi.org/10.1016/j.cell.2012.06.047.

Tao, X., & Lipsky, P. E. (2000). The Chinese anti-inflammatory and immunosuppressive herbal remedy Tripterygium wilfordii Hook F. *Rheumatic Diseases Clinics of North America, 26*, 29–50. viii.

Trepel, J., Mollapour, M., Giaccone, G., & Neckers, L. (2010). Targeting the dynamic HSP90 complex in cancer. *Nature Reviews. Cancer, 10*, 537–549. http://dx.doi.org/10.1038/nrc2887.

Uddin, S. J., et al. (2007). Gedunin, a limonoid from *Xylocarpus granatum*, inhibits the growth of CaCo-2 colon cancer cell line in vitro. *Phytotherapy Research, 21*, 757–761. http://dx.doi.org/10.1002/ptr.2159.

Vartholomaiou, E., Echeverría, P. C., & Picard, D. (2016). Unusual suspects in the twilight zone between the Hsp90 interactome and carcinogenesis. *Advances in Cancer Research, 129*, 1–30. http://dx.doi.org/10.1016/bs.acr.2015.08.001.

Vasko, R. C., et al. (2010). Mechanistic studies of sansalvamide A-amide: An allosteric modulator of Hsp90. *ACS Medicinal Chemistry Letters, 1*, 4–8. http://dx.doi.org/10.1021/ml900003t.

Walter, S., & Buchner, J. (2002). Molecular chaperones—Cellular machines for protein folding. *Angewandte Chemie, International Edition, 41*, 1098–1113. http://dx.doi.org/10.1002/1521-3773(20020402)41:7<1098::AID-ANIE1098>3.0.CO;2-9.

Wandinger, S. K., Richter, K., & Buchner, J. (2008). The Hsp90 chaperone machinery. *Journal of Biological Chemistry, 283*, 18473–18477. http://dx.doi.org/10.1074/jbc.R800007200.

Wani, M. C., Taylor, H. L., Wall, M. E., Coggon, P., & McPhail, A. T. (1971). Plant antitumor agents. VI. Isolation and structure of taxol, a novel antileukemic and antitumor agent from *Taxus brevifolia*. *Journal of the American Chemical Society, 93*, 2325–2327. http://dx.doi.org/10.1021/ja00738a045.

Wei, W., et al. (2014). Novel celastrol derivatives inhibit the growth of hepatocellular carcinoma patient-derived xenografts. *Oncotarget, 5*, 5819–5831. http://dx.doi.org/10.1016/j.bbagen.2014.06.008.

Westerheide, S. D., et al. (2004). Celastrols as inducers of the heat shock response and cytoprotection. *Journal of Biological Chemistry, 279*, 56053–56060. http://dx.doi.org/10.1074/jbc.M409267200.

Whitesell, L., Bagatell, R., & Falsey, R. (2003). The stress response: Implications for the clinical development of hsp90 inhibitors. *Current Cancer Drug Targets, 3*, 349–358.

Whitesell, L., & Lindquist, S. L. (2005). HSP90 and the chaperoning of cancer. *Nature Reviews Cancer, 5*, 761–772.

Workman, P. (2004). Altered states: Selectively drugging the Hsp90 cancer chaperone. *Trends in Molecular Medicine, 10*, 47–51.

Xu, W., & Neckers, L. (2007). Targeting the molecular chaperone heat shock protein 90 provides a multifaceted effect on diverse cell signaling pathways of cancer cells. *Clinical Cancer Research, 13*, 1625–1629. http://dx.doi.org/10.1158/1078-0432.ccr-06-2966.

Yang, H., Chen, D., Cui, Q. C., Yuan, X., & Dou, Q. P. (2006). Celastrol, a triterpene extracted from the Chinese "Thunder of God Vine," is a potent proteasome inhibitor and suppresses human prostate cancer growth in nude mice. *Cancer Research, 66*, 4758–4765. http://dx.doi.org/10.1158/0008-5472.CAN-05-4529.

Yang, H., Shi, G., & Dou, Q. P. (2007). The tumor proteasome is a primary target for the natural anticancer compound withaferin a isolated from "Indian winter cherry" *Molecular Pharmacology, 71*, 426–437. http://dx.doi.org/10.1124/mol.106.030015.

Yi, T., et al. (2008). Gambogic acid inhibits angiogenesis and prostate tumor growth by suppressing vascular endothelial growth factor receptor 2 signaling. *Cancer Research, 68*, 1843–1850. http://dx.doi.org/10.1158/0008-5472.CAN-07-5944.

Yin, Z., Henry, E. C., & Gasiewicz, T. A. (2008). (−)-Epigallocatechin-3-gallate Is a novel Hsp90 inhibitor†. *Biochemistry, 48*, 336–345. http://dx.doi.org/10.1021/bi801637q.

Yin, Z., Henry, E. C., & Gasiewicz, T. A. (2009). (-)-Epigallocatechin-3-gallate is a novel Hsp90 inhibitor. *Biochemistry, 48*, 336–345. http://dx.doi.org/10.1021/bi801637q.

Yokota, Y., Bargagna-Mohan, P., Ravindranath, P. P., Kim, K. B., & Mohan, R. (2006). Development of withaferin A analogs as probes of angiogenesis. *Bioorganic & Medicinal Chemistry Letters, 16*, 2603–2607. http://dx.doi.org/10.1016/j.bmcl.2006.02.039.

Young, J. C., Agashe, V. R., Siegers, K., & Hartl, F. U. (2004). Pathways of chaperone-mediated protein folding in the cytosol. *Nature Reviews. Molecular Cell Biology, 5*, 781–791. http://dx.doi.org/10.1038/nrm1492.

Yousuf, S. K., et al. (2011). Ring A structural modified derivatives of withaferin A and the evaluation of their cytotoxic potential. *Steroids, 76*, 1213–1222. http://dx.doi.org/10.1016/j.steroids.2011.05.012.

Yu, X. M., et al. (2005). Hsp90 inhibitors identified from a library of novobiocin analogues. *Journal of the American Chemical Society, 127*, 12778–12779. http://dx.doi.org/10.1021/ja0535864.

Yu, Y., et al. (2010). Withaferin A targets heat shock protein 90 in pancreatic cancer cells. *Biochemical Pharmacology, 79*, 542–551. http://dx.doi.org/10.1016/j.bcp.2009.09.017.

Zak, F., et al. (2004). Platinum(IV) complex with adamantylamine as nonleaving amine group: Synthesis, characterization, and in vitro antitumor activity against a panel of cisplatin-resistant cancer cell lines. *Journal of Medicinal Chemistry, 47*. http://dx.doi.org/10.1021/jm030858+.

Zanphorlin, L. M., Alves, F. R., & Ramos, C. H. (2014). The effect of celastrol, a triterpene with antitumorigenic activity, on conformational and functional aspects of the human 90 kDa heat shock protein Hsp90alpha, a chaperone implicated in the stabilization of the tumor phenotype. *Biochimica et Biophysica Acta, 1840*, 3145–3152. http://dx.doi.org/10.1016/j.bbagen.2014.06.008.

Zaveri, N. T. (2006). Green tea and its polyphenolic catechins: Medicinal uses in cancer and noncancer applications. *Life Sciences, 78*, 2073–2080. http://dx.doi.org/10.1016/j.lfs.2005.12.006.

Zhang, T., et al. (2008). A novel Hsp90 inhibitor to disrupt Hsp90/Cdc37 complex against pancreatic cancer cells. *Molecular Cancer Therapeutics*, 7, 162–170. http://dx.doi.org/10.1158/1535-7163.MCT-07-0484.

Zhang, T., et al. (2009). Characterization of celastrol to inhibit hsp90 and cdc37 interaction. *Journal of Biological Chemistry, 284*, 35381–35389. http://dx.doi.org/10.1074/jbc.M109.051532.

Zhang, L., et al. (2010). Gambogic acid inhibits Hsp90 and deregulates TNF-α/NF-κB in HeLa cells. *Biochemical and Biophysical Research Communications*, *403*, 282–287. http://dx.doi.org/10.1016/j.bbrc.2010.11.018.

Zhao, H., Brandt, G. E., Galam, L., Matts, R. L., & Blagg, B. S. (2011). Identification and initial SAR of silybin: An Hsp90 inhibitor. *Bioorganic and Medicinal Chemistry Letters*, *21*, 2659–2664. http://dx.doi.org/10.1016/j.bmcl.2010.12.088.

Zhao, H., Moroni, E., Colombo, G., & Blagg, B. S. J. (2013). Identification of a new scaffold for Hsp90 C-terminal inhibition. *ACS Medicinal Chemistry Letters*, *5*, 84–88. http://dx.doi.org/10.1021/ml400404s.

Zhao, H., Reddy Kusuma, B., & Blagg, B. S. J. (2010). Synthesis and evaluation of noviose replacements on novobiocin that manifest antiproliferative activity. *ACS Medicinal Chemistry Letters*, *1*, 311–315. http://dx.doi.org/10.1021/ml100070r.

Zhao, Q., et al. (2008). Posttranscriptional regulation of the telomerase hTERT by gambogic acid in human gastric carcinoma 823 cells. *Cancer Letters*, *262*, 223–231. http://dx.doi.org/10.1016/j.canlet.2007.12.002.

Zhao, H., et al. (2011). Engineering an antibiotic to fight cancer: Optimization of the novobiocin scaffold to produce anti-proliferative agents. *Journal of Medicinal Chemistry*, *54*, 3839–3853. http://dx.doi.org/10.1021/jm200148p.

Zhao, J., et al. (2014). Triazole containing novobiocin and biphenyl amides as Hsp90 C-terminal inhibitors. *Medicinal Chemistry Communications*, *5*, 1317–1323. http://dx.doi.org/10.1039/C4MD00102H.

Zhao, H., et al. (2015). Design, synthesis and biological evaluation of biphenylamide derivatives as Hsp90 C-terminal inhibitors. *European Journal of Medicinal Chemistry*, *89*, 442–466. http://dx.doi.org/10.1016/j.ejmech.2014.10.034.

Zhou, Y. X., & Huang, Y. L. (2009). Antiangiogenic effect of celastrol on the growth of human glioma: An in vitro and in vivo study. *Chinese Medical Journal*, *122*, 1666–1673.

Zuehlke, A., & Johnson, J. L. (2010). Hsp90 and co-chaperones twist the functions of diverse client proteins. *Biopolymers*, *93*, 211–217. http://dx.doi.org/10.1002/bip.21292.

CHAPTER FOUR

Hsp90 in Cancer: Transcriptional Roles in the Nucleus

Stuart K. Calderwood*[,1], **Len Neckers**[†]
*Department of Radiation Oncology, Beth Israel Deaconess Medical Center, Harvard Medical School, Boston, Massachusetts, USA
[†]Urologic Oncology Branch, Center for Cancer Research, National Cancer Institute, Bethesda, Maryland, USA
[1]Corresponding author: e-mail address: scalderw@bidmc.harvard.edu

Contents

Abstract

Hsp90 plays a key role in fostering metabolic pathways essential in tumorigenesis through its functions as a molecular chaperone. Multiple oncogenic factors in the membrane and cytoplasm are thus protected from degradation and destruction. Here, we have considered Hsp90's role in transcription in the nucleus. Hsp90 functions both in regulating the activity of sequence-specific transcription factors such as nuclear receptors and HSF1, as well as impacting more globally acting factors that act on chromatin and RNA polymerase II. Hsp90 influences transcription by modulating histone modification mediated by its clients SMYD3 and trithorax/MLL, as well as by regulating the processivity of RNA polymerase II through negative elongation factor. It is not currently clear how the transcriptional role of Hsp90 may be influenced by the cancer milieu although recently discovered posttranslational modification of the chaperone may be involved. Dysregulation of Hsp90 may thus influence malignant processes both by modulating the function of specific transcription factors and effects on more globally acting general components of the transcriptional machinery.

Advances in Cancer Research, Volume 129
ISSN 0065-230X
http://dx.doi.org/10.1016/bs.acr.2015.08.002

1. INTRODUCTION

Hsp90 belongs to a family of proteins, initially named for their mobility on SDS-PAGE gels, now known to have at least five family members: these include two cytoplasmic members, an ER resident cousin and a mitochondrially located sibling (Felts et al., 2000; Lindquist & Craig, 1988). The current review deals with the proteins Hsp90α and Hsp90β found in the cytoplasm and nucleus that are encoded by, respectively, the *HSP90AA1* and *HSP90AB1* genes (Sreedhar, Kalmar, Csermely, & Shen, 2004). Hsp90α and Hsp90β are both molecular chaperones and take part in the later stages of protein folding, refining the structures of client proteins into functional conformations. Both proteins can associate with ATP and carry out molecular chaperone functions that require ATP hydrolysis in the folding reaction. ATP hydrolysis by Hsp90 is, in itself, quite slow but it is increased to physiologically relevant rates by association with regulatory proteins known as cochaperones that catalyze various steps in the chaperoning of client proteins into active conformations as well as determining substrate selection (Calderwood, 2013).

The duplication event that gave rise to two related Hsp90 isoforms is thought to have occurred about 500 million years ago, permitting the development of some divergent functions for the two proteins that have only recently been investigated with any intensity (Sreedhar et al., 2004). However, many of the studies reviewed here did not take into account the potential for isoform-specific functions and the Hsp90 properties we discuss are largely the aggregate of the properties of the two isoforms. Hsp90α is the more stress-inducible family member and contains arrays of heat-inducible *cis*-acting elements (heat-shock elements, HSE) in its promoter while Hsp90β functions more as a "housekeeping chaperone." HSE are binding sites for heat shock factor 1 (HSF1), a protein that responds to proteotoxic stress by oligomerizing and activating the transcription of genes whose promoters contain HSE (Wu, 1995). This response to stress is extremely rapid and robust. HSF1 can be observed associating with the *hsp70* gene within 30 s of heat shock (Bunch et al., 2014).

2. Hsp90 AND CANCER

Expression of most HSPs, as well as HSF1, is increased in cancers with a wide range of morphologies (Ciocca & Calderwood, 2005). However, there is, as yet little solid data defining the mechanisms underlying this effect, although the ubiquity and magnitude of the changes observed suggest one or more rather generally occurring mechanisms for the increased levels of HSPs (Ciocca, Arrigo, & Calderwood, 2013). A role for HSF1 in driving up levels of *HSP* transcription might be predicted *a priori* due to its prominence in activating *HSP* promoters. As the key responder to proteotoxic stress, one might envisage a mechanism whereby changes in the tumor cell proteome, including oncogene mutation and amplification, changes in ploidy and generally enhanced translation would lead to HSF1 activation (Santagata et al., 2013; Tang et al., 2005). Indeed exposure of cells to Hsp90 inhibitors leads to the depletion of a wide range of oncoproteins due to their unfolding when the chaperone is inactivated and subsequent degradation by the proteasome (Miyata, Nakamoto, & Neckers, 2013). Such findings have led to the hypothesis that cancer cells become progressively "addicted to chaperones" and that HSP levels consequently become elevated to permit the expansion of an oncogenic proteome.

One anomaly in considering this hypothesis is that HSF1 appears to be active in cancer cells despite high levels of tumor cell Hsp90 (Mendillo et al., 2012; Fig. 1). Hsp90 has generally been considered the principal negative regulator of HSF1. Studies *in vitro* involving immunodepletion of Hsp90 have demonstrated activation of the factor (Zou, Guo, Guettouche, Smith, & Voellmy, 1998). Consequently, triggering of the heat shock response by proteotoxic stresses was considered to involve the sequestration of Hsp90 by an expanding population of unfolded proteins (Zou et al., 1998). In cancer then, elevated levels of Hsp90 would be predicted to repress HSF1 and limit increases in *HSP* gene transcription and HSP levels through the mechanism of end-product inhibition. There appears however to be "something different" about Hsp90 in cancer cells; for instance, the tumor proteome seems exquisitely sensitive to Hsp90-targeting drugs (Kamal et al., 2003). One intriguing explanation for this sensitive state might involve cancer-related alterations in the growing number of reported posttranslational modifications (PTMs) to Hsp90 that might influence its intracellular properties (Miyata, 2009; Mollapour & Neckers, 2012).

As mentioned, we still do not have a clearly defined mechanism to account for the increased levels of Hsp90 and other chaperones in most cancers. Rather than the chaperone titration model described above, direct modification of HSF1 itself could explain increases in its level and activation state during oncogenesis with consequent upregulation of Hsp90 expression. (Ciocca et al., 2013). It has for instance been shown that HSF1 is

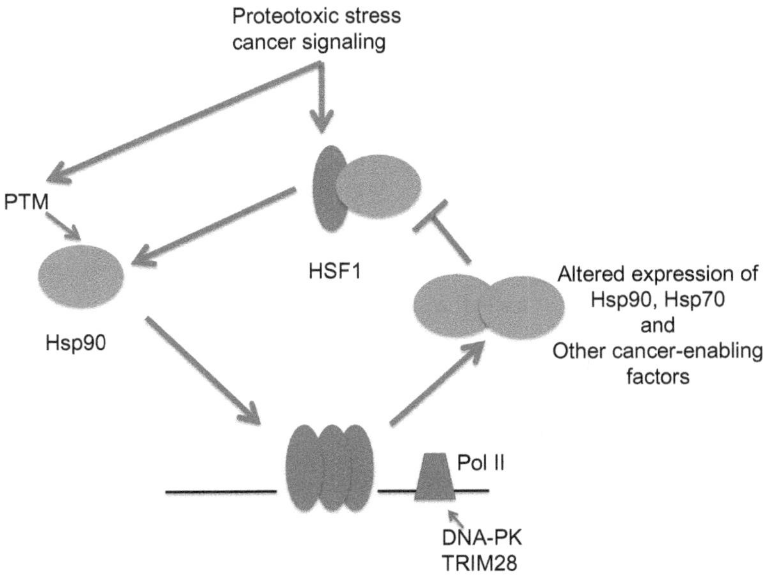

Figure 1 Hsp90 interaction with HSF1 and HSP synthesis in stress and cancer. HSF1 is depicted as being regulated by Hsp90 in a negative feedback loop. Escape from Hsp90 repression may involve modifications to HSF1 and Hsp90 itself and may include effects on the transcription factor or on more global factors including promoter proximal pausing and histone modifications. (See the color plate.)

repressed by glycogen synthase kinase 3 (GSK3), a kinase that phosphorylates HSF1 on serine 303 (Chu, Soncin, Price, Stevenson, & Calderwood, 1996; Chu, Zhong, Soncin, Stevenson, & Calderwood, 1998). Phospho-S303-HSF1 then becomes modified by SUMO (small ubiquitin-like modifier) at lysine 298, a potent repressive modification (Hietakangas et al., 2003). The SUMO moiety comes in two major varieties, encoded by the *SUMO1* and homologous *SUMO2 and 3* genes (Bettermann, Benesch, Weis, & Haybaeck, 2012; Raman, Nayak, & Muller, 2013; Subramonian et al., 2014). SUMO2/3 appears to regulate many transcription factors involved in carcinogenesis; deSUMOylation of SUMO2/3-modified factors leads to activation (Subramonian et al., 2014). One might thus propose a mechanism involving deSUMOylation of HSF1 in cancer, although no experimental evidence exists to support this hypothesis currently. Hsp90 has also been shown to be a substrate for SUMOylation and this modification may likewise modulate the properties of Hsp90 in regulating transcription (Mollapour et al., 2014).

3. Hsp90 IN THE NUCLEUS AND REGULATION OF TRANSCRIPTION

3.1 Regulation of Sequence-Specific Transcription Factors

Although Hsp90 plays significant roles in the nucleus, it is not clear that the chaperone possesses nuclear transport motifs. Indeed, Hsp90 appears to possess sequences in the C-terminus that suppress nuclear uptake (Passinen, Valkila, Manninen, Syvala, & Ylikomi, 2001). Nuclear localization may thus depend on cochaperones or other associated proteins (Longshaw, Chapple, Balda, Cheetham, & Blatch, 2004). The participation of Hsp90 in regulating transcription factor function was initially shown in studies of steroid hormone receptors (Dittmar & Pratt, 1997; Kirschke, Goswami, Southworth, Griffin, & Agard, 2014; Fig. 2, top panel). Hsp90 was shown to assemble in complexes with a number of cochaperones including p23, Hop, Hsp40, immunophilins FKBP1, and FKBP2 as well as Hsp70 in unliganded cytoplasmic glucocorticoid receptors (GR). Exposure to ligand then led to release of the GR from the chaperone complexes, recruitment of

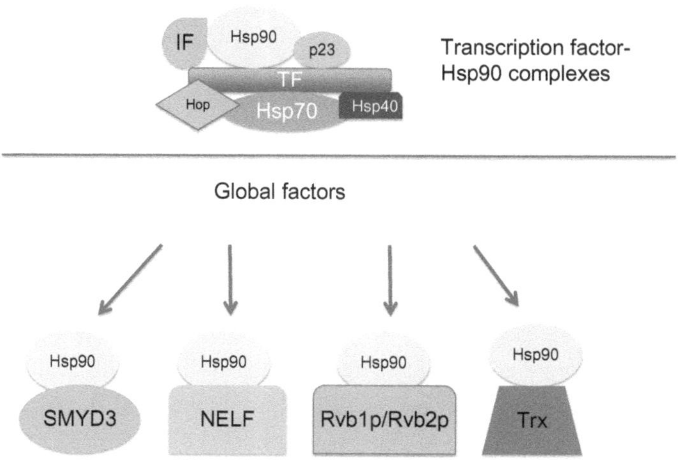

Figure 2 Participation of Hsp90 in transcription. Upper part of figure depicts Hsp90 associated with a transcription factor (TF) in a chaperone–cochaperone complex (p23, immunophilin (IF) complex—TPR protein Hop and Hsp70 and Hsp40). This type of regulation is found in nuclear receptors and HSF1. In the lower part of the figure, we show Hsp90 participating in global effects on transcriptional activation mediated by SMYD3, Trx/MLL, Rvb, and NELF involving chromatin remodeling, histone modification, and Pol II regulation. (See the color plate.)

coactivators and activation of transcription. Early studies by Picard et al. demonstrated mechanistic features of this process that have been seen repeatedly in subsequent investigations of Hsp90 client regulation (Picard et al., 1990). Binding to the chaperone complex appeared to retard *trans*-activation of gene expression mediated by GR, while disruption of such complexes inhibited hormone binding and receptor activation. Based on these and similar data, it has been suggested that the Hsp90 complex maintains the receptor in a stable inactive complex that is nonetheless poised for high-affinity hormone binding and rapid activation (Picard et al., 1990). A similar mechanism with some variation has been described for a number of other steroid hormone receptors as well as for HSF1 (Voellmy & Boellmann, 2007). Interestingly HSF1 appears to retain only a partial dependence on Hsp90 and although the chaperone can restrain transcriptional activity under uninduced conditions, free HSF1 appears to be stable in the absence of HSP binding (Zou et al., 1998).

3.2 Regulation of General Transcriptional Machinery

In addition to effects on specific transcription factors, Hsp90 seems essential for proper function of the general transcription machinery because it can mediate Pol II assembly in the cytoplasm and nuclear import of the fully assembled holoenzyme (Boulon et al., 2010). Indeed, an increasing amount of evidence for Hsp90 activities in the nucleus is accumulating which seems to imply roles beyond conventional chaperoning activities (Fig. 2, bottom panel). For instance, involvement of the chaperone in disassembly of chromatin-bound transcriptional complexes, in nucleosomal remodeling in yeast and in the activity of histone methyltransferases in mammalian cells have been reported (Bryant et al., 2008; Freeman & Yamamoto, 2002; Hamamoto et al., 2004).

There are at least two major checkpoints in transcriptional regulation that control RNA polymerase II (Pol II) activity in the nucleus and RNA synthesis. These include (1) formation of the pre-initiation complex at the transcription start site (TSS) and promoter escape of Pol II and (2) promoter proximal pausing of Pol II close to the TSS of inducible genes followed by release of Pol II for processive elongation triggered by *trans*-activators. As in the cell cycle, each of these checkpoints is regulated by members of the cell division kinase (cdk) family (Egloff & Murphy, 2008). An important target for these kinases is a region in the C-terminal domain (CTD) of the primary subunit of Pol II (RPB1) that contains multiple tandem copies of a repeating 7-amino acid sequence (Y1-S2-P3-T4-S5-P6-S7). The Pol II CTD thus presents multiple potential phosphorylation sites enabling the "hyperphosphorylation" of Pol II *in vivo* (Egloff & Murphy, 2008). The CTD does not appear to have a major role in the enzymatic properties of Pol II but instead provides a highly regulated scaffold for binding regulators of transcription. At least two of these residues, phospho-S2 and phospho-S5 appear to play decisive roles in transcription. Pol II, when phosphorylated on residue S5 is found near the TSS. This modification is mediated by a cdk7–cyclin H complex and appears to control promoter clearance. Pol II S2-phospho, the result of cdk9–cyclinT complex (also known as the positive transcription elongation factor, P-TEF-b)-mediated phosphorylation is found more frequently in the polymerase when it is associated with the gene body and 3′regions of genes. Therefore occupancy of the gene body by P-TEF-b is a characteristic of elongating Pol II species (Adelman & Lis, 2012; Egloff & Murphy, 2008). P-TEF-b has been shown to overcome the promoter proximal pausing of Pol II and to lead to RNA elongation mediated in part by Pol II S2-phospho. Regulation of transcription by P-TEF-b also involves inhibitory phosphorylation of negative regulators of elongation including the DRB sensitivity-inducing factor (DSIF) and negative elongation factor (NELF), modifications that promote release of the Pol II from the paused state (Adelman & Lis, 2012).

Cdk family molecules are among the most prominent clients of Hsp90/cochaperone Cdc37 (p50) complexes (Gray, Prince, Cheng, Stevenson, & Calderwood, 2008; Stepanova, Leng, Parker, & Harper, 1996), suggesting a potential role for such Cdk::chaperone complexes in regulating transcription. Cdk9-containing P-TEF-b complexes were shown to be critical in regulating the *Drosophila hsp70* gene (Lis, Mason, Peng, Price, & Werner, 2000). Hsp90 and Cdc37 were also shown to be essential for the assembly of the P-TEF-b complex and elongation in HIV1 transcription (Fong & Zhou, 2000; O'Keeffe, Fong, Chen, Zhou, & Zhou, 2000).

More recently it was shown, using high-resolution mapping techniques that Hsp90 is found in association with chromatin near the transcriptional start sites of multiple genes (Sawarkar, Sievers, & Paro, 2012). Chromatin-associated Hsp90 appears to have a functional role, and its inhibition by two chemically distinct inhibitors (the natural products geldanamycin and radicicol) that both target the N-domain ATP binding site led to large increases in gene transcription, apparently due to relief of Hsp90-mediated inhibition (Sawarkar et al., 2012). Highly represented among these targets were genes shown previously to be, like Hsp70, regulated by promoter proximal pausing. These included targets such as c-Myc, p53, Delta, and Notch that are well known for their dysregulation in many cancers (Sawarkar et al., 2012). As mentioned above, P-TEF-b relieves pausing and permits elongation of RNAs through phosphorylation of the CTD of Pol II at the 2-position and through modification of DSIF and NELF, suggesting potential roles for Hsp90 in these mechanisms (Adelman & Lis, 2012). Sawarkar et al. were able to show physical interactions between the NELF subunit NELF-E and Hsp90 by coimmunoprecipitation in *Drosophila* and human tissues as well as co-occupation of sites on chromatin by Hsp90 and NELF using ChIP analysis (Sawarkar et al., 2012). Hsp90 was shown to be essential for stabilizing NELF on chromatin, although not participating in NELF–Pol II interactions. Questions suggested by these data include the nature of the mechanism (s) by which Hsp90 is inactivated, and transcription permitted during pause release. One possible mechanism could include NELF phosphorylation by P-TEF-b, a change that might alter Hsp90-NELF-E affinity or direct modifications to Hsp90 during transcription. Dysregulation of Hsp90 dynamics in cancer could therefore have significant effects on the levels of transcription, particularly of *HSP* genes as well as other rapidly inducible genes that utilize promoter proximal pausing, many of which participate in the cancer phenotype (Fig. 2, bottom panel).

It should also be noted that Pol II can also accumulate in an inactive form upstream of the TSS of a number of genes induced during development in *Caenorhabditis elegans* (*C. elegans*) and this accumulation may also permit rapid and synchronized rounds of transcription (as with release of pausing) (Hsu et al., 2015; Maxwell et al., 2014). This process, known as Pol II docking or Pol II poising appears to be different in nature to pausing, with the Pol II complex binding upstream of the TSS rather than downstream as in pausing. As many of the mechanisms involved in transcription are highly conserved during evolution one would predict a role for Pol II poising in mammalian transcription. However, little evidence has accrued regarding mechanisms that govern this process, and thus a role for Hsp90 in this process is currently unknown.

3.3 Regulation of Chromatin Modifications

Another key role for Hsp90 in the nucleus appears to be interaction with Trithorax (Trx) (Tariq, Nussbaumer, Chen, Beisel, & Paro, 2009). Trx is a member of the TrxG group with homology to the mammalian mixed leukemia lineage (MLL) family of histone methyltransferases that ensure the activity of target enhancers and promoters through maintenance of, respectively, histone3 lysine4 methyl (H3K4Me1) and histone3 lysine4 trimethyl (H3K4Me3) levels (Hu et al., 2013; Morgan & Shilatifard, 2013). During the development of differentiated cells from pluripotent stem cells, Trx functions in opposition to the Polycomb genes that repress target genes through the repressive histone mark histone3 lysine27 trimethyl (H3K27Me3) (Xu & Rubin, 1993). Many developmental genes incorporate areas containing bivalent chromatin marks (both H3K4Me3 and H3K27Me3) and are considered to be in a poised state ready for either rapid induction or repression following environmental cues (Puri, Gala, Mishra, & Dhawan, 2015). Hsp90 functions to stabilize chromatin-associated Trx, and thus influences the activity of a range of developmentally regulated genes as well as HSP genes through its histone methyltransferase activity directed toward H3K4Me3 (Puri et al., 2015; Tariq et al., 2009) (Smith et al., 2004). Loss of Hsp90 function, and inactivation of Trx would favor Polycomb-mediated repression, permit demethylation of H3K4Me3 and lead to alterations of cell fate decisions as well as loss of stress protein transcription.

Translocations involving the Trx-related MLL gene occur in poor prognosis human leukemias and are thought to drive leukemogenesis by disrupting normal hematopoietic development. (Thirman et al., 1993). More recently it has been shown that members of the Trx/MLL family, particularly MLL2 and MLL3 are dysregulated in a large proportion of solid tumors (reviewed Ford & Dingwall, 2015). Thus, Hsp90 could exert important effects on tumorigenesis via regulation of Trx/MLL factors and these interactions may have significant future implications for cancer treatment (Trepel, Mollapour, Giaccone, & Neckers, 2010).

Hsp90 contains a C-terminal domain that binds to tetratricopeptide (TPR) domain proteins, the most notable of which is the scaffold protein Hop (Calderwood, 2013). One such protein is TTC5 (tetratricopeptide5) that is capable of binding to transcriptional effectors including the protein acetyltransferase p300 and stimulating transcription under stress conditions (Davies et al., 2011; Demonacos et al., 2004). Future studies might investigate a role for interaction of Hsp90 with this nuclear protein that contains six TPR domains and is thus potentially capable of binding to Hsp90 and nucleating large transcriptional complexes (Calderwood, 2013).

Another transcriptionally active molecule linked to Hsp90 regulation is the ATP-dependent chromatin remodeling factor Rvb1 (RuvB-like1), a key component of the Ino80 and SWR-C chromatin remodeling complexes (Ruden & Lu, 2008). In a yeast protein interactome screen, Hsp90 was shown to bind two novel cochaperones, TPR-containing protein associated with Hsp90 (Tah 1p) and Pih 1p (protein interacting with Hsp90), and these proteins mediated Hsp90 association with a Rvb1p/Rvbp2p complex (Zhao et al., 2005). Rvb family helicases are found in all eukaryotes and are important in chromatin remodeling/transcriptional activation, suggesting the potential significance of such Hsp90/Rvb1 interactions in the regulation of mammalian gene expression (Gnatovskiy, Mita, & Levy, 2013).

3.4 Regulation of Hsp90 Function in the Nucleus

What mechanisms might regulate Hsp90-mediated transcriptional control in the nucleus? Such regulation could involve the ATP-driven Hsp90 chaperone cycle and its influence by cochaperones (Calderwood, 2013; Neckers & Ivy, 2003). Indeed, p23/Hsp90 complexes have been shown to localize to nuclear receptor sites in a hormone-dependent manner and to attenuate transcriptional activities, while Cdc37/Hsp90 complexes can be found associated with P-TEF-b (Freeman & Yamamoto, 2002; Stepanova et al., 1996). Cochaperones have various roles in cancer and their relative abundance and activities would likely influence Hsp90 activities in the nucleus, as recently reviewed (Calderwood, 2013).

Another exciting possibility for regulation of Hsp90 function in the nucleus is PTMs of the chaperone itself as well as its associated proteins (Fig. 3). Hsp90 contains multiple potential phosphorylation sites that are emerging as key regulatory residues that respond to cell signaling pathways (Mollapour & Neckers, 2012). For instance, casein kinase 2 (CK2) seems

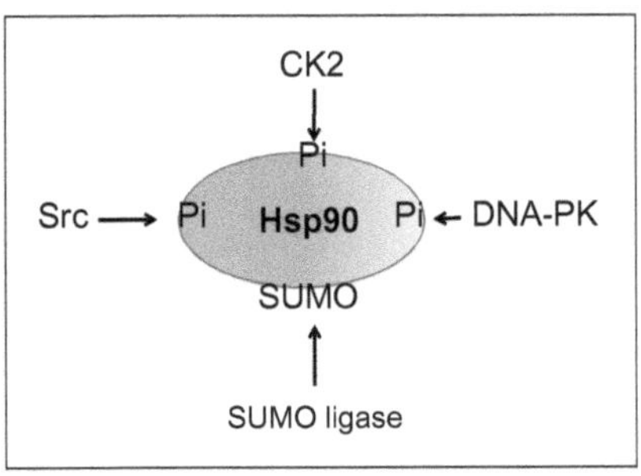

Figure 3 Posttranscriptional modifications that may influence role of Hsp90 in transcription. We depict Hsp90 phosphorylation by CK2, DNA-PK, and Src as well as its SUMOylation as known PTMs that may modulate its impact on transcriptional events. (See the color plate.)

intimately linked to Hsp90 function: CK2 is one of the major Hsp90 binding proteins, requires chaperoning by this HSP, and can phosphorylate Hsp90 at multiple sites as well as the cochaperones CDC37/p50 and FKBP52 and the transcription factor HSF1 (Miyata, 2009; Mollapour, Tsutsumi, Kim, Trepel, & Neckers, 2011; Mollapour, Tsutsumi, Truman, et al., 2011; Soncin et al., 2003). CK2 may be involved in regulating transcription factor activity/stability. For instance, phosphorylation of Hsp90α on serine 231 by CK2 destabilized association of the chaperone with the aryl hydrocarbon receptor (a nuclear receptor family protein) and diminished transcription (Ogiso et al., 2004). CK2 has been shown to be active in the nucleus, is known to play a significant role in cancer signaling and its intimate involvement with Hsp90 suggests a potential role for Hsp90–CK2 interactions in cancer (Filhol & Cochet, 2009; Filhol, Giacosa, Wallez, & Cochet, 2015; Miyata, 2009).

Although Hsp90 has been shown to interact with the Pol II complex at multiple promoter proximal paused genes through binding to NELF-E, the mechanisms involved in relief of Hsp90-associated pausing and resumption of productive elongation after transcriptional stimulus are not clear (Sawarkar et al., 2012). A potentially analogous situation has recently been reported showing that tripartate motif containing 28 (TRIM28) can stabilize Pol II pausing when bound to the ORFs of paused genes such as *hspa1b*, and then, on receipt of a stimulus such as heat shock can permit elongation (Bunch & Calderwood, 2015; Bunch et al., 2014). Conversion of TRIM28 to a factor permissive for transcriptional elongation involved the nuclear kinase DNA-dependent protein kinase (DNA-PK) which phosphorylates the protein and inactivates its *trans*-repressor function, normally transmitted through its polySUMOylated C-terminal region (Bunch & Calderwood, 2015). Interestingly, this is a mechanism "borrowed" from the DNA repair field as TRIM28 phosphorylation by DNA-PK, ATM or ATR plays a key role in repair of DNA double-strand breaks through its interaction with chromatin (Lemaitre & Soutoglou, 2014).

There is considerable current interest in novel transcriptional regulatory pathways that incorporate some of the mechanisms characterized previously in DNA maintenance and repair: roles for topoisomerase I, topoisomerase II, Ku70, Ku80, and DNA-PK have been reported (Ju & Rosenfeld, 2006; Puc et al., 2015). A number of years ago, Hsp90α was shown to be phosphorylated *in vitro* on two N-terminal sites (T5 and T7) by DNA-PK (Lees-Miller & Anderson, 1989). A recent study found that DNA damage-induced phosphorylation of these residues in cells, and this phosphorylation was

DNA-PK dependent (Solier et al., 2012). It would be intriguing to learn whether such modifications might be related to global roles for Hsp90 in transcription, such as its influence on the elongation pathway checkpoint by association with NELF or its effects on histone modification through interaction with Trx factors. Threonines 5 and 7 are found only in Hsp90α, and Hsp90α silencing affects DNA-PK stability and function that cannot be rescued by Hsp90β (Solier et al., 2012), suggesting a possible Hsp90 isoform-specific role in transcriptional regulation. In contrast, both Hsp90 isoforms can be phosphorylated by Src, a modification that appears to enhance molecular chaperone activity (Duval, Le Boeuf, Huot, & Gratton, 2007). While one intuitively regards nonreceptor tyrosine kinases such as Src as effectors of plasma membrane signaling events, these kinases play important roles in the nucleus. For instance, triggering of DNA repair leads to phosphorylation of TRIM28 by Src on chromatin and reverses gene silencing events normally mediated through the factor heterochromatin protein 1γ (HP1γ) (Kubota et al., 2013).

In addition to phosphorylation, another PTM often encountered in transcriptional activators is modification by SUMO, a modification that usually inhibits transcription due to recruitment of repressor complexes (Duval et al., 2007). Hsp90 has been shown to be SUMOylated in yeast and mammalian cells, an event that leads to recruitment of the cochaperone Aha1 and reduction in its efficiency in chaperoning some clients (Mollapour et al., 2014).

4. CONCLUSIONS

Although traditionally considered a cytosolic chaperone, recent work is revealing intriguing new roles for Hsp90 in the nucleus and in regulation of transcription. These include facilitation of transcription initiated through the binding of sequence-specific transcription factors including nuclear receptors, HSF1, and others. Regulation of nuclear client protein activity by Hsp90 involves a series of interactions (Picard et al., 1990; Sawarkar et al., 2012). In most cases, Hsp90 binds to clients in complexes containing cochaperones that stabilize the client while maintaining it in an inactive state. Release of clients from such complexes leads to a burst of activity followed by loss of such activity, presumably due to unfolding and degradation of the free client protein. The participation of Hsp90 in transcription, however, been shown to extend beyond individual client transcription factors. Global roles for the chaperone are emerging in (1) regulating elongation through NELF-E, (2) modulating epigenetic regulation through Trx/MLL binding, (3) modifying histone H3K4 methylation through association with SMYD3, and (4) affecting chromatin remodeling by associating with Rvb1/Rvb2 (Hamamoto et al., 2004; Ruden & Lu, 2008; Sawarkar et al., 2012; Tariq et al., 2009; Zhao et al., 2005; Fig. 2, bottom panel). One can thus envisage a scenario in which these "nuclear" activities of Hsp90 could influence tumorigenesis through genome-wide effects on transcription.

In addition, Hsp90 appears to assume an "altered state" in many cancers, an as yet poorly understood condition in which the chaperone is more sensitive to ATP-competitive inhibitors (Kamal et al., 2003; Moulick et al., 2011). Likely mechanisms for such altered drug sensitivity include PTMs of Hsp90 and its cochaperones (Mollapour et al., 2014; Mollapour & Neckers, 2012; Fig. 3). These and perhaps other PTMs may also influence the transcription-modifying activities of Hsp90 in cancer described here.

The investigation of Hsp90's role(s) in the nucleus is currently at an early stage and much effort will be required to establish the full significance of Hsp90 in transcription. Future studies that seem particularly compelling include the following:

(1) Many of the studies reviewed here were carried out in model organisms including yeast, *C. elegans*, and *Drosophila*. In most cases, it remains to be determined whether Hsp90 plays similar roles in mammalian cells in regulating processes such as Pol II pausing, chromatin modification by Trx and chromatin remodeling by Rvb1p/Rvbp2p containing complexes as discussed above. While the high level of conservation of many of the components of gene transcription would suggest similar functions will be seen in mammals, it will be crucial to learn whether such is really the case. If so, how important are they in mammalian biology and how do these functions become altered in cancers. In the course of this work, we anticipate discovery of other new and unexpected roles for Hsp90 in transcription.

(2) A priority also should be placed on establishing the differential/shared roles of Hsp90α and Hsp90β in the globally acting pathways of transcriptional regulation mentioned above. This will enable us to understand the regulatory roles of these two molecules more precisely as well as help guide future development of Hsp90-targeted drugs.

(3) Another major area of investigation, likely to be highly productive, will be determining the signal transduction pathways and PTMs that regulate Hsp90 activity in transcription in response to environmental cues.

A number of important PTMs have been characterized, as reviewed above, and roles for these regulatory modifications in switching transcription on or off are anticipated.

(4) Of major relevance for the current discussion are the questions, (a) What is the importance of transcriptional regulation by Hsp90 in cancer and how is it deregulated during tumor progression? (b) Finally, could the Hsp90-dependent transcriptome be selectively targeted by clinically tolerable small molecules that might become useful anticancer drugs in the future?

ACKNOWLEDGMENTS

We thank the Department of Radiation Oncology, BIDMC for support and encouragement. We are particularly thankful to Jane Trepel for valuable discussions. In addition, we thank Heeyoun Bunch, Tom Prince, and Ayesha Murshid.

This work was supported in part by NIH research grants RO-1CA047407, R01CA119045, and RO-1CA094397 (S.K.C.) and by funds from the Intramural Research Program of the National Cancer Institute (L.N., projects Z01 BC011032-01 and Z01 SC010074-12).

REFERENCES

Adelman, K., & Lis, J. T. (2012). Promoter-proximal pausing of RNA polymerase II: Emerging roles in metazoans. *Nature Reviews Genetics*, *13*, 720–731.

Bettermann, K., Benesch, M., Weis, S., & Haybaeck, J. (2012). SUMOylation in carcinogenesis. *Cancer Letters*, *316*, 113–125.

Boulon, S., Pradet-Balade, B., Verheggen, C., Molle, D., Boireau, S., Georgieva, M., et al. (2010). HSP90 and its R2TP/Prefoldin-like cochaperone are involved in the cytoplasmic assembly of RNA polymerase II. *Molecular Cell*, *39*, 912–924.

Bryant, G. O., Prabhu, V., Floer, M., Wang, X., Spagna, D., Schreiber, D., et al. (2008). Activator control of nucleosome occupancy in activation and repression of transcription. *PLoS Biology*, *6*, 2928–2939.

Bunch, H., & Calderwood, S. K. (2015). TRIM28 as a novel transcriptional elongation factor. *BMC Molecular Biology*, *16*, 14.

Bunch, H., Zheng, X., Burkholder, A., Dillon, S. T., Motola, S., Birrane, G., et al. (2014). TRIM28 regulates RNA polymerase II promoter-proximal pausing and pause release. *Nature Structural & Molecular Biology*, *21*, 876–883.

Calderwood, S. K. (2013). Molecular cochaperones: Tumor growth and cancer treatment. *Scientifica*, *2013*, 13. 217513.

Chu, B., Soncin, F., Price, B. D., Stevenson, M. A., & Calderwood, S. K. (1996). Sequential phosphorylation by mitogen-activated protein kinase and glycogen synthase kinase 3 represses transcriptional activation by heat shock factor-1. *The Journal of Biological Chemistry*, *271*, 30847–30857.

Chu, B., Zhong, R., Soncin, F., Stevenson, M. A., & Calderwood, S. K. (1998). Transcriptional activity of heat shock factor 1 at 37 degrees C is repressed through phosphorylation on two distinct serine residues by glycogen synthase kinase 3 and protein kinases Calpha and Czeta. *The Journal of Biological Chemistry*, *273*, 18640–18646.

Ciocca, D. R., Arrigo, A. P., & Calderwood, S. K. (2013). Heat shock proteins and heat shock factor 1 in carcinogenesis and tumor development: An update. *Archives of Toxicology, 87*, 19–48.

Ciocca, D. R., & Calderwood, S. K. (2005). Heat shock proteins in cancer: Diagnostic, prognostic, predictive, and treatment implications. *Cell Stress & Chaperones, 10*, 86–103.

Davies, L., Paraskevopoulou, E., Sadeq, M., Symeou, C., Pantelidou, C., Demonacos, C., et al. (2011). Regulation of glucocorticoid receptor activity by a stress responsive transcriptional cofactor. *Molecular Endocrinology, 25*, 58–71.

Demonacos, C., Krstic-Demonacos, M., Smith, L., Xu, D., O'Connor, D. P., Jansson, M., et al. (2004). A new effector pathway links ATM kinase with the DNA damage response. *Nature Cell Biology, 6*, 968–976.

Dittmar, K. D., & Pratt, W. B. (1997). Folding of the glucocorticoid receptor by the reconstituted Hsp90-based chaperone machinery. The initial hsp90.p60.hsp70-dependent step is sufficient for creating the steroid binding conformation. *The Journal of Biological Chemistry, 272*, 13047–13054.

Duval, M., Le Boeuf, F., Huot, J., & Gratton, J. P. (2007). Src-mediated phosphorylation of Hsp90 in response to vascular endothelial growth factor (VEGF) is required for VEGF receptor-2 signaling to endothelial NO synthase. *Molecular Biology of the Cell, 18*, 4659–4668.

Egloff, S., & Murphy, S. (2008). Cracking the RNA polymerase II CTD code. *Trends in Genetics, 24*, 280–288.

Felts, S. J., Owen, B. A., Nguyen, P., Trepel, J., Donner, D. B., & Toft, D. O. (2000). The hsp90-related protein TRAP1 is a mitochondrial protein with distinct functional properties. *The Journal of Biological Chemistry, 275*, 3305–3312.

Filhol, O., & Cochet, C. (2009). Protein kinase CK2 in health and disease: Cellular functions of protein kinase CK2: a dynamic affair. *Cellular and Molecular Life Sciences, 66*, 1830–1839.

Filhol, O., Giacosa, S., Wallez, Y., & Cochet, C. (2015). Protein kinase CK2 in breast cancer: The CK2beta regulatory subunit takes center stage in epithelial plasticity. *Cellular and Molecular Life Sciences, 72*, 3305–3322.

Fong, Y. W., & Zhou, Q. (2000). Relief of two built-In autoinhibitory mechanisms in P-TEFb is required for assembly of a multicomponent transcription elongation complex at the human immunodeficiency virus type 1 promoter. *Molecular and Cellular Biology, 20*, 5897–5907.

Ford, D. J., & Dingwall, A. K. (2015). The cancer COMPASS: Navigating the functions of MLL complexes in cancer. *Cancer Genetics, 208*, 178–191.

Freeman, B. C., & Yamamoto, K. R. (2002). Disassembly of transcriptional regulatory complexes by molecular chaperones. *Science, 296*, 2232–2235.

Gnatovskiy, L., Mita, P., & Levy, D. E. (2013). The human RVB complex is required for efficient transcription of type I interferon-stimulated genes. *Molecular and Cellular Biology, 33*, 3817–3825.

Gray, P. J., Jr., Prince, T., Cheng, J., Stevenson, M. A., & Calderwood, S. K. (2008). Targeting the oncogene and kinome chaperone CDC37. *Nature Reviews Cancer, 8*, 491–495.

Hamamoto, R., Furukawa, Y., Morita, M., Iimura, Y., Silva, F. P., Li, M., et al. (2004). SMYD3 encodes a histone methyltransferase involved in the proliferation of cancer cells. *Nature Cell Biology, 6*, 731–740.

Hietakangas, V., Ahlskog, J. K., Jakobsson, A. M., Hellesuo, M., Sahlberg, N. M., Holmberg, C. I., et al. (2003). Phosphorylation of serine 303 is a prerequisite for the stress-inducible SUMO modification of heat shock factor 1. *Molecular and Cellular Biology, 23*, 2953–2968.

Hsu, H. T., Chen, H. M., Yang, Z., Wang, J., Lee, N. K., Burger, A., et al. (2015). TRANSCRIPTION. Recruitment of RNA polymerase II by the pioneer transcription factor PHA-4. *Science, 348*, 1372–1376.

Hu, D., Gao, X., Morgan, M. A., Herz, H. M., Smith, E. R., & Shilatifard, A. (2013). The MLL3/MLL4 branches of the COMPASS family function as major histone H3K4 monomethylases at enhancers. *Molecular and Cellular Biology, 33*, 4745–4754.

Ju, B. G., & Rosenfeld, M. G. (2006). A breaking strategy for topoisomerase IIbeta/PARP-1-dependent regulated transcription. *Cell Cycle, 5*, 2557–2560.

Kamal, A., Thao, L., Sensintaffar, J., Zhang, L., Boehm, M. F., Fritz, L. C., et al. (2003). A high-affinity conformation of Hsp90 confers tumour selectivity on Hsp90 inhibitors. *Nature, 425*, 407–410.

Kirschke, E., Goswami, D., Southworth, D., Griffin, P. R., & Agard, D. A. (2014). Glucocorticoid receptor function regulated by coordinated action of the Hsp90 and Hsp70 chaperone cycles. *Cell, 157*, 1685–1697.

Kubota, S., Fukumoto, Y., Aoyama, K., Ishibashi, K., Yuki, R., Morinaga, T., et al. (2013). Phosphorylation of KRAB-associated protein 1 (KAP1) at Tyr-449, Tyr-458, and Tyr-517 by nuclear tyrosine kinases inhibits the association of KAP1 and heterochromatin protein 1alpha (HP1alpha) with heterochromatin. *The Journal of Biological Chemistry, 288*, 17871–17883.

Lees-Miller, S. P., & Anderson, C. W. (1989). The human double-stranded DNA-activated protein kinase phosphorylates the 90-kDa heat-shock protein, hsp90 alpha at two NH2-terminal threonine residues. *The Journal of Biological Chemistry, 264*, 17275–17280.

Lemaitre, C., & Soutoglou, E. (2014). Double strand break (DSB) repair in heterochromatin and heterochromatin proteins in DSB repair. *DNA Repair, 19*, 163–168.

Lindquist, S., & Craig, E. A. (1988). The heat-shock proteins. *Annual Review of Genetics, 22*, 631–677.

Lis, J. T., Mason, P., Peng, J., Price, D. H., & Werner, J. (2000). P-TEFb kinase recruitment and function at heat shock loci. *Genes & Development, 14*, 792–803.

Longshaw, V. M., Chapple, J. P., Balda, M. S., Cheetham, M. E., & Blatch, G. L. (2004). Nuclear translocation of the Hsp70/Hsp90 organizing protein mSTI1 is regulated by cell cycle kinases. *Journal of Cell Science, 117*, 701–710.

Maxwell, C. S., Kruesi, W. S., Core, L. J., Kurhanewicz, N., Waters, C. T., Lewarch, C. L., et al. (2014). Pol II docking and pausing at growth and stress genes in C. elegans. *Cell Reports, 6*, 455–466.

Mendillo, M. L., Santagata, S., Koeva, M., Bell, G. W., Hu, R., Tamimi, R. M., et al. (2012). HSF1 drives a transcriptional program distinct from heat shock to support highly malignant human cancers. *Cell, 150*, 549–562.

Miyata, Y. (2009). Protein kinase CK2 in health and disease: CK2: The kinase controlling the Hsp90 chaperone machinery. *Cellular and Molecular Life Sciences, 66*, 1840–1849.

Miyata, Y., Nakamoto, H., & Neckers, L. (2013). The therapeutic target Hsp90 and cancer hallmarks. *Current Pharmaceutical Design, 19*, 347–365.

Mollapour, M., Bourboulia, D., Beebe, K., Woodford, M. R., Polier, S., Hoang, A., et al. (2014). Asymmetric Hsp90 N domain SUMOylation recruits Aha1 and ATP-competitive inhibitors. *Molecular Cell, 53*, 317–329.

Mollapour, M., & Neckers, L. (2012). Post-translational modifications of Hsp90 and their contributions to chaperone regulation. *Biochimica et Biophysica Acta, 1823*, 648–655.

Mollapour, M., Tsutsumi, S., Kim, Y. S., Trepel, J., & Neckers, L. (2011a). Casein kinase 2 phosphorylation of Hsp90 threonine 22 modulates chaperone function and drug sensitivity. *Oncotarget, 2*, 407–417.

Mollapour, M., Tsutsumi, S., Truman, A. W., Xu, W., Vaughan, C. K., Beebe, K., et al. (2011b). Threonine 22 phosphorylation attenuates Hsp90 interaction with cochaperones and affects its chaperone activity. *Molecular Cell, 41*, 672–681.

Morgan, M. A., & Shilatifard, A. (2013). Drosophila SETs its sights on cancer: Trr/MLL3/4 COMPASS-like complexes in development and disease. *Molecular and Cellular Biology*, *33*, 1698–1701.

Moulick, K., Ahn, J. H., Zong, H., Rodina, A., Cerchietti, L., Gomes DaGama, E. M., et al. (2011). Affinity-based proteomics reveal cancer-specific networks coordinated by Hsp90. *Nature Chemical Biology*, *7*, 818–826.

Neckers, L., & Ivy, S. P. (2003). Heat shock protein 90. *Current Opinion in Oncology*, *15*, 419–424.

Ogiso, H., Kagi, N., Matsumoto, E., Nishimoto, M., Arai, R., Shirouzu, M., et al. (2004). Phosphorylation analysis of 90 kDa heat shock protein within the cytosolic arylhydrocarbon receptor complex. *Biochemistry*, *43*, 15510–15519.

O'Keeffe, B., Fong, Y., Chen, D., Zhou, S., & Zhou, Q. (2000). Requirement for a kinase-specific chaperone pathway in the production of a Cdk9/cyclin T1 heterodimer responsible for P-TEFb-mediated tat stimulation of HIV-1 transcription. *The Journal of Biological Chemistry*, *275*, 279–287.

Passinen, S., Valkila, J., Manninen, T., Syvala, H., & Ylikomi, T. (2001). The C-terminal half of Hsp90 is responsible for its cytoplasmic localization. *European Journal of Biochemistry / FEBS*, *268*, 5337–5342.

Picard, D., Khursheed, B., Garabedian, M. J., Fortin, M. G., Lindquist, S., & Yamamoto, K. R. (1990). Reduced levels of hsp90 compromise steroid receptor action in vivo. *Nature*, *348*, 166–168.

Puc, J., Kozbial, P., Li, W., Tan, Y., Liu, Z., Suter, T., et al. (2015). Ligand-dependent enhancer activation regulated by topoisomerase-I activity. *Cell*, *160*, 367–380.

Puri, D., Gala, H., Mishra, R., & Dhawan, J. (2015). High-wire act: The poised genome and cellular memory. *The FEBS Journal*, *282*, 1675–1691.

Raman, N., Nayak, A., & Muller, S. (2013). The SUMO system: A master organizer of nuclear protein assemblies. *Chromosoma*, *122*, 475–485.

Ruden, D. M., & Lu, X. (2008). Hsp90 affecting chromatin remodeling might explain transgenerational epigenetic inheritance in Drosophila. *Current Genomics*, *9*, 500–508.

Santagata, S., Mendillo, M. L., Tang, Y. C., Subramanian, A., Perley, C. C., Roche, S. P., et al. (2013). Tight coordination of protein translation and HSF1 activation supports the anabolic malignant state. *Science*, *341*, 1238303.

Sawarkar, R., Sievers, C., & Paro, R. (2012). Hsp90 globally targets paused RNA polymerase to regulate gene expression in response to environmental stimuli. *Cell*, *149*, 807–818.

Smith, S. T., Petruk, S., Sedkov, Y., Cho, E., Tillib, S., Canaani, E., et al. (2004). Modulation of heat shock gene expression by the TAC1 chromatin-modifying complex. *Nature Cell Biology*, *6*, 162–167.

Solier, S., Kohn, K. W., Scroggins, B., Xu, W., Trepel, J., Neckers, L., et al. (2012). Heat shock protein 90alpha (HSP90alpha), a substrate and chaperone of DNA-PK necessary for the apoptotic response. *Proceedings of the National Academy of Sciences of the United States of America*, *109*, 12866–12872.

Soncin, F., Zhang, X., Chu, B., Wang, X., Asea, A., Ann Stevenson, M., et al. (2003). Transcriptional activity and DNA binding of heat shock factor-1 involve phosphorylation on threonine 142 by CK2. *Biochemical and Biophysical Research Communications*, *303*, 700–706.

Sreedhar, A. S., Kalmar, E., Csermely, P., & Shen, Y. F. (2004). Hsp90 isoforms: Functions, expression and clinical importance. *FEBS Letters*, *562*, 11–15.

Stepanova, L., Leng, X., Parker, S. B., & Harper, J. W. (1996). Mammalian p50Cdc37 is a protein kinase-targeting subunit of Hsp90 that binds and stabilizes Cdk4. *Genes and Development*, *10*, 1491–1502.

Subramonian, D., Raghunayakula, S., Olsen, J. V., Beningo, K. A., Paschen, W., & Zhang, X. D. (2014). Analysis of changes in SUMO-2/3 modification during breast cancer progression and metastasis. *Journal of Proteome Research*, *13*, 3905–3918.

Tang, D., Khaleque, M. A., Jones, E. L., Theriault, J. R., Li, C., Wong, W. H., et al. (2005). Expression of heat shock proteins and heat shock protein messenger ribonucleic acid in human prostate carcinoma in vitro and in tumors in vivo. *Cell Stress & Chaperones, 10*, 46–58.

Tariq, M., Nussbaumer, U., Chen, Y., Beisel, C., & Paro, R. (2009). Trithorax requires Hsp90 for maintenance of active chromatin at sites of gene expression. *Proceedings of the National Academy of Sciences of the United States of America, 106*, 1157–1162.

Thirman, M. J., Gill, H. J., Burnett, R. C., Mbangkollo, D., McCabe, N. R., Kobayashi, H., et al. (1993). Rearrangement of the MLL gene in acute lymphoblastic and acute myeloid leukemias with 11q23 chromosomal translocations. *The New England Journal of Medicine, 329*, 909–914.

Trepel, J., Mollapour, M., Giaccone, G., & Neckers, L. (2010). Targeting the dynamic HSP90 complex in cancer. *Nature Reviews Cancer, 10*, 537–549.

Voellmy, R., & Boellmann, F. (2007). Chaperone regulation of the heat shock protein response. *Advances in Experimental Medicine and Biology, 594*, 89–99.

Wu, C. (1995). Heat shock transcription factors: Structure and regulation. *Annual Review of Cell and Developmental Biology, 11*, 441–469.

Xu, T., & Rubin, G. M. (1993). Analysis of genetic mosaics in developing and adult Drosophila tissues. *Development, 117*, 1223–1237.

Zhao, R., Davey, M., Hsu, Y. C., Kaplanek, P., Tong, A., Parsons, A. B., et al. (2005). Navigating the chaperone network: An integrative map of physical and genetic interactions mediated by the hsp90 chaperone. *Cell, 120*, 715–727.

Zou, J., Guo, Y., Guettouche, T., Smith, D. F., & Voellmy, R. (1998). Repression of heat shock transcription factor HSF1 activation by HSP90 (HSP90 complex) that forms a stress-sensitive complex with HSF1. *Cell, 94*, 471–480.

CHAPTER FIVE

Hsp90 as a "Chaperone" of the Epigenome: Insights and Opportunities for Cancer Therapy

Jennifer S. Isaacs[1]
Department of Cell and Molecular Pharmacology, Medical University of South Carolina, Hollings Cancer Center, Charleston, South Carolina, USA
[1]Corresponding author: e-mail address: isaacsj@musc.edu

Contents

Abstract

The cellular functions of Hsp90 have historically been attributed to its ability to chaperone client proteins involved in signal transduction. Although numerous stimuli and the signaling cascades they activate contribute to cancer progression, many of these pathways ultimately require transcriptional effectors to elicit tumor-promoting effects. Despite this obvious connection, the majority of studies evaluating Hsp90 function in malignancy have focused upon its regulation of cytosolic client proteins, and particularly members of receptor and/or kinase families. However, in recent years, Hsp90 has emerged as a pivotal orchestrator of nuclear events. Discovery of an expanding repertoire of Hsp90 clients has illuminated a vital role for Hsp90 in overseeing nuclear events and influencing gene transcription. Hence, this chapter will cast a spotlight upon several regulatory themes involving Hsp90-dependent nuclear functions. Highlighted topics include a summary of chaperone-dependent regulation of key transcription factors (TFs) and epigenetic effectors in malignancy, as well as a discussion of how the complex interplay among a subset of these TFs and epigenetic regulators may generate feed-forward loops that further support cancer progression. This chapter will also highlight less recognized indirect mechanisms whereby Hsp90-supported signaling may impinge upon epigenetic regulation. Finally, the relevance of these nuclear

Advances in Cancer Research, Volume 129
ISSN 0065-230X
http://dx.doi.org/10.1016/bs.acr.2015.09.003

events is discussed within the framework of Hsp90's capacity to enable phenotypic variation and drug resistance. These newly acquired insights expanding our understanding of Hsp90 function support the collective notion that nuclear clients are major beneficiaries of Hsp90 action, and their impairment is likely responsible for many of the anticancer effects elicited by Hsp90-targeted approaches.

1. EVIDENCE FOR HSP90 AS A "CHROMATINIZED" PROTEIN

Hsp90 is well established as a guardian of proteostasis, a function largely attributed to its chaperoning of hundreds of client proteins. Cumulative studies have shown that the largest category of Hsp90 client proteins is represented by cytosolic receptors containing Hsp90-binding domains (Taipale et al., 2012), findings consistent with the predominantly cytosolic localization of Hsp90. Hsp90 has also been observed in additional cellular compartments, highlighting its extensive functional reach. For instance, mitochondrial Hsp90 plays a key role in cellular metabolism and supports processes central to cancer progression (Chae et al., 2013). Not surprisingly, Hsp90 is also found in the nucleus. Attesting to its nuclear function, Hsp90 is significantly detected as a chromatin-bound protein and is localized in close proximity to the transcriptional start site of approximately one-third of all coding genes (Sawarkar, Sievers, & Paro, 2012). This nuclear localization of Hsp90 may appear atypical for a predominantly cytosolic chaperone, especially given that Hsp90 lacks recognizable DNA-binding motifs. This apparent discrepancy can be easily reconciled by taking into account the physical interaction of Hsp90 with an expanding repertoire of nuclear proteins involved in gene transcription and epigenetic regulation (Erlejman, Lagadari, Toneatto, Piwien-Pilipuk, & Galigniana, 2014; Khurana & Bhattacharyya, 2015; Maloney et al., 2007; Sawarkar & Paro, 2013). The following sections will discuss possible mechanisms for and implications of "chromatinized" Hsp90.

1.1 Regulation of Transcription Factor Clients

1.1.1 Hsp90-Dependent Regulation of Steroid Hormone Receptors

Perhaps the best-known function of Hsp90 is its ability to chaperone kinase clients, thereby regulating kinase stability and function in response to diverse signaling cues. Hsp90 may play a similar, if not identical, role in its regulation of clients possessing transcriptional functions. Steroid hormone receptors are among the earliest identified nuclear Hsp90 clients (Pratt & Toft, 1997; Segnitz & Gehring, 1997). Hsp90 and its co-chaperone p23 are recruited to the DNA-bound glucocorticoid receptor (Freeman & Yamamoto, 2002), a ligand-dependent transcription factor (TF) of the steroid hormone receptor family. Among other members of this family, the androgen receptor (AR) and the estrogen receptor (ER) have particular relevance for prostate and breast cancer, respectively. AR was identified as an Hsp90-interacting protein over two decades ago (Marivoet, Van Dijck, Verhoeven, & Heyns, 1992; Veldscholte, Berrevoets, Brinkmann, Grootegoed, & Mulder, 1992), and subsequently, confirmed as a client protein impacted by Hsp90 blockade (Vanaja, Mitchell, Toft, & Young, 2002). Of clinical relevance, Hsp90 inhibition suppresses AR gene activation, the latter a major contributor to the emergence of lethal castrate-resistant prostate cancer because castrate-resistant PC (CRPC) continues to rely on AR signaling even in the absence of hormone stimulation (Augello, Den, & Knudsen, 2014). Hsp90 inhibitors suppress prostate cancer growth *in vivo* (Solit et al., 2002) and delay the CRPC phenotype (Lamoureux et al., 2011; O'Malley et al., 2012). Along a similar theme, breast cancer cells become resistant to anti-estrogen therapy (Hart et al., 2015). The ER is an Hsp90 interacting client protein (Renoir, Radanyi, Faber, & Baulieu, 1990), and Hsp90 cooperates with associated co-chaperones to regulate ER-dependent transcriptional activity (Knoblauch & Garabedian, 1999). Akin to prostate cancer models, impairment of ER function by targeting Hsp90 reduces xenograft tumor growth in tamoxifen-treated hormone-refractory cancers (Beliakoff et al., 2003) and mitigates the emergence of resistance to endocrine therapy at sub-proteotoxic concentrations (Whitesell et al., 2014). Hence, the ability of Hsp90 to regulate the assembly of functional receptor complexes and the subsequent activation of these TFs in response to hormone signaling is an important facet in the development of these common cancers.

1.1.2 Hsp90 and Inflammatory Mediators: Nuclear Factor κB and STAT3

In addition to hormone receptors, Hsp90 modulates the activity of a plethora of TFs, as recently reviewed (Erlejman et al., 2014; Khurana & Bhattacharyya, 2015), and depicted schematically in Fig. 1. Of this group, nuclear factor κB (NF-κB) and signal transducer and activator of transcription 3 (STAT3) play prominent roles in inflammatory pathways associated with oncogenic transformation and cancer progression (Basseres & Baldwin, 2006; Ben-Neriah & Karin, 2011; Li, Grivennikov, & Karin, 2011). STATs

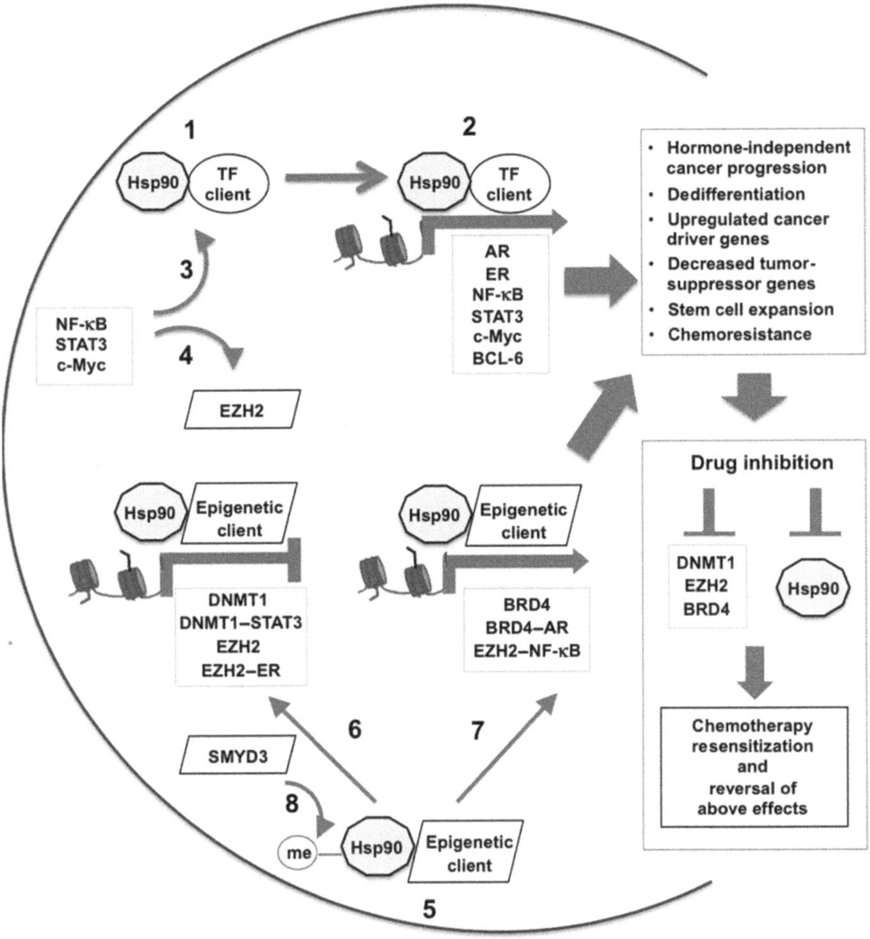

Figure 1 Hsp90 orchestrates a complex interplay of nuclear events by chaperoning diverse transcription factors and epigenetic modifiers. Large circle encompasses nuclear events impacted by Hsp90 function. (1) Hsp90 interacts with free transcription factor (TF) client proteins within the nuclear compartment. (2) A subset of Hsp90–TF complexes is also found in a chromatin-tethered state to directly modulate transcriptional events. The box encloses a subset of Hsp90 client TFs described in the text. (3) A cohort of Hsp90 client TFs is capable of directly enhancing *HSP90* gene expression. (4) This same subset of Hsp90 client TFs is also capable of upregulating expression of the epigenetic effector, and Hsp90 client, EZH2. (5) Hsp90 interacts with other epigenetic client proteins within the nuclear compartment. (6) Hsp90, in tandem with a subset of chromatin-tethered epigenetic clients (enclosed in box), may repress gene expression, most notably the expression of tumor-suppressor genes. (7) Hsp90, in tandem with a subset of chromatin-tethered epigenetic clients (enclosed in box), may promote gene expression, typically resulting in overexpression of oncogenic driver genes. (8) Complex bidirectional regulatory nodes exist. For example, Hsp90 client epigenetic proteins, such

represent a class of TFs activated upon tyrosine phosphorylation, typically in response to cytokine stimulation (Ihle, 1996). STAT3 expression and activity are critical for a number of cancer types, including pancreatic adenocarcinoma, multiple myeloma (MM), and prostate (Ammirante, Luo, Grivennikov, Nedospasov, & Karin, 2010; Chatterjee et al., 2007; Lang et al., 2007). Moreover, STAT3 has a general role in promoting tumor resistance to chemotherapy, in part via its ability to support cancer stem cell expansion (Tan, Putoczki, Stylli, & Luwor, 2014; Yu, Lee, Herrmann, Buettner, & Jove, 2014). STAT3 has been identified as an Hsp90 client (Shah, Patel, Fried, & Sehgal, 2002), and STAT3 impairment following Hsp90 inhibition may contribute to the efficacy of targeting Hsp90 in MM (Richardson et al., 2011).

NF-κB is a pivotal TF regulating the expression of proinflammatory cytokines and chemokines, including IL-6, an important pro-oncogenic factor in epithelial cancers such as breast and prostate (Sasser et al., 2007). NF-κB is indispensible for transformation in a number of cancer models (Iliopoulos, Hirsch, & Struhl, 2009), and NF-κB functions as a molecular link between inflammation and tumor progression (DiDonato, Mercurio, & Karin, 2012). NF-κB may be stimulated by a variety of signaling inputs, but typically requires prior activation of the upstream serine/threonine kinases IKK alpha and beta (Regnier et al., 1997; Woronicz, Gao, Cao, Rothe, & Goeddel, 1997). Hsp90 is a major facilitator of NF-κB activation via its association with and activation of the IKK complex (Chen, Cao, & Goeddel, 2002). More recently, NF-κB subunits have been identified as Hsp90 client proteins (Moulick et al., 2011), implicating an additional mode of direct chaperone modulation. Hsp90 inhibitors are active against tumor models dependent upon NF-κB survival signaling, such as those driven by tumor necrosis factor-alpha (TNF-α) (Wang et al., 2006). Thus, the ability to disable this critical survival signal is likely integral to the efficacy of Hsp90 inhibitor therapy.

In a majority of cancers, STAT3 and NF-κB coregulate one another, sustain a positive feedback regulatory loop (Fleming et al., 2015; Iliopoulos

---

as SMYD3, promote Hsp90 protein methylation, and modulate Hsp90 function. Collectively, these Hsp90-dependent nuclear events support a variety of key tumor-supportive pathways critical for cancer progression and therapeutic resistance. Intriguingly, pharmacological blockade of several of Hsp90's epigenetic clients restores drug sensitivity to formerly refractory tumor cells. This resensitization effect is also broadly reported to occur in response to Hsp90 inhibition, supporting the existence of functional cooperativity between these pathways. (See the color plate.)

et al., 2009), and function cooperatively to drive malignancy (He & Karin, 2011; Iliopoulos et al., 2009). Intriguingly, NF-κB upregulates Hsp90 transcription via direct promoter association (Ammirante et al., 2008; Fig. 1). Not surprisingly, Hsp90 is also a STAT3 target gene (Stephanou, Isenberg, Akira, Kishimoto, & Latchman, 1998), suggesting the likelihood of a positively reinforcing feed-forward pathway. Hsp90 expression is elevated in MM, and tumor cells demonstrating an NF-κB-mediated IL-6-activated STAT3 pathway are particularly responsive to Hsp90 inhibition (Kolosenko, Grander, & Tamm, 2014), indicating that this pathway may be clinically amenable to therapeutic targeting with Hsp90 inhibitors. Of note, both NF-κB and STAT3 regulate extensive microRNA circuits critical for linking and sustaining inflammation associated with cancer (Iliopoulos et al., 2009; Iliopoulos, Jaeger, Hirsch, Bulyk, & Struhl, 2010). The recruitment of downstream microRNA regulatory mechanisms provides a further impetus for cotargeting these TFs through Hsp90 inhibition as a treatment strategy.

1.1.3 Oncogenic Drivers: MYC

Overexpression of MYC is observed in a majority of human cancers. The MYC protein contributes to uncontrolled proliferation and dysregulated metabolism (Li & Simon, 2013), among other hallmarks of cancer (Cho et al., 2015; Gabay, Li, & Felsher, 2014). MYC elicits these effects by virtue of its ability to amplify broad transcriptional responses controlling many facets of cell behavior (Chappell & Dalton, 2013; Conacci-Sorrell, McFerrin, & Eisenman, 2014). The abundance of MYC protein is a key factor in its ability to integrate diverse signaling inputs and coordinate cellular responses (Patel, Loboda, Showe, Showe, & McMahon, 2004). Proteasomal turnover of the intrinsically unstable MYC protein plays an essential role in its functional restraint, and conversely, stabilizing mutations of MYC are prevalent in cancer as are genomic alterations that drive its overexpression. Hsp90 interacts with MYC and protects it from proteolytic destruction (Carystinos, Kandouz, Alaoui-Jamali, & Batist, 2003; Paul, Ahmed, Bhowmik, Deb, & Ghosh, 2013). *MYC*-amplified tumors exhibit a robust dependence upon MYC activity, as exemplified in MM biology, wherein MYC transcriptional activity is essential for disease maintenance (Delmore et al., 2011; Mertz et al., 2011). Hsp90 has also been shown to upregulate *MYC* at the transcriptional level (Sawarkar et al., 2012), suggesting the likelihood that Hsp90 enhances oncogenic MYC activity at multiple levels of control.

1.1.4 Relation Between TFs and Chromatin-Bound Hsp90

Most of the aforementioned studies have evaluated Hsp90–client associations within the experimental context of soluble cell lysate-derived fractions. While these studies collectively solidify a significant role for Hsp90 in regulating nuclear activities, they do not elucidate the precise mechanisms by which Hsp90 chaperones these nuclear clients to impact their transcriptional function. As a result of this uncertainty, several pertinent questions remain in regards to the relative contribution of the following mechanisms to overall regulation: (1) Hsp90 regulates soluble, non-DNA-bound forms of nuclear clients, (2) Hsp90 modulates the nuclear translocation and consequent activity of clients, (3) Hsp90 directly interacts with chromatin-bound forms of nuclear clients, and (4) Hsp90 acts in a dynamic fashion upon its nuclear clients at various points in their life cycle, in multiple cellular compartments and when soluble and chromatin bound (see Fig. 1). Although relatively few studies have evaluated and/or validated Hsp90's recruitment to promoter sites, precedent for this direct regulatory mode clearly exists. As mentioned earlier, Hsp90 associates with DNA-bound nuclear receptor complexes to modulate their assembly, disassembly, and activity (Freeman & Yamamoto, 2002). Along a similar theme, signaling cues promote recruitment of Hsp90 and NF-κB effectors to the TNF-α promoter, in tandem with the unliganded ER (Cvoro et al., 2006).

Even fewer studies have evaluated whether the presence of chromatinized Hsp90 at specific loci is dependent upon its interaction with chromatin-tethered TF clients. In support of this paradigm, Hsp90 found in association with DNA-bound MYC-containing transcriptional complexes regulates the expression of MYC target genes (Carystinos et al., 2003). Chromatinized Hsp90 has also been reported in B-cell lymphomas, the most common type of non-Hodgkin lymphoma. The Hsp90 client BCL6 plays a major role in this pathology (Ci, Polo, & Melnick, 2008), and Hsp90 is present at the promoter sites of BCL6 target genes (Cerchietti et al., 2009), implicating a role for Hsp90 as a direct mediator of chromatin-bound BCL6 stability and/or activity. Further, chromatin-bound Hsp90 is recruited to the promoters of STAT5 target genes (Moulick et al., 2011). Most recently, it was demonstrated that Hsp90 globally impacts the transcription of stimulus-dependent genes by regulating RNA polymerase II pausing via interaction with and stabilization of the negative elongation factor complex, the latter an Hsp90 client (Sawarkar et al., 2012). Further studies integrating ChIP and re-ChIP approaches will be required to validate the presence of, and requirement for, Hsp90–client interaction at chromatin-bound loci. These studies will help elucidate how stimuli can influence Hsp90 interaction with soluble and chromatinized populations of relevant nuclear Hsp90 client proteins.

An equally important query pertains to whether the interaction of Hsp90 with nuclear proteins necessarily implies the existence of a classical chaperone–client relationship. Certainly, the chaperone activity of Hsp90 is essential for at least a subset of its chromatin-bound functions (Cerchietti et al., 2009; Sawarkar et al., 2012). However, Hsp90 may be present in protein complexes in the absence of the prototypic chaperone functions associated with modulation of client protein conformation and stability. Such an atypical regulatory mode was reported for Hsp90's interaction with STAT3 and STAT5 (Moulick et al., 2011). These findings implicate an expanded role for Hsp90 in serving as a scaffold for the assembly of protein complexes, a putative role that may be particularly relevant for chromatinized Hsp90 functions.

1.2 Hsp90 Interaction with Epigenetic Effectors

Gene activation requires extensive alterations in chromatin structure. While the transcriptional functions of Hsp90 are becoming well established, emerging reports indicate that Hsp90 may also play a widespread role in gene regulation as an environmentally sensitive regulator of chromatin remodeling. Early reports have demonstrated an interaction of Hsp90 with histone proteins (Csermely & Kahn, 1991) and pointed to a role for Hsp90 as a facilitator of histone–DNA binding (Csermely, Kajtar, Hollosi, Oikarinen, & Somogyi, 1994). Chromatin-bound epigenetic functions of Hsp90 likely play an essential role in enabling Hsp90 to coordinate widespread changes in gene transcription. In support of this notion, early reports in *Drosophila* documented Hsp90 interaction with specific chromatin domains engaged in active gene transcription following heat shock (Morcillo, Diez, Carbajal, & Tanguay, 1993). The conserved function for Hsp90 as an epigenetic modifier has tremendous implications for its role in supporting the pathology of cancer as illustrated by the following examples.

1.2.1 Hsp90 Regulation of Trithorax and Trx Homologs

Within the expanding universe of chromatin remodelers, members of the Polycomb (PcG) and trithorax (Trx) family are among the most ancient and evolutionarily conserved. PcG and Trx are key catalytic components of epigenetic complexes controlling cell-lineage specification during normal development. In general, PcG and Trx play opposing roles within the context of gene regulation, with PcG repressing, and Trx activating developmental genes (Francis & Kingston, 2001; Mills, 2010; Schuettengruber, Chourrout, Vervoort, Leblanc, & Cavalli, 2007). The posttranslational modification (PTM) of histones is an essential step for subsequent modulation of chromatin structure and transcription. In this regard, Trx deposits an activating mark on H3 lysine residues (H3K4me3) (Milne et al., 2002), whereas PcG silences gene expression by depositing the repressive H3K27 trimethylation mark (Cao et al., 2002; Czermin et al., 2002; Muller et al., 2002). Of note, mutations of Trx family members in *Drosophila* generate phenotypes similar to those resulting from alterations in Hsp90-coding genes (Sollars et al., 2003). These examples provide strong support for an Hsp90–Trx axis which functions to limit variation within, or "canalize" normal developmental processes. Hence, in the "abnormal development" of cancers, such epigenetic effectors also likely serve as crucial collaborators of Hsp90 in buffering the phenotypic manifestations of increasing genetic variation/mutational load during malignant progression.

Collectively, these findings highlight the complex ways in which phenotypic variability fueling morphological evolution may arise, either during natural selection or within the context of cancer evolution in patients, following functional impairment of either chaperone activity or cooperating accessory proteins within this axis, a concept further elaborated by Jarosz, this volume. Importantly, Trx has been identified as an Hsp90 client, and Hsp90 and Trx colocalize at the promoters of Trx target genes (Tariq, Nussbaumer, Chen, Beisel, & Paro, 2009), thereby reinforcing a mechanistic basis for the functional linkage of an Hsp90–Trx axis, and further validating bona fide chromatin-centric epigenetic functions of Hsp90.

Drosophila Trx is a member of the SET1 (suppressor of variegation, enhancer of zeste, and Trx) domain family of H3K4 methyltransferases. Although Trx gene products do not have a clearly conserved functional role, their human orthologs are frequently deregulated in cancers (Kingston & Tamkun, 2014). Humans have six related SET family members, including MLL1 (mixed lineage leukemia protein-1), which plays an essential role during early development and hematopoiesis. MLL1 was shown to be an Hsp90 client (Tariq et al., 2009), and MLL1 also facilitates the heat-shock response induced by Hsp90 inhibition (Chen et al., 2014). MLL1 is a fusion partner in a large number of pathogenetic chromosomal rearrangements found in myeloid and lymphoblastic leukemias (Mohan et al., 2010) and also drives critical gene networks in solid tumors (Gallo et al., 2013; Takeda et al., 2013). Additional Trx homolog members MLL2 and MLL3 are also frequently mutated and rearranged in cancer (Morin et al., 2011; Parsons et al., 2011).

Although a role for Hsp90 in the regulation of MLL family members remains to be clarified, emerging reports indicate that Hsp90 interacts with a group of epigenetic modifiers containing the shared SET domain. Both SMYD2 and SMYD3, members of the SET and MYND (Myeloid-Nervy-DEAF1) domain-containing histone methyltransferases, are reported to interact with Hsp90 (Abu-Farha et al., 2008; Hamamoto et al., 2004). SMYD proteins control gene expression and the DNA damage response. They are of growing therapeutic interest due to their expanding role in malignancy (Spellmon, Holcomb, Trescott, Sirinupong, & Yang, 2015). These enzymes methylate histone and non-histone targets, and interestingly, Hsp90 was also identified as a substrate for SMYD-directed methylation (Abu-Farha et al., 2011; Hamamoto et al., 2004; Fig. 1). Hsp90 enhances the catalytic activity of SMYD2 (Abu-Farha et al., 2008), while SMYD3-directed Hsp90 methylation promotes Hsp90–co-chaperone interaction (Abu-Farha et al., 2011). These findings highlight the existence of a complex bidirectional regulatory mode between Hsp90 and epigenetic effectors. Additional studies are warranted to better define the extent of this bidirectional relationship within the context of the expanded cohort of SET domain-containing epigenetic modifiers.

1.2.2 Hsp90-Dependent Regulation of the Polycomb Methyltransferase EZH2

Similar to Trx, PcG homologs are conserved among species. In humans, the Polycomb-repressive complex 2 (PRC2) initiates gene silencing by depositing the repressive histone H3K27 trimethylation "mark." PRC2 epigenetically controls seminal biological processes including development, cell fate, and notably, tumor progression (Margueron & Reinberg, 2011; Sparmann & van Lohuizen, 2006). The methyltransferase enhancer of zeste homolog 2 (EZH2) is the catalytic component of PRC2. EZH2 is overexpressed in a variety of malignancies, including breast and prostate cancers, a trend that correlates with poor prognosis (Kleer et al., 2003; Varambally et al., 2002). Recent studies have demonstrated the potent anticancer effects of small-molecule inhibitors of EZH2 (McCabe et al., 2012).

EZH2 was recently identified as an Hsp90 client protein (Fiskus et al., 2009). Although it remains unclear how Hsp90 globally regulates EZH2 chromatin recruitment and associated histone methylation, a very recent report showed that the *Drosophila* Polycomb protein was required for gene repression during heat stress (Li, Lyu, et al., 2015), supporting the premise that heat stress-associated proteins such as Hsp90 likely cooperate with Polycomb effectors. Given that EZH2 is an attractive therapeutic target in diverse cancers (McCabe & Creasy, 2014; Shen & Vakoc, 2015) and Hsp90 inhibition is expected to have a major impact upon EZH2 function, it is likely that impairment of EZH2 contributes to the efficacy of Hsp90-targeted therapies (Fig. 1). Interestingly, EZH2 blockade sensitizes cancers to chemotherapy (Yamaguchi & Hung, 2014), an effect phenocopied by Hsp90-targeted agents (Acquaviva et al., 2014; Ku et al., 2014; Mitsiades et al., 2006; Miyajima et al., 2013; Sang et al., 2013). Activating point mutations of EZH2, resulting in enhanced EZH2 catalytic activity, have been found in subsets of B-cell neoplasms (Shen & Vakoc, 2015). It will be of interest to evaluate the Hsp90 dependence of these mutant proteins, given the well-known preference of the Hsp90 chaperone for mutant clients relative to matched nonmutant counterparts. Inversely, the cancer-promoting functions of EZH2 need to be balanced with recent findings that EZH2 may also possess tumor-repressive functions in certain contexts (Hock, 2012). Thus, Hsp90 inhibitors that repress EZH2 function may have deleterious consequences within specific cancer subtypes.

These reports highlight some of the complexity inherent to epigenomic targeting by Hsp90 inhibition. An additional source of complexity arises from the finding that active and repressive marks may coexist on target genes. This bivalency, particularly noted at developmental genes, is also observed in embryonic stem cells (Bernstein et al., 2006). Epigenetic bivalency is also a hallmark of cellular plasticity and is associated with activation of EMT (epithelial to mesenchymal transition) and increased "stemness" (Malouf et al., 2013). Pathological reactivation of the developmental genetic program known as EMT is associated with increased malignant potential and metastasis in diverse solid tumors (Thiery, Acloque, Huang, & Nieto, 2009; Tsai & Yang, 2013). While EZH2 is implicated as a driver of EMT events in cancer (Cao et al., 2008; Herranz et al., 2008; Tiwari et al., 2013; Tong et al., 2012), it remains to be determined how interference with Hsp90 function may alter the role of EZH2 in epigenetic bivalency and cellular plasticity associated with malignancy.

1.2.3 Cross Talk Between Hsp90 and DNA Methylation

The cytosine methylation of DNA, executed by the DNA methyltransferases (DNMTs), plays an essential role in gene repression and establishment of epigenetic memory (Bird, 2002). The loss of DNA methylation causes widespread epigenetic deregulation (Jackson-Grusby et al., 2001). In mammalian cells, PcG-mediated gene repression acts in a cooperative manner with DNA methylation. EZH2 can directly recruit DMNTs to target genes (Reynolds et al., 2006; Vire et al., 2006), and PcG-marked genes are major targets for DNMT activity (Schlesinger et al., 2007). The ability of EZH2 to interact with and recruit DNMT1 provides a mechanism by which EZH2 can mark genes for subsequent *de novo* DNA-dependent methylation (Schlesinger et al., 2007).

Hsp90 inhibition also targets DNMT1 for proteolytic destruction (Fiskus et al., 2009). Among members of the DNMT family, DNMT1 plays a prominent role in a number of processes, including somatic and stem cell viability, tissue regeneration, and homeostasis (Li, Bestor, & Jaenisch, 1992; Liao et al., 2015; Sen, Reuter, Webster, Zhu, & Khavari, 2010). In cancer, DNMTs cooperate to silence tumor-suppressive genes (Rhee et al., 2002), and cancer cells are particularly sensitive to the loss of DNMT1 (Chen et al., 2007). Small-molecule inhibitors of DNMTs such as decitabine have shown promise in a variety of cancers and resensitize tumor cells previously refractory to chemotherapeutic agents (Oronsky et al., 2014). Interestingly, in leukemia cells, additive cell killing was elicited by cotreatment with Hsp90 inhibitors and decitabine (Fiskus et al., 2009). It is currently unknown whether other DNMT family members are similarly susceptible to Hsp90-dependent regulation. Moreover, given that Hsp90 targeting concomitantly reduced expression of both EZH2 and DNMT1 (Fiskus et al., 2009), it remains unknown whether Hsp90 binds each of these proteins in isolation, or whether Hsp90 may act as a scaffold to bridge their interaction. However, these findings offer strong support for the notion that Hsp90 plays a pivotal role in regulating gene repression, at least in part via its modulation of epigenetic effectors.

1.2.4 Hsp90-Dependent Regulation of BRD4

Although most druggable epigenetic targets in cancer therapy possess catalytic activity, newer agents are capable of inactivating epigenetic targets lacking overt catalytic function. As an example of this class of proteins, bromodomain and extraterminal (BET) proteins serve as adaptors and recruit transcriptional regulatory complexes to acetylated chromatin. The conserved bromodomain motif is a chromatin-interaction module that recognizes acetylated lysine residues on histone tails, thus the designation of BET proteins as chromatin "readers" (Arrowsmith, Bountra, Fish, Lee, &

Schapira, 2012). One member of this class, BRD4, has recently been identified as a potent oncogenic factor amenable to targeted inhibition by specific small molecules such as JQ1 and I-BET (Filippakopoulos et al., 2010; Nicodeme et al., 2010). BRD4 couples histone acetylation to Pol II-mediated transcriptional elongation and is typically associated with active promoter regions (Jang et al., 2005; LeRoy, Rickards, & Flint, 2008).

BRD4 has recently been identified as an Hsp90 client (Zhou et al., 2015). This is a particularly interesting development given the unique role of BRD4 in the formation of super-enhancer domains, a restricted set of hyperacetylated chromatin regions associated with of active transcription. Super-enhancers play prominent roles in controlling cell identity during development as well as regulating expression of driver oncogenes in cancer (Hnisz et al., 2013; Loven et al., 2013). Super-enhancers promote the expression of genes that feature prominently in MM, such as MYC. Super-enhancer domains are exquisitely sensitive to BRD4 inhibition, resulting in the preferential loss of BRD4 association and concomitant-diminished expression of downstream driver genes (Dawson et al., 2011; Delmore et al., 2011; Loven et al., 2013; Mertz et al., 2011; Zuber et al., 2011). BET inhibitors can help overcome resistance to kinase inhibitor therapy (Stratikopoulos et al., 2015), and recent reports document an expanded role for BRD4 in a variety of solid tumors (Cho et al., 2014; Shi & Vakoc, 2014; Shi et al., 2014). It will be invaluable to discern whether Hsp90 globally affects BRD4 action and assists the BRD4-mediated transcriptional addictions associated with specific malignancies. This concept is particularly intriguing in that Hsp90 inhibitors possess pronounced anticancer activity against myeloma and leukemias (Moulick et al., 2011; Okawa et al., 2009). In light of the broad repertoire of Hsp90-dependent epigenetic clients also moonlighting as transcriptional drivers of these cancers, the mechanism(s) underlying the striking activity of Hsp90 inhibitors against these tumors may need to be reconsidered.

1.2.5 Hsp90-Dependent Regulation of Additional Chromatin Remodelers

The constellation of Hsp90-interacting epigenetic modifiers continues to expand (Sawarkar & Paro, 2013). Recent reports identify the hypoxia-regulated histone demethylase KDM3A as an Hsp90 client protein (Kasioulis et al., 2014). KDM3A plays a role in hypoxia-driven effects on tumorigenicity (Krieg et al., 2010) and in hormone-dependent cancers and other solid tumors (Mahajan, Lawrence, Lawrence, & Mahajan,

2014; Osawa et al., 2013; Parrish, Sechler, Winn, & Jedlicka, 2015; Sowalsky et al., 2015; Wade et al., 2015). Unbiased proteomic screens have uncovered additional Hsp90 targets, such as PRMT5 (Maloney et al., 2007), an arginine methyltransferase upregulated by MYC (Koh et al., 2015) that regulates RNA splicing, development, and pluripotency (Stopa, Krebs, & Shechter, 2015; Tee et al., 2010). Interestingly, PRMT5 also methylates cytosolic targets to drive malignant signaling (Hsu et al., 2011; Wei et al., 2013), highlighting the functional duality of these effectors with respect to histone and non-histone targets. This duality has also been observed with other epigenetic modifiers identified as Hsp90 targets, as will be discussed in the following section. These findings suggest that beyond traditional chaperone–client mechanisms, Hsp90-dependent modulation of methyltransferases and other enzymes that shape the epigenome may play an indirect but nonetheless important accessory role in calibrating the activity and/or stability of a subset of Hsp90 client proteins.

2. INDIRECT NETWORK-BASED EFFECTS OF HSP90 CLIENTS ON EPIGENETIC REGULATION

Although the prior section highlighted classical relationships involving Hsp90 interaction with a client protein "X" to elicit a direct effect "Y," the "real-life" biological output of chaperone activity is likely much more complex, encompassing indirect effects and interconnected circuitry. The following sections will touch on perhaps some less obvious aspects of this complexity.

2.1 Hsp90 TF Clients Play Pivotal Roles in Epigenetic Regulation

This section will illustrate some interconnected examples wherein known Hsp90-regulated transcriptional pathways are likely to cooperate with chromatin modifiers that have well-established roles in cancer development and progression. As described in Section 1.2.2, EZH2 is a primary facilitator of cancer progression in many solid tumors. Indeed, malignant progression may select for EZH2 upregulation, a notion supported by the observation that several TF clients also function as drivers of EZH2 transcription (Fig. 1). For instance, both NF-κB and STAT3 bind consensus sites within the EZH2 promoter to upregulate EZH2 transcription (Lin et al., 2013; Yeh et al., 2009). Hsp90 modulates NF-κB activity and enhances both STAT3 protein expression and phosphorylation (Lin et al., 2013), the latter of which directs STAT3 nuclear localization and DNA binding (Darnell, Kerr, & Stark, 1994; Ihle, 1996; Zhang, Blenis, Li, Schindler, & Chen-Kiang, 1995). This is a particularly interesting dynamic given the bidirectional regulation and functional cooperativity of these inflammatory mediators. Of clinical significance, expression of EZH2 protein and activated phospho-STAT3 is increased in cancer tissues, a trend that positively correlates with metastasis (Lin et al., 2013). EZH2 is also regulated by the MYC oncoprotein, either by virtue of direct promoter interaction, or through MYC-mediated repression of EZH2-inhibitory microRNAs (Koh et al., 2011).

In addition to regulation of promoter activity, TF clients may play important roles in recruiting epigenetic modifiers such as EZH2 to target loci. This recruitment function has recently been described for STAT5, which recruits EZH2 to mediate transcriptional repression (Mandal et al., 2011). Importantly, the choice of TF-binding partner appears to play a major role in directing EZH2 function. While an NF-κB–EZH2 interaction contributes to breast cancer, it was recently shown that these factors engage in complex functional cross talk (Lee et al., 2011). In less aggressive ER-positive breast cancers, EZH2 binds to ER to repress NF-κB target genes. Inversely, in more aggressive ER-negative basal-like breast cancers, EZH2 functions as an activator and associates with the NF-κB subunits RelA and RelB to drive target genes associated with poor outcome. Moreover, EZH2 and NF-κB coregulate additional oncogenic pathways to regulate breast stem cell expansion (Gonzalez et al., 2014). These examples emphasize the importance of considering context-dependent consequences of targeting EZH2 in malignancy.

TF clients also interact with a diverse array of epigenetic complexes. Recently, AR was shown to cooperate with BRD4 to regulate AR target gene expression and to promote the CRPC phenotype (Asangani et al., 2014). These findings implicate AR as a functional "super-enhancer" within this context. Given that both AR and BRD4 are Hsp90 client proteins, it is conceivable that Hsp90-targeted approaches may provide additive or synergistic effects with bromodomain inhibition. Similarly, STAT3 partners with DNMT1 to regulate DNA methylation and transcriptional silencing of both tumor-suppressor genes and therapeutic targets in breast cancer (Lee et al., 2012), illustrating another instance in which Hsp90 inhibition may have utility in cotargeting collaborating epigenetic effectors. As a final example, PRMT5 is essential for MYC-mediated transformation (Li, Chitnis, et al., 2015), and a PRMT5-MYC complex was shown to be critical for regulation of RNA splicing and lymphomagenesis, the former of which is a dysregulated trait integral to haematopoietic malignancies (Koh et al., 2015).

2.2 Kinase–Chromatin Cross Talk

The generally suppressive function of Hsp90 inhibitors upon client kinase activity and consequent attenuation of downstream oncogenic signaling is largely credited for the anticancer properties of these agents (Caldas-Lopes et al., 2009; Moulick et al., 2011). Although impaired signaling is arguably a major component of Hsp90 inhibitor action, it is also quite possible that at least a subset of the anti-neoplastic properties of Hsp90 inhibitors are the result of downstream effects upon a cohort of transcriptional and epigenetic regulators. Although Hsp90 regulates myriad-signaling pathways, for simplicity, we will primarily focus on AKT and extracellular signal-regulated kinase (ERK), key downstream effectors for numerous client kinases. In addition to Hsp90-dependent activation of AKT and ERK by virtue of upstream kinase signaling, reports have also demonstrated a more direct role for Hsp90-dependent regulation of AKT protein (Basso et al., 2002). Hsp90 also supports the activity of multiple upstream activating kinases of the ERK signaling cascade (Grammatikakis, Lin, Grammatikakis, Tsichlis, & Cochran, 1999; Schulte, An, & Neckers, 1997), and recent reports indicate that ERK may even be a direct client protein (Moulick et al., 2011).

2.2.1 AKT Signaling and Epigenetic Modulation

AKT activation regulates numerous signaling pathways integral to cancer progression, metastasis, and treatment failure (Pereira, Horta, Mateus, & Videira, 2015; Rodrik-Outmezguine et al., 2011). Given the dynamic nature of chromatin remodeling in response to cellular cues, it is not surprising that epigenetic effectors may also serve as AKT substrates. AKT-dependent regulation of EZH2 has emerged as one striking example of this regulatory node. Although EZH2 is typically categorized as an epigenetic repressor, recent findings indicate that PTMs of EZH2 play a major role in altering the properties and/or functions of this protein (Yamaguchi & Hung, 2014). Emerging findings indicate that the AKT-dependent phosphorylation of EZH2 at Ser21 dramatically changes EZH2's substrate specificity from histone targets to non-histone substrates. This functional switch reconciles the seemingly contradictory finding that although EZH2 protein levels are statistically increased during prostate cancer progression, a trend especially pronounced in CRPC, the corresponding inhibitory

H3K27me3 mark is not concordantly increased in these same tissue specimens (Xu et al., 2012). Mechanistically, the EZH2-Ser21 modification converts EZH2 into a transcriptional coactivator that subsequently cooperates with AR (Fig. 2). Intriguingly, EZH2's methyltransferase activity is required for this acquired activator function, indicating that EZH2 can "mark" alternate, non-histone substrates. AR is a likely member of this methylated cohort, as levels of methylated AR diminish in accordance with EZH2 blockade (Xu et al., 2012).

AKT-directed signaling also functions as a requisite stimulus for driving an EZH2-STAT3 oncogenic axis in cancers. EZH2-Ser21 catalyzes STAT3 methylation, which robustly enhances STAT3 activity via subsequent STAT3 tyrosine phosphorylation (pY705) (Kim et al., 2013; Fig. 2). This AKT-EZH2-STAT3 pathway is especially critical for maintaining the cancer stem cell population in glioblastoma multiforme (GBM), one of the most aggressive and lethal malignancies. Glioma stem cells, representing a highly malignant subpopulation universally found in GBM, are drivers of glioma propagation, tumor recurrence, and the established ability of GBM to evade conventional therapy (Chen et al., 2012; Lathia, Mack, Mulkearns-Hubert, Valentim, & Rich, 2015). The frequent coelevation of EZH2 and STAT3 levels in cancers, coupled with the sustained activation of AKT by myriad-signaling pathways, highlights the importance, and likely prevalence of this axis in a variety of malignancy-supporting scenarios. It will be of great interest in the future to dissect the respective roles of Hsp90 as a regulator of each protein within this tripartite client complex. Moreover, AKT binds to other histone sites in response to stimuli (Lee et al., 2015), supporting the notion that additional epigenetic targets and functions will be revealed. Most recently, AKT was shown to play a global role in regulating histone acetylation in a number of cancer types, including GBM (Lee et al., 2014).

2.2.2 Impact of ERK Signaling Upon Epigenetic Modifiers

Sustained ERK activation is a prominent abnormality in a large number of cancers. In addition to upstream signaling cascades, ERK is constitutively activated by genetic events, such mutation of BRAF kinase, an aberration most frequently observed in colorectal cancer and malignant melanoma (Davies et al., 2002; Hodis et al., 2012). ERK reactivation also commonly occurs as a feedback response to therapeutic intervention, which impairs clinical responses to anticancer strategies (Ahronian et al., 2015; Johannessen et al., 2010; Lai et al., 2014; Wagle et al., 2014). Mutant BRAF is a well-confirmed Hsp90 client protein (da Rocha Dias et al., 2005), and

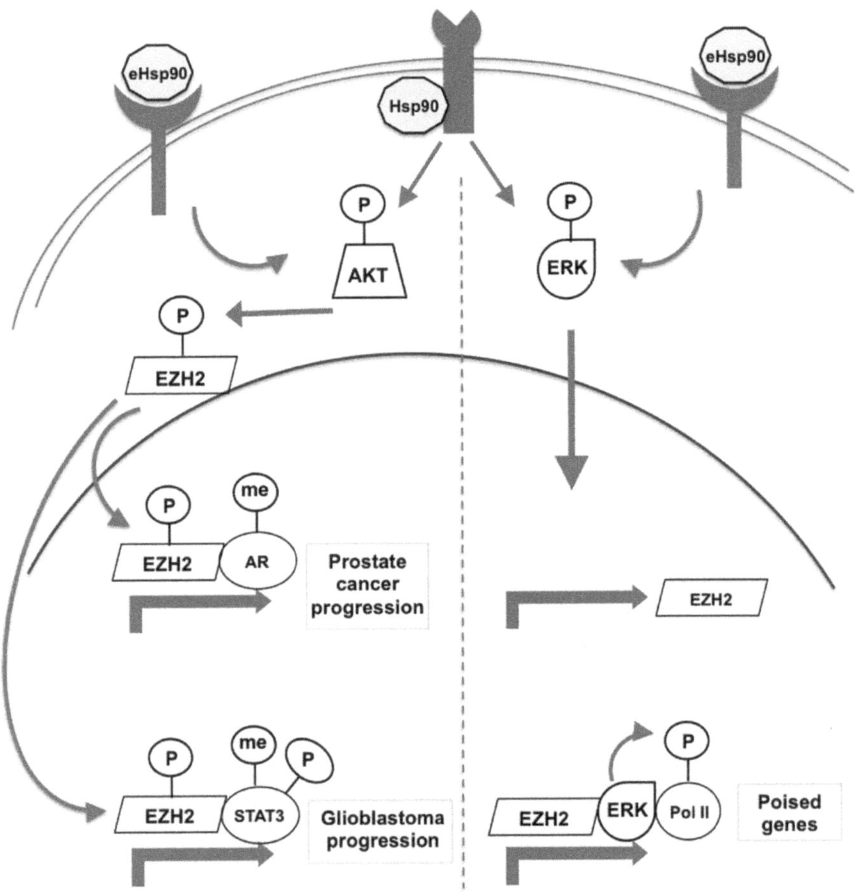

Figure 2 Hsp90 is a central mediator of kinase–chromatin cross talk. (Left of dashed line) Hsp90 chaperones numerous cytosolic receptor tyrosine kinases (RTKs) to support downstream signaling events, such as AKT activation. Hsp90 may also assist in the activation of AKT via client interaction (not shown). Activated AKT promotes EZH2 phosphorylation, an event that diminishes EZH2's affinity for histone substrates. AKT-phosphorylated EZH2 then preferentially associates with a cohort of TFs, depicted here as androgen receptor (AR) or STAT3. EZH2 also methylates its interacting partners to regulate gene transcription and cancer progression. Extracellular Hsp90 (eHsp90) may also signal through membrane proteins and receptors to drive AKT activation, although the epigenetic effects of this specific signaling portal have not yet been elucidated. (Right of dashed line) Hsp90-dependent signal transduction may also lead to ERK activation. ERK activates additional intermediates to promote the transcriptional upregulation of EZH2. ERK, which contains a DNA-binding domain, may also be corecruited with EZH2 at transcriptionally paused "or poised" genes, where it regulates polymerase II (Pol II) phosphorylation. Extracellular Hsp90 may similarly increase ERK activity, and subsequently, contribute to ERK-mediated EZH2 transcriptional events. Although not shown in this depiction, it is also possible that Hsp90 is coordinately recruited to these client-occupied DNA-binding sites.

Hsp90 inhibition has been shown to delay or overcome the emergence BRAF inhibitor resistance and to provide combinatorial benefit (Acquaviva et al., 2014; Paraiso et al., 2012; Smyth et al., 2014). While many of these studies have focused on the signaling-related functions of ERK, this kinase also plays pleiotropic roles within the context of epigenetic regulation. ERK activates intermediate substrates that facilitate EZH2 transcription, resulting in elevated EZH2 expression in cancer cells (Fujii et al., 2011; Fig. 2). ERK also contains a DNA-binding domain and is extensively associated with chromatin in embryonic stem cells, where it is recruited to active promoter regions that regulate targets promoting survival, proliferation, and pluripotency (Goke, Chan, Yan, Vingron, & Ng, 2013). Intriguingly, ERK can direct the deposition of EZH2 onto chromatin, which facilitates EZH2 recruitment at developmental loci, and enhances phosphorylation of RNA Pol II to establish poised chromatin features at these sites (Tee, Shen, Oksuz, Narendra, & Reinberg, 2014; Fig. 2). These studies clearly demonstrate the profound ability of ERK to impinge on chromatin regulation and further highlight the importance of kinase–chromatin interactions in the regulation of gene transcription.

ERK has also been shown to phosphorylate a number of histone-modifying proteins, indicating that it regulates a vast array of chromatin-dependent functions (Zhou et al., 2000), including epigenetic gene silencing during malignant EMT (Ichikawa et al., 2015). ERK kinase activity is also associated with gene hypermethylation and silencing, leading to genomic instability and a mutator phenotype especially prevalent in BRAF-driven colorectal cancers (Fang, Ou, Hutchinson, & Green, 2014). Similarly, ERK enforces the hypermethylation of tumor-suppressor genes in cancers, a function critical for Ras–ERK-mediated cellular transformation (Wu & Brenner, 2014). Not surprisingly, combination strategies comprised of ERK-targeted agents in tandem with chromatin-directed therapeutics are showing promise in overcoming treatment resistance (Johannessen et al., 2013). Of note, ERK signaling upregulates Hsp90 transcription through a variety of mechanisms (Li et al., 2010), suggesting that Hsp90 inhibition may have utility in enhancing the anticancer activity of combination ERK/chromatin-targeted regimens.

2.3 Cross Talk Between Extracellular Hsp90-Dependent Signaling and Nuclear Events

No discussion of Hsp90's relevance to cancer would be complete without including a commentary on its activity within the extracellular compartment. Now well accepted, Hsp90 executes discrete functions within distinct cellular compartments, including the cytosol, nucleus, and mitochondria. It is therefore not surprising that Hsp90 can be found extracellularly, especially in the tumor microenvironment. We, and others, have shown that extracellular Hsp90 (eHsp90) promotes tumor cell motility and invasion, and supports metastasis (Eustace et al., 2004; Gopal et al., 2011; Hance et al., 2012; Lagarrigue et al., 2010; Sidera, Gaitanou, Stellas, Matsas, & Patsavoudi, 2008; Stellas, Karameris, & Patsavoudi, 2007; Tsutsumi et al., 2008; Wang et al., 2009), and as reviewed by Jay's group, this volume. Although chaperone activity may contribute to eHsp90 function in malignancy, eHsp90 is also capable of eliciting signaling events in a manner uncoupled from this prototypic function (Cheng et al., 2008; Gopal et al., 2011). This eHsp90-stimulated pathway may be mediated by the multifunctional prosurvival receptor LDL receptor-related protein 1 (LRP1) (Fig. 2). This demonstration of an eHsp90–LRP1 axis clearly illustrates the ability of eHsp90 to participate in ligand-dependent signaling in response to environmental cues.

Of particular relevance to the discussion in Section 2.2, eHsp90 signaling events stimulate the activity of both AKT and ERK kinases (Gopal et al., 2011; Hance et al., 2012; Tsen et al., 2013). Although the potential cross talk between eHsp90 and epigenetic events has not been well studied, we recently reported that eHsp90–ERK signaling plays a major role in regulating EZH2 expression and activity in prostate cancer, the latter critical for tumor cell invasion and EMT activation (Nolan, Franco, Hance, Hayward, & Isaacs, 2015; Fig. 2). Tumor eHsp90 also appears be a conserved regulator of matrix metalloproteinase (MMP) activity (Eustace et al., 2004; Lagarrigue et al., 2010; Song et al., 2010; Stellas, El Hamidieh, & Patsavoudi, 2010), a critical proteolytic hub promoting cancer invasion and progression. MMP activation is known to robustly stimulate downstream signaling events, including ERK activation (Lue et al., 2011) and as we show (Hance et al., 2012), sustained activation of ERK may represent a prominent facet of eHsp90 function. Moreover, several reports indicate that eHsp90 stimulates EGFR signaling (Sidera et al., 2008; Thuringer et al., 2011), supporting the notion that multiple mechanisms collaborate to enforce eHsp90-dependent ERK activation. The signaling effectors impacted by eHsp90 are not limited to AKT and ERK. Signaling driven by eHsp90 may also elicit NF-κB activation (Chen et al., 2010), an event capable of promoting context-dependent EMT activation (Chen, Chen, Chen, Lee, & Huang, 2013). Finally, we have shown that eHsp90 activates the

NF-κB–STAT3 axis in cancer-associated stromal cells (Bohonowych et al., 2014). Collectively, these reports indicate that eHsp90-dependent signaling events contribute to the AKT, ERK, and NF-κB–STAT pathways frequently hyperactivated in cancers. Hence, future studies are warranted to more fully explore the potential cross talk between eHsp90-dependent signaling events and epigenetic regulation.

3. THE PERSPECTIVE: REVISITING Hsp90 FUNCTION IN CANCER THERAPY

Although "canonical" activities of Hsp90 have been extensively investigated, much less has been established regarding the functions of nuclear, and in particular, chromatinized Hsp90. This shortcoming is especially problematic, in light of the probability that the myriad nuclear functions of Hsp90 are likely key components of Hsp90's pro-tumorigenic action. Conversely, a clear understanding of how these functions and effectors operate in the context of specific types of cancer may allow for the assembly of synergistic combination drug regimens. Rational combination of "epigenetic drugs" with Hsp90 inhibitors would have the potential to lessen the dose-related toxicities of both classes of agent while simultaneously improving overall anticancer efficacy. Support for this concept is provided by the various examples cited above, where Hsp90 inhibitors or drugs that modify the epigenome were shown to resensitize tumor cells to conventional chemotherapeutic agents. Future studies are warranted to better define the global and context-dependent effects of Hsp90 inhibition on key transcriptional and epigenetic modifiers, and how these effects may ripple out to impact primary cancer drivers and/or perturb other regulatory mechanisms, such as noncoding microRNAs. To validate Hsp90 as a bona fide epigenetic modifier, it will also be necessary to determine whether chromatin-bound client proteins retain their sensitivity to Hsp90 blockade, relative to their nonchromatin-bound counterparts. Further investigation is also needed to determine whether specific PTMs of Hsp90 or of one or more of its interacting co-chaperones may help "chromatinize" Hsp90. In support of such possibilities, an earlier report indicated that histone binding increases Hsp90 phosphorylation (Csermely & Kahn, 1991). Clearly, a more comprehensive understanding of these dynamic, intricately related events may ultimately enable more precise targeting of this distinct Hsp90 population.

Considerably more work will be needed to fully understand the complex, potentially hierarchical effects of Hsp90 upon chromatin structure. Given that early studies implicated Hsp90 in chromatin condensation (Csermely et al., 1994), it would not be surprising if Hsp90 promoted formation of the large chromatin-bound complexes that mediate long-distance interactions and DNA looping to broadly regulate chromatin structure. In fact, the Hsp90 co-chaperone Aha1 interacts with condensin and cohesion-binding proteins (Sun, Hartson, & Matts, 2015), portending a role for Hsp90 in regulation of higher-order chromatin structure. This possibility is further supported by a recent report demonstrating that heat stress induces extensive remodeling of chromatin topology and facilitates promoter–enhancer interactions (Li, Lyu, et al., 2015). The potential role of Hsp90 as a global effector of chromatin structure will also need to be revisited within the clinical context of chemo- and radiotherapy response. For instance, chromatin condensation has recently been shown to confer protection from radiation-induced genomic damage by mechanisms that require further investigation (Wang et al., 2014). Indeed, from a broad perspective, the provocative findings described in this review are really just a beginning, but they provide a strong impetus to develop a revised, more comprehensive, more sophisticated view of Hsp90 function, wherein the classical signaling-related functions of cytosolic Hsp90 are integrated with the myriad nuclear activities that are now emerging for this ubiquitous chaperone.

REFERENCES

Abu-Farha, M., Lambert, J. P., Al-Madhoun, A. S., Elisma, F., Skerjanc, I. S., & Figeys, D. (2008). The tale of two domains: Proteomics and genomics analysis of SMYD2, a new histone methyltransferase. *Molecular & Cellular Proteomics: MCP*, *7*, 560–572.

Abu-Farha, M., Lanouette, S., Elisma, F., Tremblay, V., Butson, J., Figeys, D., et al. (2011). Proteomic analyses of the SMYD family interactomes identify HSP90 as a novel target for SMYD2. *Journal of Molecular Cell Biology*, *3*, 301–308.

Acquaviva, J., Smith, D. L., Jimenez, J. P., Zhang, C., Sequeira, M., He, S., et al. (2014). Overcoming acquired BRAF inhibitor resistance in melanoma via targeted inhibition of Hsp90 with ganetespib. *Molecular Cancer Therapeutics*, *13*, 353–363.

Ahronian, L. G., Sennott, E. M., Van Allen, E. M., Wagle, N., Kwak, E. L., Faris, J. E., et al. (2015). Clinical acquired resistance to RAF inhibitor combinations in BRAF-mutant colorectal cancer through MAPK pathway alterations. *Cancer Discovery*, *5*, 358–367.

Ammirante, M., Luo, J. L., Grivennikov, S., Nedospasov, S., & Karin, M. (2010). B-cell-derived lymphotoxin promotes castration-resistant prostate cancer. *Nature*, *464*, 302–305.

Ammirante, M., Rosati, A., Gentilella, A., Festa, M., Petrella, A., Marzullo, L., et al. (2008). The activity of hsp90 alpha promoter is regulated by NF-kappa B transcription factors. *Oncogene*, *27*, 1175–1178.

Arrowsmith, C. H., Bountra, C., Fish, P. V., Lee, K., & Schapira, M. (2012). Epigenetic protein families: A new frontier for drug discovery. *Nature Reviews. Drug Discovery, 11*, 384–400.

Asangani, I. A., Dommeti, V. L., Wang, X., Malik, R., Cieslik, M., Yang, R., et al. (2014). Therapeutic targeting of BET bromodomain proteins in castration-resistant prostate cancer. *Nature, 510*, 278–282.

Augello, M. A., Den, R. B., & Knudsen, K. E. (2014). AR function in promoting metastatic prostate cancer. *Cancer Metastasis Reviews, 33*, 399–411.

Basseres, D. S., & Baldwin, A. S. (2006). Nuclear factor-kappaB and inhibitor of kappaB kinase pathways in oncogenic initiation and progression. *Oncogene, 25*, 6817–6830.

Basso, A. D., Solit, D. B., Chiosis, G., Giri, B., Tsichlis, P., & Rosen, N. (2002). Akt forms an intracellular complex with heat shock protein 90 (Hsp90) and Cdc37 and is destabilized by inhibitors of Hsp90 function. *The Journal of Biological Chemistry, 277*, 39858–39866.

Beliakoff, J., Bagatell, R., Paine-Murrieta, G., Taylor, C. W., Lykkesfeldt, A. E., & Whitesell, L. (2003). Hormone-refractory breast cancer remains sensitive to the antitumor activity of heat shock protein 90 inhibitors. *Clinical Cancer Research: An Official Journal of the American Association for Cancer Research, 9*, 4961–4971.

Ben-Neriah, Y., & Karin, M. (2011). Inflammation meets cancer, with NF-kappaB as the matchmaker. *Nature Immunology, 12*, 715–723.

Bernstein, B. E., Mikkelsen, T. S., Xie, X., Kamal, M., Huebert, D. J., Cuff, J., et al. (2006). A bivalent chromatin structure marks key developmental genes in embryonic stem cells. *Cell, 125*, 315–326.

Bird, A. (2002). DNA methylation patterns and epigenetic memory. *Genes & Development, 16*, 6–21.

Bohonowych, J. E., Hance, M. W., Nolan, K. D., Defee, M., Parsons, C. H., & Isaacs, J. S. (2014). Extracellular Hsp90 mediates an NF-kappaB dependent inflammatory stromal program: Implications for the prostate tumor microenvironment. *The Prostate, 74*, 395–407.

Caldas-Lopes, E., Cerchietti, L., Ahn, J. H., Clement, C. C., Robles, A. I., Rodina, A., et al. (2009). Hsp90 inhibitor PU-H71, a multimodal inhibitor of malignancy, induces complete responses in triple-negative breast cancer models. *Proceedings of the National Academy of Sciences of the United States of America, 106*, 8368–8373.

Cao, R., Wang, L., Wang, H., Xia, L., Erdjument-Bromage, H., Tempst, P., et al. (2002). Role of histone H3 lysine 27 methylation in Polycomb-group silencing. *Science, 298*, 1039–1043.

Cao, Q., Yu, J., Dhanasekaran, S. M., Kim, J. H., Mani, R. S., Tomlins, S. A., et al. (2008). Repression of E-cadherin by the polycomb group protein EZH2 in cancer. *Oncogene, 27*, 7274–7284.

Carystinos, G. D., Kandouz, M., Alaoui-Jamali, M. A., & Batist, G. (2003). Unexpected induction of the human connexin 43 promoter by the ras signaling pathway is mediated by a novel putative promoter sequence. *Molecular Pharmacology, 63*, 821–831.

Cerchietti, L. C., Lopes, E. C., Yang, S. N., Hatzi, K., Bunting, K. L., Tsikitas, L. A., et al. (2009). A purine scaffold Hsp90 inhibitor destabilizes BCL-6 and has specific antitumor activity in BCL-6-dependent B cell lymphomas. *Nature Medicine, 15*, 1369–1376.

Chae, Y. C., Angelin, A., Lisanti, S., Kossenkov, A. V., Speicher, K. D., Wang, H., et al. (2013). Landscape of the mitochondrial Hsp90 metabolome in tumours. *Nature Communications, 4*, 2139.

Chappell, J., & Dalton, S. (2013). Roles for MYC in the establishment and maintenance of pluripotency. *Cold Spring Harbor Perspectives in Medicine, 3*, a014381.

Chatterjee, M., Jain, S., Stuhmer, T., Andrulis, M., Ungethum, U., Kuban, R. J., et al. (2007). STAT3 and MAPK signaling maintain overexpression of heat shock

proteins 90alpha and beta in multiple myeloma cells, which critically contribute to tumor-cell survival. *Blood*, *109*, 720–728.
Chen, G., Cao, P., & Goeddel, D. V. (2002). TNF-induced recruitment and activation of the IKK complex require Cdc37 and Hsp90. *Molecular Cell*, *9*, 401–410.
Chen, W. S., Chen, C. C., Chen, L. L., Lee, C. C., & Huang, T. S. (2013). Secreted heat shock protein 90alpha (HSP90alpha) induces nuclear factor-kappaB-mediated TCF12 protein expression to down-regulate E-cadherin and to enhance colorectal cancer cell migration and invasion. *The Journal of Biological Chemistry*, *288*, 9001–9010.
Chen, Y., Chen, J., Yu, J., Yang, G., Temple, E., Harbinski, F., et al. (2014). Identification of mixed lineage leukemia 1(MLL1) protein as a coactivator of heat shock factor 1(HSF1) protein in response to heat shock protein 90 (HSP90) inhibition. *The Journal of Biological Chemistry*, *289*, 18914–18927.
Chen, T., Hevi, S., Gay, F., Tsujimoto, N., He, T., Zhang, B., et al. (2007). Complete inactivation of DNMT1 leads to mitotic catastrophe in human cancer cells. *Nature Genetics*, *39*, 391–396.
Chen, J. S., Hsu, Y. M., Chen, C. C., Chen, L. L., Lee, C. C., & Huang, T. S. (2010). Secreted heat shock protein 90alpha induces colorectal cancer cell invasion through CD91/LRP-1 and NF-kappaB-mediated integrin alphaV expression. *The Journal of Biological Chemistry*, *285*, 25458–25466.
Chen, J., Li, Y., Yu, T. S., McKay, R. M., Burns, D. K., Kernie, S. G., et al. (2012). A restricted cell population propagates glioblastoma growth after chemotherapy. *Nature*, *488*, 522–526.
Cheng, C. F., Fan, J., Fedesco, M., Guan, S., Li, Y., Bandyopadhyay, B., et al. (2008). Transforming growth factor alpha (TGFalpha)-stimulated secretion of HSP90alpha: Using the receptor LRP-1/CD91 to promote human skin cell migration against a TGFbeta-rich environment during wound healing. *Molecular and Cellular Biology*, *28*, 3344–3358.
Cho, H., Herzka, T., Zheng, W., Qi, J., Wilkinson, J. E., Bradner, J. E., et al. (2014). RapidCaP, a novel GEM model for metastatic prostate cancer analysis and therapy, reveals myc as a driver of Pten-mutant metastasis. *Cancer Discovery*, *4*, 318–333.
Cho, M. H., Park, J. H., Choi, H. J., Park, M. K., Won, H. Y., Park, Y. J., et al. (2015). DOT1L cooperates with the c-Myc-p300 complex to epigenetically derepress CDH1 transcription factors in breast cancer progression. *Nature Communications*, *6*, 7821.
Ci, W., Polo, J. M., & Melnick, A. (2008). B-cell lymphoma 6 and the molecular pathogenesis of diffuse large B-cell lymphoma. *Current Opinion in Hematology*, *15*, 381–390.
Conacci-Sorrell, M., McFerrin, L., & Eisenman, R. N. (2014). An overview of MYC and its interactome. *Cold Spring Harbor Perspectives in Medicine*, *4*, a014357.
Csermely, P., & Kahn, C. R. (1991). The 90-kDa heat shock protein (hsp-90) possesses an ATP binding site and autophosphorylating activity. *The Journal of Biological Chemistry*, *266*, 4943–4950.
Csermely, P., Kajtar, J., Hollosi, M., Oikarinen, J., & Somogyi, J. (1994). The 90 kDa heat shock protein (hsp90) induces the condensation of the chromatin structure. *Biochemical and Biophysical Research Communications*, *202*, 1657–1663.
Cvoro, A., Tzagarakis-Foster, C., Tatomer, D., Paruthiyil, S., Fox, M. S., & Leitman, D. C. (2006). Distinct roles of unliganded and liganded estrogen receptors in transcriptional repression. *Molecular Cell*, *21*, 555–564.
Czermin, B., Melfi, R., McCabe, D., Seitz, V., Imhof, A., & Pirrotta, V. (2002). Drosophila enhancer of Zeste/ESC complexes have a histone H3 methyltransferase activity that marks chromosomal Polycomb sites. *Cell*, *111*, 185–196.
da Rocha Dias, S., Friedlos, F., Light, Y., Springer, C., Workman, P., & Marais, R. (2005). Activated B-RAF is an Hsp90 client protein that is targeted by the anticancer drug 17-allylamino-17-demethoxygeldanamycin. *Cancer Research*, *65*, 10686–10691.

Darnell, J. E., Jr., Kerr, I. M., & Stark, G. R. (1994). Jak-STAT pathways and transcriptional activation in response to IFNs and other extracellular signaling proteins. *Science*, *264*, 1415–1421.

Davies, H., Bignell, G. R., Cox, C., Stephens, P., Edkins, S., Clegg, S., et al. (2002). Mutations of the BRAF gene in human cancer. *Nature*, *417*, 949–954.

Dawson, M. A., Prinjha, R. K., Dittmann, A., Giotopoulos, G., Bantscheff, M., Chan, W. I., et al. (2011). Inhibition of BET recruitment to chromatin as an effective treatment for MLL-fusion leukaemia. *Nature*, *478*, 529–533.

Delmore, J. E., Issa, G. C., Lemieux, M. E., Rahl, P. B., Shi, J., Jacobs, H. M., et al. (2011). BET bromodomain inhibition as a therapeutic strategy to target c-Myc. *Cell*, *146*, 904–917.

DiDonato, J. A., Mercurio, F., & Karin, M. (2012). NF-kappaB and the link between inflammation and cancer. *Immunological Reviews*, *246*, 379–400.

Erlejman, A. G., Lagadari, M., Toneatto, J., Piwien-Pilipuk, G., & Galigniana, M. D. (2014). Regulatory role of the 90-kDa-heat-shock protein (Hsp90) and associated factors on gene expression. *Biochimica et Biophysica Acta*, *1839*, 71–87.

Eustace, B. K., Sakurai, T., Stewart, J. K., Yimlamai, D., Unger, C., Zehetmeier, C., et al. (2004). Functional proteomic screens reveal an essential extracellular role for hsp90 alpha in cancer cell invasiveness. *Nature Cell Biology*, *6*, 507–514.

Fang, M., Ou, J., Hutchinson, L., & Green, M. R. (2014). The BRAF oncoprotein functions through the transcriptional repressor MAFG to mediate the CpG Island Methylator phenotype. *Molecular Cell*, *55*, 904–915.

Filippakopoulos, P., Qi, J., Picaud, S., Shen, Y., Smith, W. B., Fedorov, O., et al. (2010). Selective inhibition of BET bromodomains. *Nature*, *468*, 1067–1073.

Fiskus, W., Buckley, K., Rao, R., Mandawat, A., Yang, Y., Joshi, R., et al. (2009). Panobinostat treatment depletes EZH2 and DNMT1 levels and enhances decitabine mediated de-repression of JunB and loss of survival of human acute leukemia cells. *Cancer Biology & Therapy*, *8*, 939–950.

Fleming, J. D., Giresi, P. G., Lindahl-Allen, M., Krall, E. B., Lieb, J. D., & Struhl, K. (2015). STAT3 acts through pre-existing nucleosome-depleted regions bound by FOS during an epigenetic switch linking inflammation to cancer. *Epigenetics & Chromatin*, *8*, 7.

Francis, N. J., & Kingston, R. E. (2001). Mechanisms of transcriptional memory. *Nature Reviews. Molecular Cell Biology*, *2*, 409–421.

Freeman, B. C., & Yamamoto, K. R. (2002). Disassembly of transcriptional regulatory complexes by molecular chaperones. *Science*, *296*, 2232–2235.

Fujii, S., Tokita, K., Wada, N., Ito, K., Yamauchi, C., Ito, Y., et al. (2011). MEK-ERK pathway regulates EZH2 overexpression in association with aggressive breast cancer subtypes. *Oncogene*, *30*, 4118–4128.

Gabay, M., Li, Y., & Felsher, D. W. (2014). MYC activation is a hallmark of cancer initiation and maintenance. *Cold Spring Harbor Perspectives in Medicine*, *4*(6). pii: a014241.

Gallo, M., Ho, J., Coutinho, F. J., Vanner, R., Lee, L., Head, R., et al. (2013). A tumorigenic MLL-homeobox network in human glioblastoma stem cells. *Cancer Research*, *73*, 417–427.

Goke, J., Chan, Y. S., Yan, J., Vingron, M., & Ng, H. H. (2013). Genome-wide kinase-chromatin interactions reveal the regulatory network of ERK signaling in human embryonic stem cells. *Molecular Cell*, *50*, 844–855.

Gonzalez, M. E., Moore, H. M., Li, X., Toy, K. A., Huang, W., Sabel, M. S., et al. (2014). EZH2 expands breast stem cells through activation of NOTCH1 signaling. *Proceedings of the National Academy of Sciences of the United States of America*, *111*, 3098–3103.

Gopal, U., Bohonowych, J. E., Lema-Tome, C., Liu, A., Garrett-Mayer, E., Wang, B., et al. (2011). A novel extracellular Hsp90 mediated co-receptor function for LRP1 regulates EphA2 dependent glioblastoma cell invasion. *PLoS One*, *6*, e17649.

Grammatikakis, N., Lin, J. H., Grammatikakis, A., Tsichlis, P. N., & Cochran, B. H. (1999). p50(cdc37) acting in concert with Hsp90 is required for Raf-1 function. *Molecular and Cellular Biology*, *19*, 1661–1672.

Hamamoto, R., Furukawa, Y., Morita, M., Iimura, Y., Silva, F. P., Li, M., et al. (2004). SMYD3 encodes a histone methyltransferase involved in the proliferation of cancer cells. *Nature Cell Biology*, *6*, 731–740.

Hance, M. W., Dole, K., Gopal, U., Bohonowych, J. E., Jezierska-Drutel, A., Neumann, C. A., et al. (2012). Secreted Hsp90 is a novel regulator of the epithelial to mesenchymal transition (EMT) in prostate cancer. *The Journal of Biological Chemistry*, *287*, 37732–37744.

Hart, C. D., Migliaccio, I., Malorni, L., Guarducci, C., Biganzoli, L., & Di Leo, A. (2015). Challenges in the management of advanced, ER-positive, HER2-negative breast cancer. *Nature Reviews. Clinical Oncology*, *12*(9), 541–552.

He, G., & Karin, M. (2011). NF-kappaB and STAT3—Key players in liver inflammation and cancer. *Cell Research*, *21*, 159–168.

Herranz, N., Pasini, D., Diaz, V. M., Franci, C., Gutierrez, A., Dave, N., et al. (2008). Polycomb complex 2 is required for E-cadherin repression by the Snail1 transcription factor. *Molecular and Cellular Biology*, *28*, 4772–4781.

Hnisz, D., Abraham, B. J., Lee, T. I., Lau, A., Saint-Andre, V., Sigova, A. A., et al. (2013). Super-enhancers in the control of cell identity and disease. *Cell*, *155*, 934–947.

Hock, H. (2012). A complex Polycomb issue: The two faces of EZH2 in cancer. *Genes & Development*, *26*, 751–755.

Hodis, E., Watson, I. R., Kryukov, G. V., Arold, S. T., Imielinski, M., Theurillat, J. P., et al. (2012). A landscape of driver mutations in melanoma. *Cell*, *150*, 251–263.

Hsu, J. M., Chen, C. T., Chou, C. K., Kuo, H. P., Li, L. Y., Lin, C. Y., et al. (2011). Crosstalk between Arg 1175 methylation and Tyr 1173 phosphorylation negatively modulates EGFR-mediated ERK activation. *Nature Cell Biology*, *13*, 174–181.

Ichikawa, K., Kubota, Y., Nakamura, T., Weng, J. S., Tomida, T., Saito, H., et al. (2015). MCRIP1, an ERK substrate, mediates ERK-induced gene silencing during epithelial-mesenchymal transition by regulating the co-repressor CtBP. *Molecular Cell*, *58*, 35–46.

Ihle, J. N. (1996). STATs: Signal transducers and activators of transcription. *Cell*, *84*, 331–334.

Iliopoulos, D., Hirsch, H. A., & Struhl, K. (2009). An epigenetic switch involving NF-kappaB, Lin28, Let-7 MicroRNA, and IL6 links inflammation to cell transformation. *Cell*, *139*, 693–706.

Iliopoulos, D., Jaeger, S. A., Hirsch, H. A., Bulyk, M. L., & Struhl, K. (2010). STAT3 activation of miR-21 and miR-181b-1 via PTEN and CYLD are part of the epigenetic switch linking inflammation to cancer. *Molecular Cell*, *39*, 493–506.

Jackson-Grusby, L., Beard, C., Possemato, R., Tudor, M., Fambrough, D., Csankovszki, G., et al. (2001). Loss of genomic methylation causes p53-dependent apoptosis and epigenetic deregulation. *Nature Genetics*, *27*, 31–39.

Jang, M. K., Mochizuki, K., Zhou, M., Jeong, H. S., Brady, J. N., & Ozato, K. (2005). The bromodomain protein Brd4 is a positive regulatory component of P-TEFb and stimulates RNA polymerase II-dependent transcription. *Molecular Cell*, *19*, 523–534.

Johannessen, C. M., Boehm, J. S., Kim, S. Y., Thomas, S. R., Wardwell, L., Johnson, L. A., et al. (2010). COT drives resistance to RAF inhibition through MAP kinase pathway reactivation. *Nature*, *468*, 968–972.

Johannessen, C. M., Johnson, L. A., Piccioni, F., Townes, A., Frederick, D. T., Donahue, M. K., et al. (2013). A melanocyte lineage program confers resistance to MAP kinase pathway inhibition. *Nature*, *504*, 138–142.

Kasioulis, I., Syred, H. M., Tate, P., Finch, A., Shaw, J., Seawright, A., et al. (2014). Kdm3a lysine demethylase is an Hsp90 client required for cytoskeletal rearrangements during spermatogenesis. *Molecular Biology of the Cell, 25*, 1216–1233.

Khurana, N., & Bhattacharyya, S. (2015). Hsp90, the concertmaster: Tuning transcription. *Frontiers in Oncology, 5*, 100.

Kim, E., Kim, M., Woo, D. H., Shin, Y., Shin, J., Chang, N., et al. (2013). Phosphorylation of EZH2 activates STAT3 signaling via STAT3 methylation and promotes tumorigenicity of glioblastoma stem-like cells. *Cancer Cell, 23*, 839–852.

Kingston, R. E., & Tamkun, J. W. (2014). Transcriptional regulation by trithorax-group proteins. *Cold Spring Harbor Perspectives in Biology, 6*, a019349.

Kleer, C. G., Cao, Q., Varambally, S., Shen, R., Ota, I., Tomlins, S. A., et al. (2003). EZH2 is a marker of aggressive breast cancer and promotes neoplastic transformation of breast epithelial cells. *Proceedings of the National Academy of Sciences of the United States of America, 100*, 11606–11611.

Knoblauch, R., & Garabedian, M. J. (1999). Role for Hsp90-associated cochaperone p23 in estrogen receptor signal transduction. *Molecular and Cellular Biology, 19*, 3748–3759.

Koh, C. M., Bezzi, M., Low, D. H., Ang, W. X., Teo, S. X., Gay, F. P., et al. (2015). MYC regulates the core pre-mRNA splicing machinery as an essential step in lymphomagenesis. *Nature, 523*, 96–100.

Koh, C. M., Iwata, T., Zheng, Q., Bethel, C., Yegnasubramanian, S., & De Marzo, A. M. (2011). Myc enforces overexpression of EZH2 in early prostatic neoplasia via transcriptional and post-transcriptional mechanisms. *Oncotarget, 2*, 669–683.

Kolosenko, I., Grander, D., & Tamm, K. P. (2014). IL-6 activated JAK/STAT3 pathway and sensitivity to Hsp90 inhibitors in multiple myeloma. *Current Medicinal Chemistry, 21*, 3042–3047.

Krieg, A. J., Rankin, E. B., Chan, D., Razorenova, O., Fernandez, S., & Giaccia, A. J. (2010). Regulation of the histone demethylase JMJD1A by hypoxia-inducible factor 1 alpha enhances hypoxic gene expression and tumor growth. *Molecular and Cellular Biology, 30*, 344–353.

Ku, S., Lasorsa, E., Adelaiye, R., Ramakrishnan, S., Ellis, L., & Pili, R. (2014). Inhibition of Hsp90 augments docetaxel therapy in castrate resistant prostate cancer. *PLoS One, 9*, e103680.

Lagarrigue, F., Dupuis-Coronas, S., Ramel, D., Delsol, G., Tronchere, H., Payrastre, B., et al. (2010). Matrix metalloproteinase-9 is upregulated in nucleophosmin-anaplastic lymphoma kinase-positive anaplastic lymphomas and activated at the cell surface by the chaperone heat shock protein 90 to promote cell invasion. *Cancer Research, 70*, 6978–6987.

Lai, A. Z., Cory, S., Zhao, H., Gigoux, M., Monast, A., Guiot, M. C., et al. (2014). Dynamic reprogramming of signaling upon met inhibition reveals a mechanism of drug resistance in gastric cancer. *Science Signaling, 7*, ra38.

Lamoureux, F., Thomas, C., Yin, M. J., Kuruma, H., Fazli, L., Gleave, M. E., et al. (2011). A novel HSP90 inhibitor delays castrate-resistant prostate cancer without altering serum PSA levels and inhibits osteoclastogenesis. *Clinical Cancer Research: An Official Journal of the American Association for Cancer Research, 17*, 2301–2313.

Lang, S. A., Moser, C., Gaumann, A., Klein, D., Glockzin, G., Popp, F. C., et al. (2007). Targeting heat shock protein 90 in pancreatic cancer impairs insulin-like growth factor-I receptor signaling, disrupts an interleukin-6/signal-transducer and activator of transcription 3/hypoxia-inducible factor-1alpha autocrine loop, and reduces orthotopic tumor growth. *Clinical Cancer Research: An Official Journal of the American Association for Cancer Research, 13*, 6459–6468.

Lathia, J. D., Mack, S. C., Mulkearns-Hubert, E. E., Valentim, C. L., & Rich, J. N. (2015). Cancer stem cells in glioblastoma. *Genes & Development, 29*, 1203–1217.

Lee, J. V., Carrer, A., Shah, S., Snyder, N. W., Wei, S., Venneti, S., et al. (2014). Akt-dependent metabolic reprogramming regulates tumor cell histone acetylation. *Cell Metabolism*, *20*, 306–319.

Lee, J. H., Kang, B. H., Jang, H., Kim, T. W., Choi, J., Kwak, S., et al. (2015). AKT phosphorylates H3-threonine 45 to facilitate termination of gene transcription in response to DNA damage. *Nucleic Acids Research*, *43*, 4505–4516.

Lee, S. T., Li, Z., Wu, Z., Aau, M., Guan, P., Karuturi, R. K., et al. (2011). Context-specific regulation of NF-kappaB target gene expression by EZH2 in breast cancers. *Molecular Cell*, *43*, 798–810.

Lee, H., Zhang, P., Herrmann, A., Yang, C., Xin, H., Wang, Z., et al. (2012). Acetylated STAT3 is crucial for methylation of tumor-suppressor gene promoters and inhibition by resveratrol results in demethylation. *Proceedings of the National Academy of Sciences of the United States of America*, *109*, 7765–7769.

LeRoy, G., Rickards, B., & Flint, S. J. (2008). The double bromodomain proteins Brd2 and Brd3 couple histone acetylation to transcription. *Molecular Cell*, *30*, 51–60.

Li, E., Bestor, T. H., & Jaenisch, R. (1992). Targeted mutation of the DNA methyltransferase gene results in embryonic lethality. *Cell*, *69*, 915–926.

Li, Y., Chitnis, N., Nakagawa, H., Kita, Y., Natsugoe, S., Yang, Y., et al. (2015). PRMT5 is required for lymphomagenesis triggered by multiple oncogenic drivers. *Cancer Discovery*, *5*, 288–303.

Li, N., Grivennikov, S. I., & Karin, M. (2011). The unholy trinity: Inflammation, cytokines, and STAT3 shape the cancer microenvironment. *Cancer Cell*, *19*, 429–431.

Li, L., Lyu, X., Hou, C., Takenaka, N., Nguyen, H. Q., Ong, C. T., et al. (2015). Widespread rearrangement of 3D chromatin organization underlies polycomb-mediated stress-induced silencing. *Molecular Cell*, *58*, 216–231.

Li, W., Miao, X., Qi, Z., Zeng, W., Liang, J., & Liang, Z. (2010). Hepatitis B virus X protein upregulates HSP90alpha expression via activation of c-Myc in human hepatocarcinoma cell line, HepG2. *Virology Journal*, 7, 45.

Li, B., & Simon, M. C. (2013). Molecular pathways: Targeting MYC-induced metabolic reprogramming and oncogenic stress in cancer. *Clinical Cancer Research: An Official Journal of the American Association for Cancer Research*, *19*, 5835–5841.

Liao, J., Karnik, R., Gu, H., Ziller, M. J., Clement, K., Tsankov, A. M., et al. (2015). Targeted disruption of DNMT1, DNMT3A and DNMT3B in human embryonic stem cells. *Nature Genetics*, *47*, 469–478.

Lin, Y. W., Ren, L. L., Xiong, H., Du, W., Yu, Y. N., Sun, T. T., et al. (2013). Role of STAT3 and vitamin D receptor in EZH2-mediated invasion of human colorectal cancer. *The Journal of Pathology*, *230*, 277–290.

Loven, J., Hoke, H. A., Lin, C. Y., Lau, A., Orlando, D. A., Vakoc, C. R., et al. (2013). Selective inhibition of tumor oncogenes by disruption of super-enhancers. *Cell*, *153*, 320–334.

Lue, H. W., Yang, X., Wang, R., Qian, W., Xu, R. Z., Lyles, R., et al. (2011). LIV-1 promotes prostate cancer epithelial-to-mesenchymal transition and metastasis through HB-EGF shedding and EGFR-mediated ERK signaling. *PLoS One*, *6*, e27720.

Mahajan, K., Lawrence, H. R., Lawrence, N. J., & Mahajan, N. P. (2014). ACK1 tyrosine kinase interacts with histone demethylase KDM3A to regulate the mammary tumor oncogene HOXA1. *The Journal of Biological Chemistry*, *289*, 28179–28191.

Maloney, A., Clarke, P. A., Naaby-Hansen, S., Stein, R., Koopman, J. O., Akpan, A., et al. (2007). Gene and protein expression profiling of human ovarian cancer cells treated with the heat shock protein 90 inhibitor 17-allylamino-17-demethoxygeldanamycin. *Cancer Research*, *67*, 3239–3253.

Malouf, G. G., Taube, J. H., Lu, Y., Roysarkar, T., Panjarian, S., Estecio, M. R., et al. (2013). Architecture of epigenetic reprogramming following Twist1-mediated epithelial-mesenchymal transition. *Genome Biology*, *14*, R144.

Mandal, M., Powers, S. E., Maienschein-Cline, M., Bartom, E. T., Hamel, K. M., Kee, B. L., et al. (2011). Epigenetic repression of the Igk locus by STAT5-mediated recruitment of the histone methyltransferase Ezh2. *Nature Immunology*, *12*, 1212–1220.

Margueron, R., & Reinberg, D. (2011). The Polycomb complex PRC2 and its mark in life. *Nature*, *469*, 343–349.

Marivoet, S., Van Dijck, P., Verhoeven, G., & Heyns, W. (1992). Interaction of the 90-kDa heat shock protein with native and in vitro translated androgen receptor and receptor fragments. *Molecular and Cellular Endocrinology*, *88*, 165–174.

McCabe, M. T., & Creasy, C. L. (2014). EZH2 as a potential target in cancer therapy. *Epigenomics*, *6*, 341–351.

McCabe, M. T., Ott, H. M., Ganji, G., Korenchuk, S., Thompson, C., Van Aller, G. S., et al. (2012). EZH2 inhibition as a therapeutic strategy for lymphoma with EZH2-activating mutations. *Nature*, *492*, 108–112.

Mertz, J. A., Conery, A. R., Bryant, B. M., Sandy, P., Balasubramanian, S., Mele, D. A., et al. (2011). Targeting MYC dependence in cancer by inhibiting BET bromodomains. *Proceedings of the National Academy of Sciences of the United States of America*, *108*, 16669–16674.

Mills, A. A. (2010). Throwing the cancer switch: Reciprocal roles of polycomb and trithorax proteins. *Nature Reviews. Cancer*, *10*, 669–682.

Milne, T. A., Briggs, S. D., Brock, H. W., Martin, M. E., Gibbs, D., Allis, C. D., et al. (2002). MLL targets SET domain methyltransferase activity to Hox gene promoters. *Molecular Cell*, *10*, 1107–1117.

Mitsiades, C. S., Mitsiades, N. S., McMullan, C. J., Poulaki, V., Kung, A. L., Davies, F. E., et al. (2006). Antimyeloma activity of heat shock protein-90 inhibition. *Blood*, *107*, 1092–1100.

Miyajima, N., Tsutsumi, S., Sourbier, C., Beebe, K., Mollapour, M., Rivas, C., et al. (2013). The HSP90 inhibitor ganetespib synergizes with the MET kinase inhibitor crizotinib in both crizotinib-sensitive and -resistant MET-driven tumor models. *Cancer Research*, *73*, 7022–7033.

Mohan, M., Herz, H. M., Takahashi, Y. H., Lin, C., Lai, K. C., Zhang, Y., et al. (2010). Linking H3K79 trimethylation to Wnt signaling through a novel Dot1-containing complex (DotCom). *Genes & Development*, *24*, 574–589.

Morcillo, G., Diez, J. L., Carbajal, M. E., & Tanguay, R. M. (1993). HSP90 associates with specific heat shock puffs (hsr omega) in polytene chromosomes of Drosophila and Chironomus. *Chromosoma*, *102*, 648–659.

Morin, R. D., Mendez-Lago, M., Mungall, A. J., Goya, R., Mungall, K. L., Corbett, R. D., et al. (2011). Frequent mutation of histone-modifying genes in non-Hodgkin lymphoma. *Nature*, *476*, 298–303.

Moulick, K., Ahn, J. H., Zong, H., Rodina, A., Cerchietti, L., Gomes DaGama, E. M., et al. (2011). Affinity-based proteomics reveal cancer-specific networks coordinated by Hsp90. *Nature Chemical Biology*, *7*, 818–826.

Muller, J., Hart, C. M., Francis, N. J., Vargas, M. L., Sengupta, A., Wild, B., et al. (2002). Histone methyltransferase activity of a Drosophila Polycomb group repressor complex. *Cell*, *111*, 197–208.

Nicodeme, E., Jeffrey, K. L., Schaefer, U., Beinke, S., Dewell, S., Chung, C. W., et al. (2010). Suppression of inflammation by a synthetic histone mimic. *Nature*, *468*, 1119–1123.

Nolan, K. D., Franco, O. E., Hance, M. W., Hayward, S. W., & Isaacs, J. S. (2015). Tumor-secreted Hsp90 subverts polycomb function to drive prostate tumor growth and invasion. *The Journal of Biological Chemistry*, *290*, 8271–8282.

Okawa, Y., Hideshima, T., Steed, P., Vallet, S., Hall, S., Huang, K., et al. (2009). SNX-2112, a selective Hsp90 inhibitor, potently inhibits tumor cell growth, angiogenesis, and osteoclastogenesis in multiple myeloma and other hematologic tumors by abrogating signaling via Akt and ERK. *Blood*, *113*, 846–855.

O'Malley, K. J., Langmann, G., Ai, J., Ramos-Garcia, R., Vessella, R. L., & Wang, Z. (2012). Hsp90 inhibitor 17-AAG inhibits progression of LuCaP35 xenograft prostate tumors to castration resistance. *The Prostate*, *72*, 1117–1123.

Oronsky, B., Oronsky, N., Scicinski, J., Fanger, G., Lybeck, M., & Reid, T. (2014). Rewriting the epigenetic code for tumor resensitization: A review. *Translational Oncology*, *7*, 626–631.

Osawa, T., Tsuchida, R., Muramatsu, M., Shimamura, T., Wang, F., Suehiro, J., et al. (2013). Inhibition of histone demethylase JMJD1A improves anti-angiogenic therapy and reduces tumor-associated macrophages. *Cancer Research*, *73*, 3019–3028.

Paraiso, K. H., Haarberg, H. E., Wood, E., Rebecca, V. W., Chen, Y. A., Xiang, Y., et al. (2012). The HSP90 inhibitor XL888 overcomes BRAF inhibitor resistance mediated through diverse mechanisms. *Clinical Cancer Research: An Official Journal of the American Association for Cancer Research*, *18*, 2502–2514.

Parrish, J. K., Sechler, M., Winn, R. A., & Jedlicka, P. (2015). The histone demethylase KDM3A is a microRNA-22-regulated tumor promoter in Ewing Sarcoma. *Oncogene*, *34*, 257–262.

Parsons, D. W., Li, M., Zhang, X., Jones, S., Leary, R. J., Lin, J. C., et al. (2011). The genetic landscape of the childhood cancer medulloblastoma. *Science*, *331*, 435–439.

Patel, J. H., Loboda, A. P., Showe, M. K., Showe, L. C., & McMahon, S. B. (2004). Analysis of genomic targets reveals complex functions of MYC. *Nature Reviews. Cancer*, *4*, 562–568.

Paul, I., Ahmed, S. F., Bhowmik, A., Deb, S., & Ghosh, M. K. (2013). The ubiquitin ligase CHIP regulates c-Myc stability and transcriptional activity. *Oncogene*, *32*, 1284–1295.

Pereira, L., Horta, S., Mateus, R., & Videira, M. A. (2015). Implications of Akt2/Twist crosstalk on breast cancer metastatic outcome. *Drug Discovery Today*, *20*(9), 1152–1158.

Pratt, W. B., & Toft, D. O. (1997). Steroid receptor interactions with heat shock protein and immunophilin chaperones. *Endocrine Reviews*, *18*, 306–360.

Regnier, C. H., Song, H. Y., Gao, X., Goeddel, D. V., Cao, Z., & Rothe, M. (1997). Identification and characterization of an IkappaB kinase. *Cell*, *90*, 373–383.

Renoir, J. M., Radanyi, C., Faber, L. E., & Baulieu, E. E. (1990). The non-DNA-binding heterooligomeric form of mammalian steroid hormone receptors contains a hsp90-bound 59-kilodalton protein. *The Journal of Biological Chemistry*, *265*, 10740–10745.

Reynolds, P. A., Sigaroudinia, M., Zardo, G., Wilson, M. B., Benton, G. M., Miller, C. J., et al. (2006). Tumor suppressor p16INK4A regulates polycomb-mediated DNA hypermethylation in human mammary epithelial cells. *The Journal of Biological Chemistry*, *281*, 24790–24802.

Rhee, I., Bachman, K. E., Park, B. H., Jair, K. W., Yen, R. W., Schuebel, K. E., et al. (2002). DNMT1 and DNMT3b cooperate to silence genes in human cancer cells. *Nature*, *416*, 552–556.

Richardson, P. G., Mitsiades, C. S., Laubach, J. P., Lonial, S., Chanan-Khan, A. A., & Anderson, K. C. (2011). Inhibition of heat shock protein 90 (HSP90) as a therapeutic strategy for the treatment of myeloma and other cancers. *British Journal of Haematology*, *152*, 367–379.

Rodrik-Outmezguine, V. S., Chandarlapaty, S., Pagano, N. C., Poulikakos, P. I., Scaltriti, M., Moskatel, E., et al. (2011). mTOR kinase inhibition causes feedback-dependent biphasic regulation of AKT signaling. *Cancer Discovery*, *1*, 248–259.

Sang, J., Acquaviva, J., Friedland, J. C., Smith, D. L., Sequeira, M., Zhang, C., et al. (2013). Targeted inhibition of the molecular chaperone Hsp90 overcomes ALK inhibitor resistance in non-small cell lung cancer. *Cancer Discovery*, *3*, 430–443.

Sasser, A. K., Sullivan, N. J., Studebaker, A. W., Hendey, L. F., Axel, A. E., & Hall, B. M. (2007). Interleukin-6 is a potent growth factor for ER-alpha-positive human breast cancer. *FASEB Journal: Official Publication of the Federation of American Societies for Experimental Biology*, *21*, 3763–3770.

Sawarkar, R., & Paro, R. (2013). Hsp90@chromatin.nucleus: An emerging hub of a networker. *Trends in Cell Biology*, *23*, 193–201.

Sawarkar, R., Sievers, C., & Paro, R. (2012). Hsp90 globally targets paused RNA polymerase to regulate gene expression in response to environmental stimuli. *Cell*, *149*, 807–818.

Schlesinger, Y., Straussman, R., Keshet, I., Farkash, S., Hecht, M., Zimmerman, J., et al. (2007). Polycomb-mediated methylation on Lys27 of histone H3 pre-marks genes for de novo methylation in cancer. *Nature Genetics*, *39*, 232–236.

Schuettengruber, B., Chourrout, D., Vervoort, M., Leblanc, B., & Cavalli, G. (2007). Genome regulation by polycomb and trithorax proteins. *Cell*, *128*, 735–745.

Schulte, T. W., An, W. G., & Neckers, L. M. (1997). Geldanamycin-induced destabilization of Raf-1 involves the proteasome. *Biochemical and Biophysical Research Communications*, *239*, 655–659.

Segnitz, B., & Gehring, U. (1997). The function of steroid hormone receptors is inhibited by the hsp90-specific compound geldanamycin. *The Journal of Biological Chemistry*, *272*, 18694–18701.

Sen, G. L., Reuter, J. A., Webster, D. E., Zhu, L., & Khavari, P. A. (2010). DNMT1 maintains progenitor function in self-renewing somatic tissue. *Nature*, *463*, 563–567.

Shah, M., Patel, K., Fried, V. A., & Sehgal, P. B. (2002). Interactions of STAT3 with caveolin-1 and heat shock protein 90 in plasma membrane raft and cytosolic complexes. Preservation of cytokine signaling during fever. *The Journal of Biological Chemistry*, *277*, 45662–45669.

Shen, C., & Vakoc, C. R. (2015). Gain-of-function mutation of chromatin regulators as a tumorigenic mechanism and an opportunity for therapeutic intervention. *Current Opinion in Oncology*, *27*, 57–63.

Shi, J., & Vakoc, C. R. (2014). The mechanisms behind the therapeutic activity of BET bromodomain inhibition. *Molecular Cell*, *54*, 728–736.

Shi, J., Wang, Y., Zeng, L., Wu, Y., Deng, J., Zhang, Q., et al. (2014). Disrupting the interaction of BRD4 with diacetylated Twist suppresses tumorigenesis in basal-like breast cancer. *Cancer Cell*, *25*, 210–225.

Sidera, K., Gaitanou, M., Stellas, D., Matsas, R., & Patsavoudi, E. (2008). A critical role for HSP90 in cancer cell invasion involves interaction with the extracellular domain of HER-2. *The Journal of Biological Chemistry*, *283*, 2031–2041.

Smyth, T., Paraiso, K. H., Hearn, K., Rodriguez-Lopez, A. M., Munck, J. M., Haarberg, H. E., et al. (2014). Inhibition of HSP90 by AT13387 delays the emergence of resistance to BRAF inhibitors and overcomes resistance to dual BRAF and MEK inhibition in melanoma models. *Molecular Cancer Therapeutics*, *13*, 2793–2804.

Solit, D. B., Zheng, F. F., Drobnjak, M., Munster, P. N., Higgins, B., Verbel, D., et al. (2002). 17-Allylamino-17-demethoxygeldanamycin induces the degradation of androgen receptor and HER-2/neu and inhibits the growth of prostate cancer xenografts. *Clinical Cancer Research: An Official Journal of the American Association for Cancer Research*, *8*, 986–993.

Sollars, V., Lu, X., Xiao, L., Wang, X., Garfinkel, M. D., & Ruden, D. M. (2003). Evidence for an epigenetic mechanism by which Hsp90 acts as a capacitor for morphological evolution. *Nature Genetics*, *33*, 70–74.

Song, X., Wang, X., Zhuo, W., Shi, H., Feng, D., Sun, Y., et al. (2010). The regulatory mechanism of extracellular Hsp90{alpha} on matrix metalloproteinase-2 processing and tumor angiogenesis. *The Journal of Biological Chemistry*, *285*, 40039–40049.

Sowalsky, A. G., Xia, Z., Wang, L., Zhao, H., Chen, S., Bubley, G. J., et al. (2015). Whole transcriptome sequencing reveals extensive unspliced mRNA in metastatic castration-resistant prostate cancer. *Molecular Cancer Research: MCR*, *13*, 98–106.

Sparmann, A., & van Lohuizen, M. (2006). Polycomb silencers control cell fate, development and cancer. *Nature Reviews Cancer*, *6*, 846–856.

Spellmon, N., Holcomb, J., Trescott, L., Sirinupong, N., & Yang, Z. (2015). Structure and function of SET and MYND domain-containing proteins. *International Journal of Molecular Sciences*, *16*, 1406–1428.

Stellas, D., El Hamidieh, A., & Patsavoudi, E. (2010). Monoclonal antibody 4C5 prevents activation of MMP2 and MMP9 by disrupting their interaction with extracellular HSP90 and inhibits formation of metastatic breast cancer cell deposits. *BMC Cell Biology*, *11*, 51.

Stellas, D., Karameris, A., & Patsavoudi, E. (2007). Monoclonal antibody 4C5 immunostains human melanomas and inhibits melanoma cell invasion and metastasis. *Clinical Cancer Research: An Official Journal of the American Association for Cancer Research*, *13*, 1831–1838.

Stephanou, A., Isenberg, D. A., Akira, S., Kishimoto, T., & Latchman, D. S. (1998). The nuclear factor interleukin-6 (NF-IL6) and signal transducer and activator of transcription-3 (STAT-3) signalling pathways co-operate to mediate the activation of the hsp90beta gene by interleukin-6 but have opposite effects on its inducibility by heat shock. *The Biochemical Journal*, *330*(Pt. 1), 189–195.

Stopa, N., Krebs, J. E., & Shechter, D. (2015). The PRMT5 arginine methyltransferase: Many roles in development, cancer and beyond. *Cellular and Molecular Life Sciences: CMLS*, *72*, 2041–2059.

Stratikopoulos, E. E., Dendy, M., Szabolcs, M., Khaykin, A. J., Lefebvre, C., Zhou, M. M., et al. (2015). Kinase and BET inhibitors together clamp inhibition of PI3K signaling and overcome resistance to therapy. *Cancer Cell*, *27*, 837–851.

Sun, L., Hartson, S. D., & Matts, R. L. (2015). Identification of proteins associated with Aha1 in HeLa cells by quantitative proteomics. *Biochimica et Biophysica Acta*, *1854*, 365–380.

Taipale, M., Krykbaeva, I., Koeva, M., Kayatekin, C., Westover, K. D., Karras, G. I., et al. (2012). Quantitative analysis of HSP90-client interactions reveals principles of substrate recognition. *Cell*, *150*, 987–1001.

Takeda, S., Liu, H., Sasagawa, S., Dong, Y., Trainor, P. A., Cheng, E. H., et al. (2013). HGF-MET signals via the MLL-ETS2 complex in hepatocellular carcinoma. *The Journal of Clinical Investigation*, *123*, 3154–3165.

Tan, F. H., Putoczki, T. L., Stylli, S. S., & Luwor, R. B. (2014). The role of STAT3 signaling in mediating tumor resistance to cancer therapy. *Current Drug Targets*, *15*, 1341–1353.

Tariq, M., Nussbaumer, U., Chen, Y., Beisel, C., & Paro, R. (2009). Trithorax requires Hsp90 for maintenance of active chromatin at sites of gene expression. *Proceedings of the National Academy of Sciences of the United States of America*, *106*, 1157–1162.

Tee, W. W., Pardo, M., Theunissen, T. W., Yu, L., Choudhary, J. S., Hajkova, P., et al. (2010). Prmt5 is essential for early mouse development and acts in the cytoplasm to maintain ES cell pluripotency. *Genes & Development*, *24*, 2772–2777.

Tee, W. W., Shen, S. S., Oksuz, O., Narendra, V., & Reinberg, D. (2014). Erk1/2 activity promotes chromatin features and RNAPII phosphorylation at developmental promoters in mouse ESCs. *Cell*, *156*, 678–690.

Thiery, J. P., Acloque, H., Huang, R. Y., & Nieto, M. A. (2009). Epithelial-mesenchymal transitions in development and disease. *Cell*, *139*, 871–890.

Thuringer, D., Hammann, A., Benikhlef, N., Fourmaux, E., Bouchot, A., Wettstein, G., et al. (2011). Transactivation of the epidermal growth factor receptor by heat shock protein 90 via Toll-like receptor 4 contributes to the migration of glioblastoma cells. *The Journal of Biological Chemistry*, *286*, 3418–3428.

Tiwari, N., Tiwari, V. K., Waldmeier, L., Balwierz, P. J., Arnold, P., Pachkov, M., et al. (2013). Sox4 is a master regulator of epithelial-mesenchymal transition by controlling Ezh2 expression and epigenetic reprogramming. *Cancer Cell*, *23*, 768–783.

Tong, Z. T., Cai, M. Y., Wang, X. G., Kong, L. L., Mai, S. J., Liu, Y. H., et al. (2012). EZH2 supports nasopharyngeal carcinoma cell aggressiveness by forming a co-repressor complex with HDAC1/HDAC2 and Snail to inhibit E-cadherin. *Oncogene*, *31*, 583–594.

Tsai, J. H., & Yang, J. (2013). Epithelial-mesenchymal plasticity in carcinoma metastasis. *Genes & Development*, *27*, 2192–2206.

Tsen, F., Bhatia, A., O'Brien, K., Cheng, C. F., Chen, M., Hay, N., et al. (2013). Extracellular heat shock protein 90 signals through subdomain II and the NPVY motif of LRP-1 receptor to Akt1 and Akt2: A circuit essential for promoting skin cell migration in vitro and wound healing in vivo. *Molecular and Cellular Biology*, *33*, 4947–4959.

Tsutsumi, S., Scroggins, B., Koga, F., Lee, M. J., Trepel, J., Felts, S., et al. (2008). A small molecule cell-impermeant Hsp90 antagonist inhibits tumor cell motility and invasion. *Oncogene*, *27*, 2478–2487.

Vanaja, D. K., Mitchell, S. H., Toft, D. O., & Young, C. Y. (2002). Effect of geldanamycin on androgen receptor function and stability. *Cell Stress & Chaperones*, *7*, 55–64.

Varambally, S., Dhanasekaran, S. M., Zhou, M., Barrette, T. R., Kumar-Sinha, C., Sanda, M. G., et al. (2002). The polycomb group protein EZH2 is involved in progression of prostate cancer. *Nature*, *419*, 624–629.

Veldscholte, J., Berrevoets, C. A., Brinkmann, A. O., Grootegoed, J. A., & Mulder, E. (1992). Anti-androgens and the mutated androgen receptor of LNCaP cells: Differential effects on binding affinity, heat-shock protein interaction, and transcription activation. *Biochemistry*, *31*, 2393–2399.

Vire, E., Brenner, C., Deplus, R., Blanchon, L., Fraga, M., Didelot, C., et al. (2006). The Polycomb group protein EZH2 directly controls DNA methylation. *Nature*, *439*, 871–874.

Wade, M. A., Jones, D., Wilson, L., Stockley, J., Coffey, K., Robson, C. N., et al. (2015). The histone demethylase enzyme KDM3A is a key estrogen receptor regulator in breast cancer. *Nucleic Acids Research*, *43*, 196–207.

Wagle, N., Van Allen, E. M., Treacy, D. J., Frederick, D. T., Cooper, Z. A., Taylor-Weiner, A., et al. (2014). MAP kinase pathway alterations in BRAF-mutant melanoma patients with acquired resistance to combined RAF/MEK inhibition. *Cancer Discovery*, *4*, 61–68.

Wang, X., Ju, W., Renouard, J., Aden, J., Belinsky, S. A., & Lin, Y. (2006). 17-Allylamino-17-demethoxygeldanamycin synergistically potentiates tumor necrosis factor-induced lung cancer cell death by blocking the nuclear factor-kappaB pathway. *Cancer Research*, *66*, 1089–1095.

Wang, M., Kern, A. M., Hulskotter, M., Greninger, P., Singh, A., Pan, Y., et al. (2014). EGFR-mediated chromatin condensation protects KRAS-mutant cancer cells against ionizing radiation. *Cancer Research*, *74*, 2825–2834.

Wang, X., Song, X., Zhuo, W., Fu, Y., Shi, H., Liang, Y., et al. (2009). The regulatory mechanism of Hsp90alpha secretion and its function in tumor malignancy. *Proceedings of the National Academy of Sciences of the United States of America*, *106*, 21288–21293.

Wei, H., Wang, B., Miyagi, M., She, Y., Gopalan, B., Huang, D. B., et al. (2013). PRMT5 dimethylates R30 of the p65 subunit to activate NF-kappaB. *Proceedings of the National Academy of Sciences of the United States of America, 110*, 13516–13521.

Whitesell, L., Santagata, S., Mendillo, M. L., Lin, N. U., Proia, D. A., & Lindquist, S. (2014). HSP90 empowers evolution of resistance to hormonal therapy in human breast cancer models. *Proceedings of the National Academy of Sciences of the United States of America, 111*, 18297–18302.

Woronicz, J. D., Gao, X., Cao, Z., Rothe, M., & Goeddel, D. V. (1997). IkappaB kinase-beta: NF-kappaB activation and complex formation with IkappaB kinase-alpha and NIK. *Science, 278*, 866–869.

Wu, B. K., & Brenner, C. (2014). Suppression of TET1-dependent DNA demethylation is essential for KRAS-mediated transformation. *Cell Reports, 9*, 1827–1840.

Xu, K., Wu, Z. J., Groner, A. C., He, H. H., Cai, C., Lis, R. T., et al. (2012). EZH2 oncogenic activity in castration-resistant prostate cancer cells is Polycomb-independent. *Science, 338*, 1465–1469.

Yamaguchi, H., & Hung, M. C. (2014). Regulation and role of EZH2 in cancer. *Cancer Research and Treatment: Official Journal of Korean Cancer Association, 46*, 209–222.

Yeh, H. Y., Cheng, S. W., Lin, Y. C., Yeh, C. Y., Lin, S. F., & Soo, V. W. (2009). Identifying significant genetic regulatory networks in the prostate cancer from microarray data based on transcription factor analysis and conditional independency. *BMC Medical Genomics, 2*, 70.

Yu, H., Lee, H., Herrmann, A., Buettner, R., & Jove, R. (2014). Revisiting STAT3 signalling in cancer: New and unexpected biological functions. *Nature Reviews. Cancer, 14*, 736–746.

Zhang, X., Blenis, J., Li, H. C., Schindler, C., & Chen-Kiang, S. (1995). Requirement of serine phosphorylation for formation of STAT-promoter complexes. *Science, 267*, 1990–1994.

Zhou, X., Fan, L. X., Peters, D. J., Trudel, M., Bradner, J. E., & Li, X. (2015). Therapeutic targeting of BET bromodomain protein, Brd4, delays cyst growth in ADPKD. *Human Molecular Genetics, 24*, 3982–3993.

Zhou, X., Richon, V. M., Wang, A. H., Yang, X. J., Rifkind, R. A., & Marks, P. A. (2000). Histone deacetylase 4 associates with extracellular signal-regulated kinases 1 and 2, and its cellular localization is regulated by oncogenic Ras. *Proceedings of the National Academy of Sciences of the United States of America, 97*, 14329–14333.

Zuber, J., Shi, J., Wang, E., Rappaport, A. R., Herrmann, H., Sison, E. A., et al. (2011). RNAi screen identifies Brd4 as a therapeutic target in acute myeloid leukaemia. *Nature, 478*, 524–528.

CHAPTER SIX

Emerging Roles of Extracellular Hsp90 in Cancer

Daniel Senh Wong*, Daniel G. Jay†,1
*Graduate Program in Cellular and Molecular Physiology, Sackler School of Graduate Biomedical Sciences, Tufts University, Boston, Massachusetts, USA
†Department of Developmental, Molecular, and Chemical Biology, School of Medicine, Tufts University, Boston, Massachusetts, USA
[1]Corresponding author: e-mail address: daniel.jay@tufts.edu

Contents

Abstract

Heat shock protein 90 (Hsp90) is a highly expressed chaperone that modulates the function and stability of hundreds of cellular client proteins. In this capacity, Hsp90 impacts human health in myriad ways and it is accordingly a high-interest molecular target in the oncology setting. This interest has led to a large number of clinical trials to evaluate the potential benefit of Hsp90 inhibitors in cancer treatment and, more recently, in combination with chemotherapeutic agents. Although these studies are still ongoing, some issues have arisen, such as toxicity effects associated with administration of these agents.

We and others have identified a novel role for Hsp90 outside of cancer cells. This extracellular Hsp90 (eHsp90) was shown to be critical for the regulation of tumor invasiveness and metastasis, central processes associated with cancer lethality. Since these initial papers, a considerable cohort of studies has expanded upon this role, implicating eHsp90 in the activation of a number of proteins that support tumor cell invasion. As eHsp90 is preferentially detected on the surface of tumor cells, and within their surrounding microenvironment, it is possible that drugs capable of selectively targeting eHsp90 may exploit this differential expression. This selectivity may, in turn, enable treatment regimens with reduced target-related toxicity. This review will focus on our

Advances in Cancer Research, Volume 129
ISSN 0065-230X
http://dx.doi.org/10.1016/bs.acr.2016.01.001

current understanding of eHsp90, particularly in cancer, and we will discuss the relevance of eHsp90 as a biomarker for invasive cancer and its potential as a drug target.

1. HISTORY AND OVERVIEW OF EXTRACELLULAR Hsp90

While the intracellular roles of Hsp90 are well known and widely studied (Csermely, Schnaider, Sőti, Prohászka, & Nardai, 1998; Jackson, 2012; Jarosz, 2016; Pearl & Prodromou, 2006; Taipale, Jarosz, & Lindquist, 2010), the translocation of cytoplasmic Hsp90 to the cell surface and secretion of Hsp90 from cells to generate a pool of extracellular Hsp90 (eHsp90) is a comparatively recent discovery. It is presently unclear whether eHsp90 present at the cell surface is distinct from the secreted eHsp90 pool. An equally unresolved question is the identity of the population responsible for changes in cell behavior and phenotype. As many publications have examined eHsp90 from only one of these two compartments, the terminology eHsp90 will generally refer to both protein populations, while surface eHsp90 and secreted eHsp90 will be clarified when possible.

Given that Hsp90 is a predominantly cytoplasmic protein chaperone, the presence of secreted eHsp90 was originally thought to be the result of necrotic cell death and lysis, resulting in the release of intracellular contents into the surrounding milieu (Basu, Binder, Suto, Anderson, & Srivastava, 2000; Saito, Dai, & Ohtsuka, 2005). However, this death-dependent release conflicted with the detection of eHsp90 on the surface of viable tumor cells (Srivastava, DeLeo, & Old, 1986; Ullrich, Robinson, Law, Willingham, & Appella, 1986), on neurons in the developing nervous system (Thomaidou & Patsavoudi, 1993), and in extracellular vesicles derived from dendritic cells (Théry et al., 2001, 1999). Since these early reports, we and others have demonstrated Hsp90 on the tumor surface and in media conditioned by these tumor cells (Eustace et al., 2004; Sidera, Gaitanou, Stellas, Matsas, & Patsavoudi, 2008).

Secretion of Hsp90 was first described in vascular smooth muscle cells (VSMCs) in response to oxidative stress, via stimulation of extracellular signal-related kinase 1 and 2 (ERK1/2) signaling (Liao et al., 2000). eHsp90 was also found within exosomes from dendritic cells of the immune system, further demonstrating that its extracellular localization could be uncoupled from apoptotic processes (Théry et al., 2001, 1999). Using an unbiased functional proteomic screen that employed fluorophore-assisted light inactivation (FALI) of proteins, we identified eHsp90α as an extracellular protein required for invasion (Eustace et al., 2004). We further showed that Hsp90α, rather than Hsp90β, was preferentially secreted by HT-1080 fibrosarcoma cancer cells and also found on the surface of these cells. This differential expression of Hsp90 isoforms on the cell exterior further supports a regulated mechanism of secretion.

To address the question of eHsp90 function, we employed FALI to show that eHsp90α activated matrix metalloproteinase 2 (MMP2) and mediated cancer cell invasion (Eustace et al., 2004). It was subsequently shown that eHsp90 activated additional extracellular clients to support cell motility and invasion (Sidera et al., 2008). For example, eHsp90 interacts with the extracellular domain of the receptor tyrosine kinase human epidermal growth factor receptor-2 (HER-2, also known as ErbB-2, Neu), leading to cytoskeletal rearrangements required for cell motility and invasion (Sidera et al., 2008). This group also showed that eHsp90 could activate a second matrix metalloproteinase, MMP9 (Stellas, Hamidieh, & Patsavoudi, 2010). Multiple groups have since demonstrated that eHsp90 is accessible to a number of specific antibodies and small molecules capable of neutralizing eHsp90 activity (Barrott et al., 2013; Devarakonda, Kita, Phoenix, & Claffey, 2015; McCready, Wong, Burlison, Ying, & Jay, 2014; Stellas et al., 2010; Tsutsumi et al., 2007). Together, these studies implicate eHsp90 as a central hub for activation of extracellular proteins required for cancer invasion.

2. MECHANISM OF eHsp90 SECRETION

Hsp90 is secreted from cancer cells, and while the exact mechanism of Hsp90 secretion is not completely understood, some parts of the process have been elucidated. The canonical protein secretory pathway involves vesicular transport of soluble proteins from the endoplasmic reticulum (ER) to the Golgi apparatus and on to the plasma membrane for secretion, but this process requires an N-terminal signal sequence that is not present in Hsp90. While the Picard laboratory suggested this classical secretion might occur for an Hsp90α (*HSP90AA1-2*) isoform via a pseudo-signal sequence (Picard, 2004), we showed that this was expressed at very low levels and did not contribute significantly to the secreted eHsp90 population (McCready, Sims, Chan, & Jay, 2010). Instead, Hsp90α exits the cell via exosomes, microvesicles generated by the endocytic pathway. The importance of this mechanism was shown in a series of experiments using primary human keratinocytes or various cancer cell lines in which Brefeldin A (BFA) inhibition of ER–Golgi protein transport did not reduce Hsp90 secretion, but treatment of cells with dimethyl amiloride (DMA) to block the exosomal secretion pathway significantly reduced the amount of secreted eHsp90 (Cheng et al., 2008; McCready et al., 2010). Translocation of Hsp90 to the cell surface has been associated with increased Hsp90 secretion (Lei, Venkatakrishnan, Yu, & Kazlauskas, 2007), and a recent study showed that phospholipase C gamma (PLCγ) and protein kinase C gamma (PKCγ) are required for translocation of Hsp90 to the plasma membrane (Yang et al., 2014), but how Hsp90 is loaded onto exosomes for secretion is not understood.

Although the regulation of Hsp90 secretion is complex, the process appears to be regulated, at least in part, by a cohort of posttranslational modifications. Specifically, PKA-dependent phosphorylation of the Thr-90 residue, along with cleavage of the EEVD motif from the C-terminal TPR repeat-binding domain, leads to Hsp90 secretion (Lei et al., 2007; Wang et al., 2009). Phosphorylation and other posttranslational modifications of Hsp90 such as acetylation (Yang et al., 2008) are known to affect not only its cellular localization (Lei et al., 2007; Wang et al., 2009) but also its affinity for binding client proteins (Barrott & Haystead, 2013; Mollapour & Neckers, 2012), which may also be important for its targeting to exosomes and for exosomal cargo selection.

Hsp90 secretion may occur as a specific response to environmental stimuli. For example, Hsp90 secretion increases in response to oxidative stress (Liao et al., 2000) and hypoxia (Li et al., 2007; Sahu et al., 2012; Woodley et al., 2009). Secretion also increases following stimulation with various growth factors including transforming growth factor alpha (TGFα) (Cheng et al., 2008) and epidermal growth factor (EGF) (Yang et al., 2014). Secretion of both Hsp90α and β isoforms has been reported, but there are conflicting reports on whether both remain stable and active in the extracellular environment. The original study showing Hsp90 secretion from VSMCs identified only the Hsp90α isoform (Liao et al., 2000), which was also reported by our lab within the context of fibrosarcoma cells (Eustace et al., 2004). Also in 2004, a study investigating Hsp90 as a neuron-specific antigen in the developing nervous system showed that both Hsp90α and β isoforms were present (Sidera, Samiotaki, Yfanti, Panayotou, & Patsavoudi, 2004), while a more recent study looking at eHsp90 in cancer showed that only Hsp90α appears to stabilize and activate extracellular client proteins like MMP2, while Hsp90β is rapidly degraded

(Song et al., 2010). Because the segregation of extracellular functions between Hsp90α and Hsp90β is unsettled, we use eHsp90 to broadly refer to both isoforms outside the cell.

3. ASSOCIATION OF eHsp90 WITH EXOSOMES

While eHsp90 has been detected in exosomes secreted by human cancer cells (Hegmans et al., 2004) as well as from dendritic cells (Théry et al., 2001), it is unclear whether Hsp90 is an integral component of exosomes or merely cargo. Although some groups have reported eHsp90 as an exosomal protein (Février & Raposo, 2004; Théry et al., 2001), its association with exosomes has not been consistently observed (Gupta & Knowlton, 2007). Interestingly, exosome secretion has been shown to be a calcium-dependent process (Savina, Furlán, Vidal, & Colombo, 2003), and Hsp90 can interact with the calcium-binding protein calmodulin (Minami, Kawasaki, Suzuki, & Yahara, 1993). It has been reported that Hsp90 inhibition perturbs the endocytic pathway, changing the sorting pattern of endosomes, resulting in the accumulation of endosomes within multivesicular bodies and lysosomes, independent of cargo identity (Cortese et al., 2013). These data suggest that Hsp90 may have a role in exosome secretion, although additional studies will be required to better define this mechanism.

Current evidence indicates that eHsp90 is localized to the extracellular side of the exosomal membrane, rather than the luminal side as would be expected for a cytoplasmic protein. The typical formation of exosomes by inward budding of the plasma membrane to form endosomes, followed by inward budding of endosomal membranes to form multivesicular bodies, should capture cytoplasmic Hsp90 in the lumen of exosomes. Thus, cytoplasmic proteins should be protected from protease digestion by the exosomal membrane. However, the initial report of Hsp90 secretion from VSMCs showed that eHsp90 was susceptible to protease digestion (Liao et al., 2000). Subsequent studies using isolated exosomes have replicated this finding (McCready et al., 2010; Yang et al., 2014), suggesting that Hsp90 is on the exosomal surface rather than within the lumen of exosomes. Transmission electron micrographs of immunogold-labeled exosomes also show that most eHsp90 is present on the exosomal surface, while a fraction appears to be unbound and present between exosomes (Yang et al., 2014). The mechanism responsible for eHsp90 localization to the external side of the exosomal membrane is not known and warrants further study.

Even more perplexing, additional studies demonstrate the presence of eHsp90 in the conditioned media of cells in association with several members of its known cochaperone complexes, including Hsp70, Hop, Hsp40, and p23 (Sims, McCready, & Jay, 2011). It is unclear how this might occur, as cochaperone complexes are nonmembrane bound, and likely too large to be translocated as an intact assemblage across a lipid bilayer. No pore protein is known to move any members of this complex across plasma or endosomal membranes. Hsp90 is, however, known to translocate into the nucleus via the nuclear pore complex (Echeverría et al., 2009), so its translocation across membranes can and does occur. Another Hsp90-dependent process that may provide insight into the mechanism of its membrane translocation is antigen presentation in dendritic cells, where Hsp90 chaperoning of the peptide is required for crosspresentation (Ichiyanagi et al., 2010). However, it is hypothesized that only short, unfolded peptides are translocated in this way, and so the comparatively large Hsp90 protein and its cochaperones would not be expected to utilize this pathway without being refolded after crossing the membrane.

A number of the stress stimuli that induce Hsp90 translocation and secretion are commonly found in cancers, such as DNA damage (Yu, Harris, & Levine, 2006), increased growth factor signaling (Cheng et al., 2008; Lei et al., 2007; Li et al., 2007; Wang et al., 2009), heat stress (Clayton, Turkes, Navabi, Mason, & Tabi, 2005), hypoxia (Jayaprakash et al., 2015; Li et al., 2007), and oxidative stress (Liao et al., 2000). Stress-induced alterations in Hsp90 homeostasis are reflected in observations that Hsp90 is widely overexpressed in human cancers (Becker et al., 2004; Ferrarini, Heltai, Zocchi, & Rugarli, 1992; Pick et al., 2007; Wang, Heuvelman, Carroll, Dufield, & Masferrer, 2010) and also present at the cell surface in breast cancer (Devarakonda et al., 2015; Eustace et al., 2004; Pick et al., 2007), fibrosarcoma (Eustace et al., 2004), glioblastoma (Gopal et al., 2011), melanoma (Becker et al., 2004), and prostate cancer (Hance et al., 2012). Analysis of patient tumor biopsy samples and survival data found that high expression and membrane localization of eHsp90 are associated with decreased survival of patients with breast cancer, although the study did not distinguish between the α and β Hsp90 isoforms (Pick et al., 2007). These tumor-associated stressors and resultant changes in eHsp90 expression profiles have specific implications for eHsp90-directed signaling and for the initiation of prometastatic signaling, as detailed in the subsequent sections.

4. eHsp90-DEPENDENT SIGNALING AND PROMOTILITY FUNCTIONS

Cell surface eHsp90 interacts with low-density lipoprotein receptor-related protein (LRP1, also CD91) (Basu, Binder, Ramalingam, & Srivastava, 2001) and may be anchored to the cell surface in addition to MMP9 by CD44 (Lagarrigue et al., 2010). This eHsp90–LRP1 interaction induces a number of promotility signaling cascades, including ERK1/2 and AKT1/2 activation, molecular responses that are essential for wound healing (Cheng et al., 2008; Tsen et al., 2013; Fig. 1). The interaction between eHsp90 and LRP1 requires the extracellular subdomain II of LRP1, as well as the NPVY motif of the LRP1 cytoplasmic domain (Tsen et al., 2013). Interestingly, cytoplasmic Hsp90β stabilizes the intracellular domain of LRP1 (Jayaprakash et al., 2015), indicating cooperativity with the eHsp90–LRP1 pathway. Activation of AKT by eHsp90 in glioma leads to subsequent phosphorylation of the receptor tyrosine kinase ephrin type-A receptor 2 (EPHA2), which promotes its interaction with LRP1 to sustain AKT activation and promotility signaling (Gopal et al., 2011).

The interaction of eHsp90 with other cell surface proteins has also been proposed to contribute to the progression of malignant disease. Aberrant HER-2 activity mediates cancer cell invasion and is associated with breast cancer metastasis (Medress & Hayden Gephart, 2015). The binding of eHsp90 to the extracellular domain of HER-2 (Sidera et al., 2008) in conjunction with Cdc37 (El Hamidieh, Grammatikakis, & Patsavoudi, 2012) contributes to increased cell motility. HER-2 downstream signaling activates both the progrowth ERK pathway and the prosurvival phosphatidylinositol-3-kinase (PI3K) pathway (Calderwood, Khaleque, Sawyer, & Ciocca, 2006). HER-2 also signals through Src (Belsches-Jablonski et al., 2001), which is present at focal adhesions in a protein complex with focal adhesion kinase. This adhesion complex is critical for cell motility via its interaction with integrins, which bridge the cytoskeleton and extracellular matrix (ECM) (Mitra, Hanson, & Schlaepfer, 2005; Tsutsumi et al., 2007; Wozniak, Modzelewska, Kwong, & Keely, 2004). Normal cells are dependent upon attachment to the ECM and will die if attachment is lost (anoikis) (Frisch & Screaton, 2001). eHsp90 activity appears to be required for the interaction of Src and integrin β1 at focal adhesions that is critical for cell migration (Tsutsumi et al., 2007). In addition, eHsp90 signaling through a HER-2–Src pathway promotes anoikis

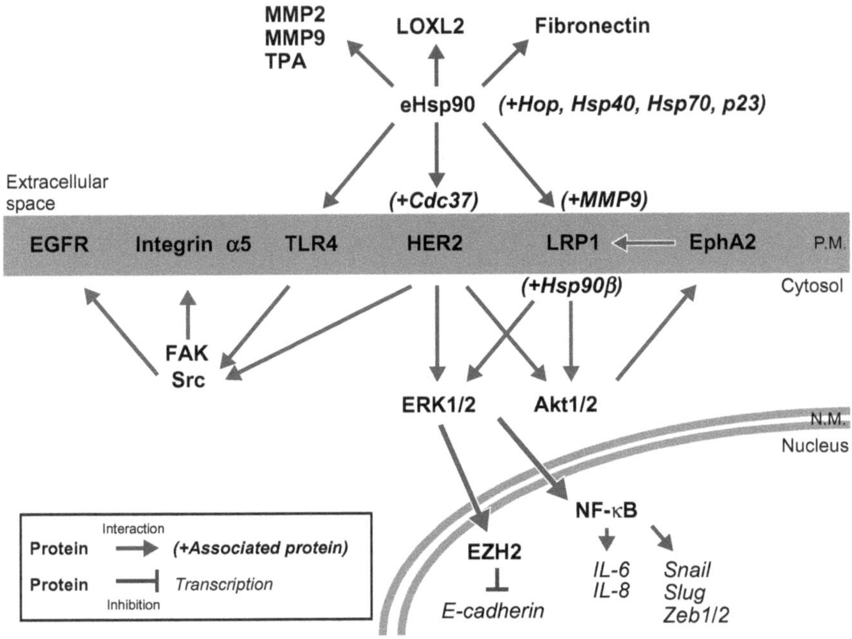

Figure 1 eHsp90 interacts with extracellular clients and surface receptors to promote cell migration and invasion in development, wound healing, and cancer. eHsp90 can be found in the extracellular space, along with some members of the cochaperone complexes that are found within cells (Hop, Hsp40, Hsp70, and p23). Interaction of eHsp90 with matrix metalloproteinases (MMP2, MMP9) and tissue plasminogen activator (TPA) facilitates cell invasion. eHsp90 has also been found to associate with lysyl oxidase-like protein 2 (LOXL2), which crosslinks collagen and alters ECM stiffness, as well as fibronectin, a component of the ECM. The binding of eHsp90 to several cell surface receptors has also been shown to induce intracellular signaling pathways. eHsp90 interacts with the Toll-like receptor 4 (TLR), signaling through the Src pathway to focal adhesion kinase (FAK) and integrin α5, which is critical for cell motility, as well as transactivation of epithelial growth factor receptor (EGFR). eHsp90 binds to the extracellular domain of HER2 in conjunction with its cochaperone Cdc37, resulting in signaling through the Src, ERK, and AKT pathways. Signaling to integrin α5 through Src allows cells to become anchorage independent. eHsp90 also binds to low-density lipoprotein receptor-related protein 1 (LRP1) at the cell surface to also signal through the ERK and AKT pathways. AKT activation leads to activation of ephrin type-A receptor 2 (EPHA2), which again activates LRP1 to sustain signaling. The cytoplasmic tail of LRP1 is stabilized by Hsp90β. eHsp90 signaling also supports an epithelial to mesenchymal transition (EMT) phenotype via ERK signaling by inducing expression of enhancer of zeste homolog 2 (EZH2) and reducing expression of the epithelial marker E-cadherin. eHsp90-activated ERK and NF-κB may also increase expression of mesenchymal markers Snail, Slug, and Zeb1/2 via NF-κB. NF-κB also induces transcription of the inflammatory cytokines IL-6 and IL-8. P.M., plasma membrane; N.M., nuclear membrane. (See the color plate.)

resistance and proliferation in an integrin α5-dependent manner (Haenssen et al., 2010). The interaction of eHsp90 and Toll-like receptor 4 (TLR4) also signals through Src to transactivate epithelial growth factor receptor, which increases cell migration (Thuringer et al., 2011; Fig. 1).

Increased migration and invasion is a characteristic of cells that have undergone epithelial–mesenchymal transition (EMT), a process by which polarized, nonmotile epithelial cells lose their polarity and become motile and invasive mesenchymal cells. This process is important in normal development and wound healing, but when dysregulated, may also support cancer metastasis. While eHsp90-dependent signaling has many diverse effects, an important consequence is activation of the EMT program, as recently shown in prostate and colorectal cancer models (Chen, Chen, Chen, Lee, & Huang, 2013; Hance et al., 2012). EMT status can be assessed by monitoring molecular markers such as a loss of E-cadherin expression and increased expression of the mesenchymal markers Snail, Slug, Zeb1, and Zeb2 (Polyak & Weinberg, 2009). Exposure of prostate cancer cells to eHsp90 promoted molecular events consistent with EMT activation, and this transition required MMP9 and ERK activity (Hance et al., 2012). eHsp90-induced activation of ERK also increased expression of enhancer of zeste homolog 2 (EZH2), a methyltransferase of the polycomb repressor complex, leading to epigenetic repression of E-cadherin (Nolan, Franco, Hance, Hayward, & Isaacs, 2015). Work in colorectal cancer cells showed that eHsp90 promotion of EMT is mediated by an LRP1–NF-κB pathway (Nagaraju et al., 2015), suggesting the importance of context in eHsp90-mediated signaling alterations (Fig. 1).

5. eHsp90-DEPENDENT ACTIONS IN MATRIX REMODELING AND CELLULAR INVASION

Remodeling of the ECM is an ongoing, dynamic process in normal tissue that is important for development, the maintenance of homeostasis, cell motility, and wound healing. Modification of the ECM is accomplished through the regulation and activation of extracellular proteins that cross-link molecules secreted by fibroblasts and other cell types to produce more of the matrix and by proteases that degrade and remodel the ECM matrix. Dysregulation of this process is associated with the development of disease (reviewed in Bonnans, Chou, & Werb, 2014). Although ECM remodeling in response to eHsp90 stimulation has not been directly established, eHsp90 is involved in the activation of several proteases known to modify the ECM

(Eustace et al., 2004; Lei et al., 2007; McCready et al., 2010, 2014; Stellas et al., 2010).

Dysregulated ECM remodeling is a characteristic of cancer cell invasion and the development of metastatic disease. eHsp90 has been shown to bind directly to fibronectin in the ECM (Hunter et al., 2014), consistent with a role for eHsp90 in ECM stability or organization. Interestingly, several of eHsp90's extracellular clients are proteases that are secreted as inactive proenzymes (zymogens) that must be proteolytically cleaved for activation (Gaffney, Solomonov, Zehorai, & Sagi, 2015; Khan & James, 1998). Several groups, including ours, have shown that eHsp90 interacts with inactive proteases, or their activating proteins, to facilitate enzyme activation and subsequent cell migration and invasion (Eustace et al., 2004; McCready et al., 2010; Sims et al., 2011; Song et al., 2010; Yu & Stamenkovic, 2000). Specifically, we demonstrated that the association of eHsp90 with MMP2 increased the abundance of activated forms of MMPs, which in turn increased tumor cell invasion (Eustace et al., 2004). MMP2 activation can be enhanced by Hsp70, Hop, and other members of Hsp90 cochaperone complexes (Sims et al., 2011) in an ATP-independent manner (Sims et al., 2011; Song et al., 2010). We have suggested that eHsp90 may serve to increase pro-MMP2 cleavage to its active form via interaction of cochaperone complexes with extracellular proteases, although this remains to be definitively established (Sims et al., 2011). Curiously, it has also been demonstrated that MMP2 processing can be regulated by eHsp90 binding without ATP hydrolysis via its ability to protect the protease from inactivation rather than by promoting its activation (Song et al., 2010).

Mass spectrometry analysis of eHsp90 complexes immunoprecipitated from media conditioned by MDA-MB-231 breast cancer cells revealed that eHsp90 also interacts with tissue plasminogen activator (TPA) (McCready et al., 2010) and lysyl oxidase-like protein 2 (LOXL2) (McCready et al., 2014). Generation of active plasmin from plasminogen by TPA elicits diverse effects, including liberation of ECM-bound growth factors, ECM remodeling, MMP activation, and cleavage of cell surface receptors (reviewed by Deryugina & Quigley, 2012). LOXL2 is a member of the lysyl oxidase family of proteins, which modify the ECM by cross-linking collagen. Cross-linking alters the stiffness of the matrix, resulting in phenotypic changes such as increased cell motility and proliferation (Barker, Bird, Lang, & Erler, 2013; Kim et al., 2014). Increased expression of LOXL2 by cancer cells has been associated with increased invasion (Moon et al., 2013) and poor clinical prognosis (Ahn et al., 2013; Barker et al., 2011).

However, the role of LOX family members is complex, as studies have also implicated LOX and LOXL2 in tumor-suppressing effects (Mayorca-Guiliani & Erler, 2013). Hence, global inhibition of these proteins may have complex and unanticipated effects on tumor progression and metastasis. This complexity may be explained by the differential functions elicited by the LOXL2 propeptide relative to the mature cleaved form. While the LOXL2 propeptide has been shown to function as a tumor suppressor by its inhibition of the MAP kinase pathway (Agra, Cidre, García-García, de la Parra, & Alonso, 2013), the cleaved and active LOXL2 may possess oncogenic activity by enhancing EMT and ECM remodeling (Barker et al., 2011; Vadasz et al., 2005).

6. SELECTIVE TARGETING OF eHsp90 TO TREAT CANCERS

Targeting Hsp90 for anticancer therapy may seem paradoxical, because while it is overexpressed in cancer cells, it is also abundant within normal cells and is required for cell survival under both basal and stress conditions. However, in many cancer cells, Hsp90 appears to serve a unique role in stabilizing proteins within an otherwise unsuitable environment or chaperoning mutant oncoproteins that exist in unstable conformations (An, Schulte, & Neckers, 2000; Azoitei et al., 2014; Whitesell & Lindquist, 2005; Whitesell et al., 2014; Workman, Burrows, Neckers, & Rosen, 2007). Thus, inhibiting Hsp90 may sensitize cancer cells to stress-induced damage and cell death. Alternatively, given the global role of Hsp90 in protein homeostasis, inhibiting its activity with cell-permeable agents carries the substantial risk of enough target-related systemic toxicity to limit therapeutic efficacy. Indeed, recent clinical trials of second-generation Hsp90 inhibitors given at maximally tolerated doses either alone or in combination with other agents have been sadly disappointing (Cambridge News, 2014; Oki et al., 2015; Spreafico et al., 2015; Synta Pharmaceutical, 2015).

It is an obvious goal to improve tumor selectivity and to consequently reduce systemic toxicity following administration of Hsp90-targeted agents. Recent studies suggest that the Hsp90 chaperone populations found in cancer cells may have unique properties that allow for enhanced drug uptake (Castro et al., 2005; Kamal et al., 2003; Moulick et al., 2011; Nayar et al., 2013; Taldone, Ochiana, Patel, & Chiosis, 2014; Zong et al., 2015), but this is a developing and unsettled area of research. Therapeutic approaches designed to blunt eHsp90 action could provide a much needed strategy for dealing with the problem of poor therapeutic index confronting current clinical development of pan-Hsp90 inhibitors, given the specific association of eHsp90 with the malignant state, and the ability of its targeting to inhibit activation of a specific cohort of extracellular clients essential for tumor cell motility, invasion, and metastasis. Moreover, the limited repertoire of eHsp90 clients may translate to a reduction in target-related drug toxicity when compared with the more global effects of conventional cell-permeable drugs, as supported by recent studies (McCready et al., 2014). However, despite this therapeutic promise, preclinical eHsp90-targeted agents have not yet entered clinical trials. Small molecule candidates developed in addition to DMAG-*N*-oxide (Tsutsumi et al., 2007) include STA-12-7191, a biotinylated ganetespib (McCready et al., 2014), and HS-27, the latter of which was shown to have potential utility for cancer cell imaging (Barrott et al., 2013). Interestingly, a recent study indicates that tumor surface eHsp90 may undergo active transport resulting in its reinternalization (Barrott et al., 2013). If validated, this mechanism may provide a further rationale to explore use of eHsp90-targeted agents in an oncology setting. The tumor-specific reinternalization of eHsp90-targeted agents would be predicted to contribute to tumor-specific drug accumulation, further enhance drug selectivity, and limit host toxicity.

Targeting eHsp90 may also be achieved by the use of antibody-based approaches. The earliest eHsp90-specific agent, monoclonal antibody 4C5, was first developed and characterized by Thomaidou and Patsavoudi over two decades ago (Stellas et al., 2010; Thomaidou, Dori, & Patsavoudi, 1995; Thomaidou & Patsavoudi, 1993). Preclinical studies of 4C5 demonstrated its ability to inhibit breast cancer metastasis (Sidera, El Hamidieh, Mamalaki, & Patsavoudi, 2011; Stellas et al., 2010). Additional groups have demonstrated the ability of Hsp90-targeted antibodies to limit tumor motility, invasion and metastasis in prostate, glioma, breast, and melanoma models (Gopal et al., 2011; Hance et al., 2012; Tsutsumi et al., 2009). Recently, a group has isolated Hsp90-specific antibodies from human breast cancer patients. One such purified antibody, HCAb2, was demonstrated to bind to cell surface eHsp90, and to inhibit breast cancer cell motility (Devarakonda et al., 2015), indicating that circulating Hsp90 antibodies may be generated in response to detection of tumor eHsp90.

Given these early results, one may predict that the detection of putative Hsp90-neutralizing autoantibodies in patient sera may be correlated with positive clinical outcome. However, studies examining anti-Hsp90 autoantibodies in patient sera have produced conflicting results. Contrary to this prediction, the presence of autoantibodies in the sera of patients has been associated with poor prognosis in breast cancer (Conroy et al., 1995; Conroy, Sasieni, Fentiman, & Latchman, 1998) and ovarian cancer (Luo, Herrera, Soosaipillai, & Diamandis, 2002), while favorable associations were observed for patients with osteosarcoma (Trieb et al., 1999). It is unclear at this time whether these differences are due to intrinsic tumor specificity or to functional differences in neutralizing activities of the antibodies toward eHsp90. The presence of Hsp90 autoantibodies has been correlated with advanced disease stage (Conroy et al., 1995, 1998; Luo et al., 2002), as well as with positive response to chemotherapy (Trieb et al., 1999), although levels do not necessarily correlate with tumor resection alone (Conroy et al., 1998). These findings suggest that this physiological antigenic response to tumor eHsp90 antigen should be further evaluated and may have the potential to be further developed within a therapeutic setting.

7. eHsp90 AS A CANCER BIOMARKER

eHsp90 may have potential as a clinical biomarker because of its preferential secretion in cancer cells, coupled with the ability to noninvasively assay eHsp90 levels via enzyme-linked immunosorbent assay (ELISA) performed on sera derived from routine blood draws. Recent studies using animal models, as well as clinical samples and patient data, have explored this idea. Proteomic analysis of serum samples from patients with hepatocellular carcinoma (HCC) revealed significantly elevated levels of serum eHsp90 in HCC samples compared to normal controls, demonstrating the potential utility of serum eHsp90 as a biomarker for HCC, although this study did not differentiate between the two Hsp90 isoforms (Sun et al., 2010). Another study used a mouse model in tandem with ELISA for Hsp90α to selectively detect eHsp90α in the plasma of tumor-bearing animals (Wang et al., 2009). This same group used a similar ELISA assay to survey plasma samples from patients with breast, liver, lung, or pancreatic cancer. Of note, serum eHsp90α levels were significantly higher in the patient groups with tumor burden, and positively correlated with tumor malignancy and metastasis (Wang et al., 2009).

An Hsp90α ELISA test for detection of plasma eHsp90 was approved for use in China in the care of patients with lung cancer (Cheng, 2013). This ELISA approach (Yantai Protgen Biotechnology Development Co. Ltd., Yantai, China) was used to examine plasma levels of Hsp90α in samples from

2247 patients with lung cancer where it detected significantly elevated plasma Hsp90α levels, the magnitude of which also correlated with more advanced disease stage as well as disease progression (Shi et al., 2014). Another ELISA study examined serum levels of Hsp90β from patients with lung tumors. Although serum Hsp90β was elevated in patients with tumors compared to individuals without tumors, the assay did not achieve sufficient specificity to accurately distinguish between benign and malignant tumors (Rong et al., 2014). The results of these studies indicate that, at least in lung cancer, plasma Hsp90α levels may be a useful diagnostic and prognostic biomarker to assess disease status and predict outcome. Testing the association of eHsp90α with clinicopathological metrics in additional cancer types will be an important next step toward the goal of determining the value of secreted Hsp90α in improving early cancer detection and/or disease progression. Toward this goal, the protein ELISA test is currently in clinical trials in China for breast, colorectal, and liver cancer (NCT02324101, NCT02324114, and NCT02324127).

8. FUTURE DIRECTIONS

Much work is needed to improve our understanding of the function of eHsp90 and its roles in supporting cancer progression. It is unclear whether eHsp90 on the surface of cells is distinct from the secreted eHsp90 present on exosomes, or if these populations represent an interchangeable pool. It is also unclear how eHsp90 translocates from the cytoplasm to the surface of exosomes and how, or if, it functions on exosomes. While the chaperone is associated with exosomes, it is not known whether it is simply cargo on these vesicles, or if Hsp90 has a role in the regulation of exosome secretion from cells or uptake by recipient cells. Moreover, how eHsp90 reaches the exosomal surface and whether it defines a specific exosomal subpopulation remain open questions. The important role of exosomes in communication between tumor cells and stroma makes addressing eHsp90's role in this process an intriguing area for investigation. It is also unclear how eHsp90 operates in the extracellular space, as many of its intracellular functions are ATP dependent, but the concentration of ATP in the extracellular environment is very low. It has been suggested that posttranslational modifications such as phosphorylation and acetylation could increase the affinity of Hsp90 for ATP and/or client proteins. Alternatively, ATP-independent chaperone activities reported for its C-terminal domain could be very important to its functioning within the extracellular environment. These open questions have great basic biological significance, and their answers will have important implications for targeting eHsp90 for cancer therapy.

Currently available data suggest that eHsp90 might provide an opportunity for developing more selective and less toxic anticancer therapies than global Hsp90 inhibitors. Among the pathways regulated by eHsp90 are developmental and wound-healing processes that are inappropriately reactivated in cancer and critical for disease progression, such as cell migration, invasion, and the EMT program. Therefore, inhibition of eHsp90 may have clinical utility in reducing or preventing metastasis to improve cancer control and improve patient outcome. If eHsp90 at the cell surface is reinternalized, it is also possible that tethering cytotoxic compounds to eHsp90-targeting molecules could improve the selectivity of conventional chemotherapeutics and reduce their toxic side effects.

The development of rapid and sensitive clinical tests that utilize plasma eHsp90 levels as a biomarker is an exciting and promising translational application that has the potential to improve cancer detection and diagnosis. While this test has only been approved in China and for just one type of cancer, we await results of the additional trials underway for breast, colorectal, and liver cancers. Finally, while clinical trials continue for pan-Hsp90 inhibitors, the importance of eHsp90 in other types of cancer, and the effects on normal tissues of selective inhibitors of eHsp90 need to be defined. Although early-stage impermeant Hsp90 inhibitors may serve as a platform for the development of antimetastatic drugs, these need to be more rigorously tested in preclinical animal models before they are ready for evaluation in cancer patients. Although no agents suitable for clinical testing are yet available, recent and emerging data collectively support the premise that eHsp90 indeed represents a functionally relevant target for cancer therapy, a notion poised to have a significant impact on future strategies to block Hsp90 function in malignancy.

REFERENCES

Agra, N., Cidre, F., García-García, L., de la Parra, J., & Alonso, J. (2013). Lysyl oxidase is downregulated by the EWS/FLI1 oncoprotein and its propeptide domain displays tumor supressor activities in Ewing sarcoma cells. *PLoS One*, *8*(6), e66281. http://doi.org/10.1371/journal.pone.0066281.

Ahn, S. G., Dong, S. M., Oshima, A., Kim, W. H., Lee, H. M., Lee, S. A., et al. (2013). LOXL2 expression is associated with invasiveness and negatively influences survival in breast cancer patients. *Breast Cancer Research and Treatment*, *141*(1), 89–99. http://doi.org/10.1007/s10549-013-2662-3.

An, W. G., Schulte, T. W., & Neckers, L. M. (2000). The heat shock protein 90 antagonist geldanamycin alters chaperone association with p210bcr-abl and v-src proteins before their degradation by the proteasome. *Cell Growth and Differentiation, 11*(7), 355–360.

Azoitei, N., Diepold, K., Brunner, C., Rouhi, A., Genze, F., Becher, A., et al. (2014). HSP90 supports tumor growth and angiogenesis through PRKD2 protein stabilization. *Cancer Research, 74*(23), 7125–7136. http://doi.org/10.1158/0008-5472.CAN-14-1017.

Barker, H. E., Bird, D., Lang, G., & Erler, J. T. (2013). Tumor-secreted LOXL2 activates fibroblasts through FAK signaling. *Molecular Cancer Research, 11*(11), 1425–1436. http://doi.org/10.1158/1541-7786.MCR-13-0033-T.

Barker, H. E., Chang, J., Cox, T. R., Lang, G., Bird, D., Nicolau, M., et al. (2011). LOXL2-mediated matrix remodeling in metastasis and mammary gland involution. *Cancer Research, 71*(5), 1561–1572. http://doi.org/10.1158/0008-5472.CAN-10-2868.

Barrott, J. J., & Haystead, T. A. J. (2013). Hsp90, an unlikely ally in the war on cancer. *FEBS Journal, 280*(6), 1381–1396. http://doi.org/10.1111/febs.12147.

Barrott, J. J., Hughes, P. F., Osada, T., Yang, X.-Y., Hartman, Z. C., Loiselle, D. R., et al. (2013). Optical and radioiodinated tethered Hsp90 inhibitors reveal selective internalization of ectopic Hsp90 in malignant breast tumor cells. *Chemistry & Biology, 20*(9), 1187–1197. http://doi.org/10.1016/j.chembiol.2013.08.004.

Basu, S., Binder, R. J., Ramalingam, T., & Srivastava, P. K. (2001). CD91 is a common receptor for heat shock proteins gp96, hsp90, hsp70, and calreticulin. *Immunity, 14*(3), 303–313. http://doi.org/10.1016/S1074-7613(01)00111-X.

Basu, S., Binder, R. J., Suto, R., Anderson, K. M., & Srivastava, P. K. (2000). Necrotic but not apoptotic cell death releases heat shock proteins, which deliver a partial maturation signal to dendritic cells and activate the NF-κB pathway. *International Immunology, 12*(11), 1539–1546. http://doi.org/10.1093/intimm/12.11.1539.

Becker, B., Multhoff, G., Farkas, B., Wild, P.-J., Landthaler, M., Stolz, W., et al. (2004). Induction of Hsp90 protein expression in malignant melanomas and melanoma metastases. *Experimental Dermatology, 13*(1), 27–32. http://doi.org/10.1111/j.0906-6705.2004.00114.x.

Belsches-Jablonski, A. P., Biscardi, J. S., Peavy, D. R., Tice, D. A., Romney, D. A., & Parsons, S. J. (2001). Src family kinases and HER2 interactions in human breast cancer cell growth and survival. *Oncogene, 20*(12), 1465–1475. http://dx.doi.org/10.1038/sj.onc.1204205.

Bonnans, C., Chou, J., & Werb, Z. (2014). Remodelling the extracellular matrix in development and disease. *Nature Reviews. Molecular Cell Biology, 15*(12), 786–801. http://doi.org/10.1038/nrm3904.

Calderwood, S. K., Khaleque, M. A., Sawyer, D. B., & Ciocca, D. R. (2006). Heat shock proteins in cancer: Chaperones of tumorigenesis. *Trends in Biochemical Sciences, 31*(3), 164–172. http://doi.org/10.1016/j.tibs.2006.01.006.

Cambridge News. (2014). *Vernalis announces Novartis has ceased development work on AUY922 cancer treatment.* Retrieved January 1, 2016, from, http://www.cambridge-news.co.uk/Vernalis-announces-Novartis-ceased-development/story-25740513-detail/story.html.

Castro, J. E., Prada, C. E., Loria, O., Kamal, A., Chen, L., Burrows, F. J., et al. (2005). ZAP-70 is a novel conditional heat shock protein 90 (Hsp90) client: Inhibition of Hsp90 leads to ZAP-70 degradation, apoptosis, and impaired signaling in chronic lymphocytic leukemia. *Blood, 106*(7), 2506–2512. http://doi.org/10.1182/blood-2005-03-1099.

Chen, W.-S., Chen, C.-C., Chen, L.-L., Lee, C.-C., & Huang, T.-S. (2013). Secreted heat shock protein 90α (HSP90α) induces nuclear factor-κB-mediated TCF12 protein expression to down-regulate E-cadherin and to enhance colorectal cancer cell migration and invasion. *Journal of Biological Chemistry, 288*(13), 9001–9010. http://doi.org/10.1074/jbc.M112.437897.

Cheng, Y. (2013). Scientists develop new test for cancer. *China Daily*, (November 20). Retrieved from, http://www.chinadaily.com.cn/china/2013-11/20/content_17117168.htm.

Cheng, C.-F., Fan, J., Fedesco, M., Guan, S., Li, Y., Bandyopadhyay, B., et al. (2008). Transforming growth factor alpha (TGFalpha)-stimulated secretion of HSP90alpha: Using the receptor LRP-1/CD91 to promote human skin cell migration against a TGFbeta-rich environment during wound healing. *Molecular and Cellular Biology*, *28*(10), 3344–3358. http://doi.org/10.1128/MCB.01287-07.

Clayton, A., Turkes, A., Navabi, H., Mason, M. D., & Tabi, Z. (2005). Induction of heat shock proteins in B-cell exosomes. *Journal of Cell Science*, *118*(16), 3631–3638. http://doi.org/10.1242/jcs.02494.

Conroy, S., Gibson, S., Brunström, G., Isenberg, D., Luqmani, Y., & Latchman, D. (1995). Autoantibodies to 90 kDa heat-shock protein in sera of breast cancer patients. *The Lancet*, *345*(8942), 126. http://doi.org/10.1016/S0140-6736(95)90090-X.

Conroy, S. E., Sasieni, P. D., Fentiman, I., & Latchman, D. S. (1998). Autoantibodies to the 90 kDa heat shock protein and poor survival in breast cancer patients. *European Journal of Cancer (Oxford, England: 1990)*, *34*(6), 942–943.

Cortese, K., Howes, M. T., Lundmark, R., Tagliatti, E., Bagnato, P., Petrelli, A., et al. (2013). The HSP90 inhibitor geldanamycin perturbs endosomal structure and drives recycling ErbB2 and transferrin to modified MVBs/lysosomal compartments. *Molecular Biology of the Cell*, *24*(2), 129–144. http://doi.org/10.1091/mbc.E12-04-0282.

Csermely, P., Schnaider, T., Sőti, C., Prohászka, Z., & Nardai, G. (1998). The 90-kDa molecular chaperone family: Structure, function, and clinical applications. A comprehensive review. *Pharmacology & Therapeutics*, *79*(2), 129–168. http://doi.org/10.1016/S0163-7258(98)00013-8.

Deryugina, E. I., & Quigley, J. P. (2012). Cell surface remodeling by plasmin: A new function for an old enzyme. *BioMed Research International*, *2012*, e564259. http://dx.doi.org/10.1155/2012/564259. http://doi.org/10.1155/2012/564259.

Devarakonda, C. V., Kita, D., Phoenix, K. N., & Claffey, K. P. (2015). Patient-derived heavy chain antibody targets cell surface HSP90 on breast tumors. *BMC Cancer*, *15*(1), 614. http://doi.org/10.1186/s12885-015-1608-z.

Echeverría, P. C., Mazaira, G., Erlejman, A., Gomez-Sanchez, C., Pilipuk, G. P., & Galigniana, M. D. (2009). Nuclear import of the glucocorticoid receptor-hsp90 complex through the nuclear pore complex is mediated by its interaction with Nup62 and importin β. *Molecular and Cellular Biology*, *29*(17), 4788–4797. http://doi.org/10.1128/MCB.00649-09.

El Hamidieh, A., Grammatikakis, N., & Patsavoudi, E. (2012). Cell surface Cdc37 participates in extracellular HSP90 mediated cancer cell invasion. *PLoS One*, 7(8), e42722. http://doi.org/10.1371/journal.pone.0042722.

Eustace, B. K., Sakurai, T., Stewart, J. K., Yimlamai, D., Unger, C., Zehetmeier, C., et al. (2004). Functional proteomic screens reveal an essential extracellular role for hsp90α in cancer cell invasiveness. *Nature Cell Biology*, *6*(6), 507–514. http://doi.org/10.1038/ncb1131.

Ferrarini, M., Heltai, S., Zocchi, M. R., & Rugarli, C. (1992). Unusual expression and localization of heat-shock proteins in human tumor cells. *International Journal of Cancer*, *51*(4), 613–619. http://doi.org/10.1002/ijc.2910510418.

Février, B., & Raposo, G. (2004). Exosomes: Endosomal-derived vesicles shipping extracellular messages. *Current Opinion in Cell Biology*, *16*(4), 415–421. http://doi.org/10.1016/j.ceb.2004.06.003.

Frisch, S. M., & Screaton, R. A. (2001). Anoikis mechanisms. *Current Opinion in Cell Biology*, *13*(5), 555–562. http://doi.org/10.1016/S0955-0674(00)00251-9.

Gaffney, J., Solomonov, I., Zehorai, E., & Sagi, I. (2015). Multilevel regulation of matrix metalloproteinases in tissue homeostasis indicates their molecular specificity in vivo. *Matrix Biology, 44–46,* 191–199. http://doi.org/10.1016/j.matbio.2015.01.012.

Gopal, U., Bohonowych, J. E., Lema-Tome, C., Liu, A., Garrett-Mayer, E., Wang, B., et al. (2011). A novel extracellular Hsp90 mediated co-receptor function for LRP1 regulates EphA2 dependent glioblastoma cell invasion. *PLoS One, 6*(3), e17649. http://doi.org/10.1371/journal.pone.0017649.

Gupta, S., & Knowlton, A. A. (2007). HSP60 trafficking in adult cardiac myocytes: Role of the exosomal pathway. *American Journal of Physiology. Heart and Circulatory Physiology, 292*(6), H3052–H3056. http://doi.org/10.1152/ajpheart.01355.2006.

Haenssen, K. K., Caldwell, S. A., Shahriari, K. S., Jackson, S. R., Whelan, K. A., Klein-Szanto, A. J., et al. (2010). ErbB2 requires integrin α5 for anoikis resistance via Src regulation of receptor activity in human mammary epithelial cells. *Journal of Cell Science, 123*(8), 1373–1382. http://doi.org/10.1242/jcs.050906.

Hance, M. W., Dole, K., Gopal, U., Bohonowych, J. E., Jezierska-Drutel, A., Neumann, C. A., et al. (2012). Secreted Hsp90 is a novel regulator of the epithelial to mesenchymal transition (EMT) in prostate cancer. *Journal of Biological Chemistry, 287*(45), 37732–37744. http://doi.org/10.1074/jbc.M112.389015.

Hegmans, J. P. J. J., Bard, M. P. L., Hemmes, A., Luider, T. M., Kleijmeer, M. J., Prins, J.-B., et al. (2004). Proteomic analysis of exosomes secreted by human mesothelioma cells. *The American Journal of Pathology, 164*(5), 1807–1815. http://doi.org/10.1016/S0002-9440(10)63739-X.

Hunter, M. C., O'Hagan, K. L., Kenyon, A., Dhanani, K. C. H., Prinsloo, E., & Edkins, A. L. (2014). Hsp90 binds directly to fibronectin (FN) and inhibition reduces the extracellular fibronectin matrix in breast cancer cells. *PLoS One, 9*(1), e86842. http://doi.org/10.1371/journal.pone.0086842.

Ichiyanagi, T., Imai, T., Kajiwara, C., Mizukami, S., Nakai, A., Nakayama, T., et al. (2010). Essential role of endogenous heat shock protein 90 of dendritic cells in antigen cross-presentation. *The Journal of Immunology, 185*(5), 2693–2700. http://doi.org/10.4049/jimmunol.1000821.

Jackson, S. (2012). *Hsp90: Structure and function* (pp. 1–86). Berlin, Heidelberg: Springer. Retrieved from, http://www.springerlink.com/content/755g44131410v6rv/abstract/.

Jarosz, D. (2016). Hsp90: A global regulator of the genotype-to-phenotype map in cancers. In J. Isaacs & L. Whitesell (Eds.), *Advances in Cancer Research, 129,* 225–247. http://www.sciencedirect.com/science/article/pii/S0065230X15000974.

Jayaprakash, P., Dong, H., Zou, M., Bhatia, A., O'Brien, K., Chen, M., et al. (2015). Hsp90α and Hsp90β together operate a hypoxia and nutrient paucity stress-response mechanism during wound healing. *Journal of Cell Science, 128*(8), 1475–1480. http://doi.org/10.1242/jcs.166363.

Kamal, A., Thao, L., Sensintaffar, J., Zhang, L., Boehm, M. F., Fritz, L. C., et al. (2003). A high-affinity conformation of Hsp90 confers tumour selectivity on Hsp90 inhibitors. *Nature, 425*(6956), 407–410. http://doi.org/10.1038/nature01913.

Khan, A. R., & James, M. N. (1998). Molecular mechanisms for the conversion of zymogens to active proteolytic enzymes. *Protein Science: A Publication of the Protein Society,* 7(4), 815–836.

Kim, B.-R., Dong, S. M., Seo, S. H., Lee, J.-H., Lee, J. M., Lee, S.-H., et al. (2014). Lysyl oxidase-like 2 (LOXL2) controls tumor-associated cell proliferation through the interaction with MARCKSL1. *Cellular Signalling, 26*(9), 1765–1773. http://doi.org/10.1016/j.cellsig.2014.05.010.

Lagarrigue, F., Dupuis-Coronas, S., Ramel, D., Delsol, G., Tronchère, H., Payrastre, B., et al. (2010). Matrix metalloproteinase-9 is upregulated in nucleophosmin-anaplastic lymphoma kinase-positive anaplastic lymphomas and activated at the cell surface by the chaperone heat shock protein 90 to promote cell invasion. *Cancer Research*, *70*(17), 6978–6987. http://doi.org/10.1158/0008-5472.CAN-10-0861.

Lei, H., Venkatakrishnan, A., Yu, S., & Kazlauskas, A. (2007). Protein kinase A-dependent translocation of Hsp90 alpha impairs endothelial nitric-oxide synthase activity in high glucose and diabetes. *The Journal of Biological Chemistry*, *282*(13), 9364–9371. http://doi.org/10.1074/jbc.M608985200.

Li, W., Li, Y., Guan, S., Fan, J., Cheng, C.-F., Bright, A. M., et al. (2007). Extracellular heat shock protein-90alpha: Linking hypoxia to skin cell motility and wound healing. *The EMBO Journal*, *26*(5), 1221–1233. http://doi.org/10.1038/sj.emboj.7601579.

Liao, D.-F., Jin, Z.-G., Baas, A. S., Daum, G., Gygi, S. P., Aebersold, R., et al. (2000). Purification and identification of secreted oxidative stress-induced factors from vascular smooth muscle cells. *Journal of Biological Chemistry*, *275*(1), 189–196. http://doi.org/10.1074/jbc.275.1.189.

Luo, L.-Y., Herrera, I., Soosaipillai, A., & Diamandis, E. P. (2002). Identification of heat shock protein 90 and other proteins as tumour antigens by serological screening of an ovarian carcinoma expression library. *British Journal of Cancer*, *87*(3), 339–343. http://doi.org/10.1038/sj.bjc.6600439.

Mayorca-Guiliani, A., & Erler, J. T. (2013). The potential for targeting extracellular LOX proteins in human malignancy. *OncoTargets and Therapy*, *6*, 1729–1735. http://doi.org/10.2147/OTT.S38110.

McCready, J., Sims, J. D., Chan, D., & Jay, D. G. (2010). Secretion of extracellular hsp90α via exosomes increases cancer cell motility: A role for plasminogen activation. *BMC Cancer*, *10*(1), 294. http://doi.org/10.1186/1471-2407-10-294.

McCready, J., Wong, D. S., Burlison, J. A., Ying, W., & Jay, D. G. (2014). An impermeant ganetespib analog inhibits extracellular Hsp90-mediated cancer cell migration that involves lysyl oxidase 2-like protein. *Cancers*, *6*(2), 1031–1046. http://doi.org/10.3390/cancers6021031.

Medress, Z., & Hayden Gephart, M. (2015). Molecular and genetic predictors of breast-to-brain metastasis: Review and case presentation. *Cureus*, 7(1), e246. http://doi.org/10.7759/cureus.246.

Minami, Y., Kawasaki, H., Suzuki, K., & Yahara, I. (1993). The calmodulin-binding domain of the mouse 90-kDa heat shock protein. *Journal of Biological Chemistry*, *268*(13), 9604–9610.

Mitra, S. K., Hanson, D. A., & Schlaepfer, D. D. (2005). Focal adhesion kinase: In command and control of cell motility. *Nature Reviews. Molecular Cell Biology*, *6*(1), 56–68. http://doi.org/10.1038/nrm1549.

Mollapour, M., & Neckers, L. (2012). Post-translational modifications of Hsp90 and their contributions to chaperone regulation. *Biochimica et Biophysica Acta (BBA)—Molecular Cell Research*, *1823*(3), 648–655. http://doi.org/10.1016/j.bbamcr.2011.07.018.

Moon, H.-J., Finney, J., Xu, L., Moore, D., Welch, D. R., & Mure, M. (2013). MCF-7 cells expressing nuclear associated lysyl oxidase-like 2 (LOXL2) exhibit an epithelial-to-mesenchymal transition (EMT) phenotype and are highly invasive in vitro. *Journal of Biological Chemistry*, *288*(42), 30000–30008. http://doi.org/10.1074/jbc.C113.502310.

Moulick, K., Ahn, J. H., Zong, H., Rodina, A., Cerchietti, L., Gomes DaGama, E. M., et al. (2011). Affinity-based proteomics reveal cancer-specific networks coordinated by Hsp90. *Nature Chemical Biology*, 7(11), 818–826. http://doi.org/10.1038/nchembio.670.

Nagaraju, G. P., Long, T.-E., Park, W., Landry, J. C., Taliaferro-Smith, L., Farris, A. B., et al. (2015). Heat shock protein 90 promotes epithelial to mesenchymal transition, invasion, and migration in colorectal cancer. *Molecular Carcinogenesis*, *54*(10), 1147–1158. http://doi.org/10.1002/mc.22185.

Nayar, U., Lu, P., Goldstein, R. L., Vider, J., Ballon, G., Rodina, A., et al. (2013). Targeting the Hsp90-associated viral oncoproteome in gammaherpesvirus-associated malignancies. *Blood*, *122*(16), 2837–2847. http://doi.org/10.1182/blood-2013-01-479972.

Nolan, K. D., Franco, O. E., Hance, M. W., Hayward, S. W., & Isaacs, J. S. (2015). Tumor-secreted Hsp90 subverts polycomb function to drive prostate tumor growth and invasion. *Journal of Biological Chemistry*, *290*(13), 8271–8282. http://doi.org/10.1074/jbc.M115.637496.

Oki, Y., Younes, A., Knickerbocker, J., Samaniego, F., Nastoupil, L., Hagemeister, F., et al. (2015). Experience with HSP90 inhibitor AUY922 in patients with relapsed or refractory non-Hodgkin lymphoma. *Haematologica*, *100*(7), e272–e274. http://doi.org/10.3324/haematol.2015.126557.

Pearl, L. H., & Prodromou, C. (2006). Structure and mechanism of the Hsp90 molecular chaperone machinery. *Annual Review of Biochemistry*, *75*(1), 271–294. http://doi.org/10.1146/annurev.biochem.75.103004.142738.

Picard, D. (2004). Hsp90 invades the outside. *Nature Cell Biology*, *6*(6), 479–480. http://doi.org/10.1038/ncb0604-479.

Pick, E., Kluger, Y., Giltnane, J. M., Moeder, C., Camp, R. L., Rimm, D. L., et al. (2007). High HSP90 expression is associated with decreased survival in breast cancer. *Cancer Research*, *67*(7), 2932–2937. http://doi.org/10.1158/0008-5472.CAN-06-4511.

Polyak, K., & Weinberg, R. A. (2009). Transitions between epithelial and mesenchymal states: Acquisition of malignant and stem cell traits. *Nature Reviews. Cancer*, *9*(4), 265–273. http://doi.org/10.1038/nrc2620.

Rong, B., Zhao, C., Liu, H., Ming, Z., Cai, X., Gao, W., et al. (2014). Identification and verification of Hsp90-beta as a potential serum biomarker for lung cancer. *American Journal of Cancer Research*, *4*(6), 874–885.

Sahu, D., Zhao, Z., Tsen, F., Cheng, C.-F., Park, R., Situ, A. J., et al. (2012). A potentially common peptide target in secreted heat shock protein-90α for hypoxia-inducible factor-1α–positive tumors. *Molecular Biology of the Cell*, *23*(4), 602–613. http://doi.org/10.1091/mbc.E11-06-0575.

Saito, K., Dai, Y., & Ohtsuka, K. (2005). Enhanced expression of heat shock proteins in gradually dying cells and their release from necrotically dead cells. *Experimental Cell Research*, *310*(1), 229–236. http://doi.org/10.1016/j.yexcr.2005.07.014.

Savina, A., Furlán, M., Vidal, M., & Colombo, M. I. (2003). Exosome release is regulated by a calcium-dependent mechanism in K562 cells. *Journal of Biological Chemistry*, *278*(22), 20083–20090. http://doi.org/10.1074/jbc.M301642200.

Shi, Y., Liu, X., Lou, J., Han, X., Zhang, L., Wang, Q., et al. (2014). Plasma levels of heat shock protein 90 alpha associated with lung cancer development and treatment responses. *Clinical Cancer Research*, *20*(23), 6016–6022. http://doi.org/10.1158/1078-0432.CCR-14-0174.

Sidera, K., El Hamidieh, A., Mamalaki, A., & Patsavoudi, E. (2011). The 4C5 cell-impermeable anti-HSP90 antibody with anti-cancer activity, is composed of a single light chain dimer. *PLoS One*, *6*(9), e23906. http://doi.org/10.1371/journal.pone.0023906.

Sidera, K., Gaitanou, M., Stellas, D., Matsas, R., & Patsavoudi, E. (2008). A critical role for HSP90 in cancer cell invasion involves interaction with the extracellular domain of HER-2. *Journal of Biological Chemistry*, *283*(4), 2031–2041. http://doi.org/10.1074/jbc.M701803200.

Sidera, K., Samiotaki, M., Yfanti, E., Panayotou, G., & Patsavoudi, E. (2004). Involvement of cell surface HSP90 in cell migration reveals a novel role in the developing nervous system. *Journal of Biological Chemistry*, *279*(44), 45379–45388. http://doi.org/10.1074/jbc.M405486200.

Sims, J. D., McCready, J., & Jay, D. G. (2011). Extracellular heat shock protein (Hsp)70 and Hsp90α assist in matrix metalloproteinase-2 activation and breast cancer cell migration and invasion. *PLoS One*, *6*(4), e18848. http://doi.org/10.1371/journal.pone.0018848.

Song, X., Wang, X., Zhuo, W., Shi, H., Feng, D., Sun, Y., et al. (2010). The regulatory mechanism of extracellular Hsp90{alpha} on matrix metalloproteinase-2 processing and tumor angiogenesis. *The Journal of Biological Chemistry*, *285*(51), 40039–40049. http://doi.org/10.1074/jbc.M110.181941.

Spreafico, A., Delord, J.-P., De Mattos-Arruda, L., Berge, Y., Rodon, J., Cottura, E., et al. (2015). A first-in-human phase I, dose-escalation, multicentre study of HSP990 administered orally in adult patients with advanced solid malignancies. *British Journal of Cancer*, *112*(4), 650–659. http://doi.org/10.1038/bjc.2014.653.

Srivastava, P. K., DeLeo, A. B., & Old, L. J. (1986). Tumor rejection antigens of chemically induced sarcomas of inbred mice. *Proceedings of the National Academy of Sciences of the United States of America*, *83*(10), 3407–3411.

Stellas, D., Hamidieh, A. E., & Patsavoudi, E. (2010). Monoclonal antibody 4C5 prevents activation of MMP2 and MMP9 by disrupting their interaction with extracellular HSP90 and inhibits formation of metastatic breast cancer cell deposits. *BMC Cell Biology*, *11*(1), 51. http://doi.org/10.1186/1471-2121-11-51.

Sun, Y., Zang, Z., Xu, X., Zhang, Z., Zhong, L., Zan, W., et al. (2010). Differential proteomics identification of HSP90 as potential serum biomarker in hepatocellular carcinoma by two-dimensional electrophoresis and mass spectrometry. *International Journal of Molecular Sciences*, *11*(4), 1423–1433. http://doi.org/10.3390/ijms11041423.

Synta Pharmaceutical. (2015). *Synta announces termination for futility of ganetespib phase 3 GALAXY-2 trial in lung cancer [Press Release]*. Retrieved from, http://www.businesswire.com/news/home/20151020006900/en/.

Taipale, M., Jarosz, D. F., & Lindquist, S. (2010). HSP90 at the hub of protein homeostasis: Emerging mechanistic insights. *Nature Reviews. Molecular Cell Biology*, *11*(7), 515–528. http://doi.org/10.1038/nrm2918.

Taldone, T., Ochiana, S. O., Patel, P. D., & Chiosis, G. (2014). Selective targeting of the stress chaperone as a therapeutic strategy. *Trends in Pharmacological Sciences*, *35*(11), 592–603. http://doi.org/10.1016/j.tips.2014.09.001.

Théry, C., Boussac, M., Véron, P., Ricciardi-Castagnoli, P., Raposo, G., Garin, J., et al. (2001). Proteomic analysis of dendritic cell-derived exosomes: A secreted subcellular compartment distinct from apoptotic vesicles. *The Journal of Immunology*, *166*(12), 7309–7318. http://doi.org/10.4049/jimmunol.166.12.7309.

Théry, C., Regnault, A., Garin, J., Wolfers, J., Zitvogel, L., Ricciardi-Castagnoli, P., et al. (1999). Molecular characterization of dendritic cell-derived exosomes selective accumulation of the heat shock protein Hsc73. *The Journal of Cell Biology*, *147*(3), 599–610. http://doi.org/10.1083/jcb.147.3.599.

Thomaidou, D., Dori, I., & Patsavoudi, E. (1995). Developmental expression and functional characterization of the 4C5 antigen in the postnatal cerebellar cortex. *Journal of Neurochemistry*, *64*(5), 1937–1944. http://doi.org/10.1046/j.1471-4159.1995.64051937.x.

Thomaidou, D., & Patsavoudi, E. (1993). Identification of a novel neuron-specific surface antigen in the developing nervous system, by monoclonal antibody 4C5. *Neuroscience*, *53*(3), 813–827. http://doi.org/10.1016/0306-4522(93)90626-Q.

Thuringer, D., Hammann, A., Benikhlef, N., Fourmaux, E., Bouchot, A., Wettstein, G., et al. (2011). Transactivation of the epidermal growth factor receptor by heat shock protein 90 via toll-like receptor 4 contributes to the migration of glioblastoma cells. *Journal of Biological Chemistry*, *286*(5), 3418–3428. http://doi.org/10.1074/jbc.M110.154823.

Trieb, K., Gerth, R., Holzer, G., Grohs, J. G., Berger, P., & Kotz, R. (1999). Antibodies to heat shock protein 90 in osteosarcoma patients correlate with response to neoadjuvant chemotherapy. *British Journal of Cancer*, *82*(1), 85–87. http://doi.org/10.1054/bjoc.1999.0881.

Tsen, F., Bhatia, A., O'Brien, K., Cheng, C.-F., Chen, M., Hay, N., et al. (2013). Extracellular heat shock protein 90 signals through subdomain II and the NPVY motif of LRP-1 receptor to Akt1 and Akt2: A circuit essential for promoting skin cell migration in vitro and wound healing in vivo. *Molecular and Cellular Biology*, *33*(24), 4947–4959. http://doi.org/10.1128/MCB.00559-13.

Tsutsumi, S., Mollapour, M., Graf, C., Lee, C.-T., Scroggins, B. T., Xu, W., et al. (2009). Hsp90 charged-linker truncation reverses the functional consequences of weakened hydrophobic contacts in the N domain. *Nature Structural & Molecular Biology*, *16*(11), 1141–1147. http://doi.org/10.1038/nsmb.1682.

Tsutsumi, S., Scroggins, B., Koga, F., Lee, M.-J., Trepel, J., Felts, S., et al. (2007). A small molecule cell-impermeant Hsp90 antagonist inhibits tumor cell motility and invasion. *Oncogene*, *27*(17), 2478–2487. http://doi.org/10.1038/sj.onc.1210897.

Ullrich, S. J., Robinson, E. A., Law, L. W., Willingham, M., & Appella, E. (1986). A mouse tumor-specific transplantation antigen is a heat shock-related protein. *Proceedings of the National Academy of Sciences of the United States of America*, *83*(10), 3121–3125.

Vadasz, Z., Kessler, O., Akiri, G., Gengrinovitch, S., Kagan, H. M., Baruch, Y., et al. (2005). Abnormal deposition of collagen around hepatocytes in Wilson's disease is associated with hepatocyte specific expression of lysyl oxidase and lysyl oxidase like protein-2. *Journal of Hepatology*, *43*(3), 499–507. http://doi.org/10.1016/j.jhep.2005.02.052.

Wang, X., Heuvelman, D. M., Carroll, J. A., Dufield, D. R., & Masferrer, J. L. (2010). Geldanamycin-induced PCNA degradation in isolated Hsp90 complex from cancer cells. *Cancer Investigation*, *28*(6), 635–641. http://doi.org/10.3109/07357901003630983.

Wang, X., Song, X., Zhuo, W., Fu, Y., Shi, H., Liang, Y., et al. (2009). The regulatory mechanism of Hsp90α secretion and its function in tumor malignancy. *Proceedings of the National Academy of Sciences of the United States of America*, *106*(50), 21288–21293. http://doi.org/10.1073/pnas.0908151106.

Whitesell, L., & Lindquist, S. L. (2005). HSP90 and the chaperoning of cancer. *Nature Reviews. Cancer*, *5*(10), 761–772. http://doi.org/10.1038/nrc1716.

Whitesell, L., Santagata, S., Mendillo, M. L., Lin, N. U., Proia, D. A., & Lindquist, S. (2014). HSP90 empowers evolution of resistance to hormonal therapy in human breast cancer models. *Proceedings of the National Academy of Sciences of the United States of America*, *111*(51), 18297–18302. http://doi.org/10.1073/pnas.1421323111.

Woodley, D. T., Fan, J., Cheng, C.-F., Li, Y., Chen, M., Bu, G., et al. (2009). Participation of the lipoprotein receptor LRP1 in hypoxia-HSP90α autocrine signaling to promote keratinocyte migration. *Journal of Cell Science*, *122*(10), 1495–1498. http://doi.org/10.1242/jcs.047894.

Workman, P., Burrows, F., Neckers, L., & Rosen, N. (2007). Drugging the cancer chaperone HSP90. *Annals of the New York Academy of Sciences*, *1113*(1), 202–216. http://doi.org/10.1196/annals.1391.012.

Wozniak, M. A., Modzelewska, K., Kwong, L., & Keely, P. J. (2004). Focal adhesion regulation of cell behavior. *Biochimica et Biophysica Acta (BBA)—Molecular Cell Research*, *1692*(2–3), 103–119. http://doi.org/10.1016/j.bbamcr.2004.04.007.

Yang, Y., Rao, R., Shen, J., Tang, Y., Fiskus, W., Nechtman, J., et al. (2008). Role of acetylation and extracellular location of heat shock protein 90α in tumor cell invasion. *Cancer Research*, *68*(12), 4833–4842. http://doi.org/10.1158/0008-5472.CAN-08-0644.

Yang, J., Song, X., Chen, Y., Lu, X., Fu, Y., & Luo, Y. (2014). PLCγ1–PKCγ signaling-mediated Hsp90α plasma membrane translocation facilitates tumor metastasis. *Traffic*, *15*(8), 861–878. http://doi.org/10.1111/tra.12179.

Yu, X., Harris, S. L., & Levine, A. J. (2006). The regulation of exosome secretion: A novel function of the p53 protein. *Cancer Research*, *66*(9), 4795–4801. http://doi.org/10.1158/0008-5472.CAN-05-4579.

Yu, Q., & Stamenkovic, I. (2000). Cell surface-localized matrix metalloproteinase-9 proteolytically activates TGF-β and promotes tumor invasion and angiogenesis. *Genes & Development*, *14*(2), 163–176. http://doi.org/10.1101/gad.14.2.163.

Zong, H., Gozman, A., Caldas-Lopes, E., Taldone, T., Sturgill, E., Brennan, S., et al. (2015). A hyperactive signalosome in acute myeloid leukemia drives addiction to a tumor-specific Hsp90 species. *Cell Reports*, *13*(10), 2159–2173. http://doi.org/10.1016/j.celrep.2015.10.073.

CHAPTER SEVEN

GRP94/gp96 in Cancer: Biology, Structure, Immunology, and Drug Development

Bill X. Wu, Feng Hong, Yongliang Zhang, Ephraim Ansa-Addo, Zihai Li[1]
Hollings Cancer Center, Department of Microbiology and Immunology, Medical University of South Carolina, Charleston, South Carolina, USA
[1]Corresponding author: e-mail address: zihai@musc.edu

Contents

Advances in Cancer Research, Volume 129
ISSN 0065-230X
http://dx.doi.org/10.1016/bs.acr.2015.09.001

Abstract

As an endoplasmic reticulum heat-shock protein 90 (HSP90) paralog, GRP94 (glucose-regulated protein 94)/gp96 (hereafter referred to as GRP94) has been shown to be an essential master chaperone for multiple receptors including Toll-like receptors, Wnt coreceptors, and integrins. Clinically, expression of GRP94 correlates with advanced stage and poor survival in a variety of cancers. Recent preclinical studies have also revealed that GRP94 expression is closely linked to cancer growth and metastasis in melanoma, ovarian cancer, multiple myeloma, lung cancer, and inflammation-associated colon cancer. Thus, GRP94 is an attractive therapeutic target in a number of malignancies. The chaperone function of GRP94 depends on its ATPase domain, which is structurally distinct from HSP90, allowing design of highly selective GRP94-targeted inhibitors. In this chapter, we discuss the biology and structure–function relationship of GRP94. We also summarize the immunological roles of GRP94 based on the studies documented over the last two decades, as these pertain to tumorigenesis and cancer progression. Finally, the structure-based rationale for the design of selective small-molecule inhibitors of GRP94 and their potential application in the treatment of cancer are highlighted.

1. INTRODUCTION

Glucose-regulated protein 94 (GRP94) (Lee, Delegeane, & Scharff, 1981), also known as gp96 (Srivastava, DeLeo, & Old, 1986), endoplasmin (Koch, Smith, Macer, Webster, & Mortara, 1986), ERp99 (Lewis, Mazzarella, & Green, 1985), and HSP90b1 (Chen, Piel, Gui, Bruford, & Monteiro, 2005), is an endoplasmic reticulum (ER) paralog of heat-shock protein (HSP90). Like other HSPs, GRP94 is induced by the accumulation of misfolded proteins (Kozutsumi, Segal, Normington, Gething, & Sambrook, 1988). It binds and hydrolyzes ATP (Dollins, Warren, Immormino, & Gewirth, 2007; Li & Srivastava, 1993), is the most abundant protein in the ER lumen, and is ubiquitously present in nucleated cells. GRP94 is thought to play critical roles in general protein quality control in the ER; however, its specific roles and modes of action have only recently been revealed. Genetic studies using knockout (KO) and gain-of-function systems surprisingly demonstrated that GRP94 plays obligatory roles in the folding of a variety of innate immune receptors such as Toll-like receptors (TLRs) and integrins, raising an intriguing question about the roles of GRP94 in the development of inflammation-associated diseases including cancer. In addition, the structure of GRP94 at atomic resolution has finally been determined, which has resolved a long-standing controversy regarding the intrinsic ATP binding and ATPase activity of GRP94. Interestingly, not only the structure of GRP94 exhibits similar, but also distinct features compared with its cytosolic counterpart HSP90, allowing development of GRP94-specific inhibitors. Indeed, purine scaffold-based selective GRP94 inhibitors have been developed (Patel et al., 2013; Taldone, Ochiana, Patel, & Chiosis, 2014), paving the way for development of selective inhibitors of GRP94 as therapeutic agents. Meanwhile, in addition to autoimmune and inflammatory diseases, molecular epidemiologic studies of cancer have consistently demonstrated significant positive association between GRP94 expression and aggressiveness. In this review, we discuss the biology and structure–function relationship of GRP94, with specific emphasis on emerging and unresolved questions about this ancient molecule. The structural basis for the design of selective GRP94 small-molecule inhibitors and their potential use in the treatment of inflammation-associated cancer are also highlighted.

2. GRP94 AS A MASTER IMMUNE CHAPERONE

Compromised immune surveillance and impaired antitumor immunity are key mechanisms in oncogenesis and cancer progression (Dunn, Bruce, Ikeda, Old, & Schreiber, 2002). The immune functions of GRP94 are well documented; for example, GRP94 was shown to be required for early B- and T-cell lymphopoiesis (Staron et al., 2010), for the innate immune defense function of macrophages and for the immune suppressive function of regulatory T cells (Tregs) (Zhang, Ansa-Addo, & Li, 2015; Zhang, Wu, et al., 2015). Through genetic and biochemical studies, GRP94 was identified as an essential chaperone for the folding of TLRs (Randow & Seed, 2001; Yang et al., 2007), integrins (Staron et al., 2010), Wnt coreceptor low-density lipoprotein receptor-related protein 6 (LRP6) (Hua et al., 2013; Liu et al., 2010), glycoprotein A repetitions predominant (GARP) (Zhang, Wu, et al., 2015), and platelet glycoprotein Ib/IX/V complex (Liu et al., 2013, 2010; Randow & Seed, 2001; Staron et al., 2011; Yang et al., 2007; Zhang, Wu, et al., 2015) (Table 1). These findings suggest that GRP94 serves as a specialized immune chaperone that controls receptors for immune function in multicellular organisms. In the following sections, we will discuss the role of GRP94-client proteins, including TLRs, integrins, and Wnt coreceptors, in the regulation of cancer immunology.

Table 1 GRP94: A Master Immune Chaperone

GRP94-Client Protein	Knockout Model	References
TLRs	GRP94-mutant leukemia pre-B cell, macrophage-specific KO mice	Randow and Seed (2001), Yang et al. (2007), and Liu et al. (2010)
Integrins	GRP94-mutant leukemia pre-B cell, hematopoietic-specific KO BMT mice	Randow and Seed (2001) and Staron et al. (2010)
LRP6	Gut epithelial-specific KO mice	Liu et al. (2013)
LRRC32 (GARP)	Treg-specific KO, platelet-specific KO mice	Zhang, Wu, et al. (2015)
Platelet glycoprotein Ib/IX/V	Hematopoietic-specific KO BMT mice	Staron et al. (2011)

2.1 GRP94 Participates in Pathogen Defense via Chaperoning TLRs

TLRs are important pathogen pattern recognition receptors for microbial products. Upon activation, TLRs initiate specific immunoactivities, via MyD88-dependent or -independent pathways (Kawai & Akira, 2010; Medzhitov, 2007). There are 13 members of the TLR family that have been identified in mammals. TLR3, TLR7, TLR8, TLR9, and TLR13 are localized primarily in endolysosomes, whereas the rest of TLRs are expressed on the cell surface. By isolating a mouse cell line deficient in bacterial products induced NF-κB-dependent responses, Randow and Seed (2001) identified a specific and restricted role for the ER chaperone GRP94 in the folding, assembly, and export of TLRs (TLR1,2,4). Yang et al. generated macrophage-specific GRP94 KO mice using the LysM-cre/flox genetic approach and analyzed macrophage responses to TLR agonists (Yang et al., 2007). They found that GRP94-deficient macrophages failed to respond to ligands for plasma membrane TLRs (TLR2, TLR4, and TLR5) and endolysosome-located TLR7 and TLR9, demonstrating that the folding and trafficking of these TLRs are dependent on GRP94. Macrophage-specific GRP94 KO mice are highly susceptible to *Listeria monocytogenes* infection, indicating the important role of GRP94 in regulating TLR responses and host defense to Listeria infection. Later, Liu et al. demonstrated that GRP94 chaperones multiple TLRs in a manner that is dependent on another ER luminal protein, CNPY3 (Liu et al., 2010). Additionally, CNPY3 interacts with GRP94 to promote substrate loading in an ATP-dependent manner. Interestingly, they also identified that TLR3 is distinct from other TLRs because its folding and maturation are independent of GRP94.

2.2 GRP94 Chaperones Multiple Integrins Which Are Important for Immune Cell Interaction

Similar to TLRs, integrins are type I transmembrane receptor dimers formed by α and β subunits. Eighteen α subunits and eight β subunits can heterodimerize to form twenty-four distinct integrin family members (Miranti & Brugge, 2002). Integrins play critical roles for the immune system in leukocyte trafficking and migration, immunological synapse formation, costimulation and phagocytosis as they mediate cell–cell, cell–extracellular matrix, and cell–pathogen interactions (Luo, Carman, & Springer, 2007). In addition to TLR deficiency in GRP94 mutant pre-B leukemia (pre-B) cells, Randow and Seed also identified other integrins, including CD11a (αL), CD18 (β2), and CD49d (α4), to be retained intracellularly in GRP94-mutant pre-B cells. These studies indicate that GRP94 chaperones integrins. Staron et al. expanded on the GRP94-client network by determining the differential expression of integrins by wild-type (WT) and GRP94 KO hematopoietic cells. They found that GRP94 is essential for the expression of a majority of integrin subunits, including α1, α2, α4, αD, αE, αL, αM, αV, and αX, within the hematopoietic system (Staron et al., 2010). Interestingly, they found that the functional requirements for TLRs and integrins in hematopoiesis are highly lineage specific as deletion of GRP94 led to selective and stage-specific defects in both T and B lymphopoiesis, but not in NK-cell development or myelopoiesis. Further work by Wu et al. identified a C-terminus loop structure within the client-binding domain (CBD) of GRP94 (amino acids 652–678) to be critical for chaperoning both integrins and TLRs (Wu et al., 2012). Intriguingly, there is substrate specificity for the binding of GRP94 to integrins and TLRs because two residues, Met658 and Met662 in the C-terminal loop are critical for GRP94 binding to integrins but not to TLRs.

2.3 GRP94 Plays a Role in Regulating Wnt Signaling Pathway by Cochaperoning LRP6 with MesD

The canonical Wnt/LRP6 signaling pathway is evolutionarily conserved and plays key roles during the development of many organ systems. The critical requirement of the Wnt/LRP6/β-catenin signaling pathway in dendritic cells (DCs) (Swafford & Manicassamy, 2015), T-cell (Xue & Zhao, 2012), and B-cell biology (Qiang & Rudikoff, 2004) has been well discussed. In an effort to explore the client network of GRP94, GRP94-client complexes were immunoprecipitated from the pre-B leukemia cell line. Unexpectedly, MesD was discovered as a GRP94-interacting molecule (Liu et al., 2013). MesD was shown previously to be a critical chaperone for the surface expression of LRP5/6. Moreover, GRP94 binds to both LRP6 and MesD to facilitate optimal interaction between the two proteins. In the absence of GRP94, the Wnt/β-catenin signaling pathway is greatly impaired (Hua et al., 2013; Liu et al., 2013). Together, these data strongly indicate that the maturation and cell surface expression of LRP6 is dependent on the coordinated action of both GRP94 and MesD. Further work demonstrated that the significance of this finding *in vivo*, by showing that conditional deletion of GRP94 results in loss of intestinal homeostasis with a subsequent increase of bacterial translocation through gut epithelium barrier. Based on the fact that WT function of the hematopoietic system was unable to rescue the KO gut phenotype, and that gut-specific deletion of GRP94 recapitulates the gut pathology of knocking out Tcf4 (a major downstream effector of Wnt signaling) in mice, it is clear that gut-intrinsic Wnt signaling significantly contributes to gut mucosal immunity. As aforementioned, the Wnt/LRP6/β-catenin signaling pathway contributes important roles in immune regulation and oncogenesis (Hua et al., 2013; Liu et al., 2013). Given this importance, further investigation of the precise mechanisms by which GRP94 chaperones LRP6 will be useful in developing a GRP94-targeted therapeutic strategy for cancer- and immune-related diseases.

2.4 GRP94 Chaperones GARP/LRRC32 and Safeguards Regulatory T-Cell Suppressive Function

Transforming growth factor beta (TGF-β) plays multiple essential roles in cancer biology (Katsuno, Lamouille, & Derynck, 2013; Li & Flavell, 2008; Massague, 2008). First, TGF-β controls diverse oncogenic processes including proliferation, differentiation, apoptosis, angiogenesis, epithelial–mesenchymal transition (EMT), and metastasis. Second, it is a master regulator that plays critical roles in enabling cancers to evade antitumor immunity. Leucine-rich repeats domain 32 (LRRC32), also known as GARP, was recently reported to be highly expressed on activated Tregs and platelets. Functional analyses verified that GARP serves as a docking protein for the surface expression of latent membrane TGF-β. Tregs with high expression of GARP show potent suppressive capacity (Tran et al., 2009; Wang et al., 2009). Interestingly, GARP shares structural similarities with TLR and GPIb, two known client proteins of GRP94. Zhang, Wu, et al. (2015) observed that both Tregs and platelets show defective expression of GARP, in respectively, Treg-specific (FoxP3-cre) and platelet-specific (PF4-cre) GRP94 KO mice. These observations indicate that GRP94 chaperones GARP, a conclusion further strengthened by demonstration of direct physical interaction and the finding that GARP cannot exit the ER in the absence of GRP94, indicative of incorrect protein folding. Because GRP94 chaperones both GARP and integrins, TGF-β bioavailability was greatly impaired in GRP94-null Treg cells, an expected result, given that GARP and integrins regulate TGF-β bioavailability (Tran et al., 2009). This knowledge should prove useful in developing therapies to control TGF-β-related pathologies such as cancers where TGF-β mediates cell invasion and metastasis, and autoimmune-diseases dependent upon the GRP94/ GARP–integrin/TGF-β axis.

2.5 GRP94 Is Essential for Platelet Function via Chaperoning Platelet Glycoprotein Ib/IX/V

Glycoprotein Ib–IX–V complex (GPIb–IX–IV) is exclusively expressed in platelets and megakaryocytes, which is responsible for Von Willebrand factor (VWF)-mediated platelet activation and aggregation. As a result, GPIb–IX–IV is involved in the life cycle of platelets as they contribute to thrombosis, inflammation, and cancer metastasis. In addition to VWF, GPIb–IX–IV can also bind to a number of other ligands in circulation, including thrombin, P-selectin, integrin αMβ2, factor XI, factor XII, high-molecular-weight kininogen, as well as a number of snake venom proteins (Li & Emsley, 2013). Genetic KO of GRP94 in mice not only led to a complete loss of the mature GPIb–IX complex in a cell-intrinsic fashion but also resulted in the macrothrombocytopenia and prolonged bleeding time characteristic of the Bernard–Soulier syndrome (Staron et al., 2011). Specifically, GRP94 interacted with GPIX, but not with other subunits of the GPIb–IX complex. In the absence of GRP94, GPIX becomes unstable and degraded via the classic endoplasmic reticulum associated degradation (ERAD) pathway (Staron et al., 2011). Interestingly, GPIb expression is not limited to platelets, as multiple cancer cell types also express the protein. GPIb–IX–IV complex itself has been considered as an oncoprotein, because its expression is associated with genomic instability and cell transformation (Li, Lu, Cohen, & Prochownik, 2008). Thus, GRP94-dependent regulation of GPIb–IX–IV expression may represent an additional mechanism by which GRP94 enables cancer development.

3. ROLES OF GRP94 IN CANCER: EXPANDING THE GRP94-CLIENT NETWORK

3.1 GRP94 and Its Cancer-Promoting Clientele

As a master ER chaperone linking protein quality control to stress and inflammation, GRP94 has been found to promote cancer via its role in folding a variety of clients (Fig. 1). Besides the above-discussed TLRs, integrins and GARP, GRP94 also controls the maturation and secretion of insulin-like growth factors (IGFs), which are important mitogenic and prosurvival factors for many cancers (Wanderling et al., 2007). A systemic reduction of IGFs was indeed observed in a GRP94 conditional deletion mouse model (Barton et al., 2012). After finding that GRP94 binds to LRP6 (Liu et al., 2013), our group further demonstrated that GRP94 deficiency in human multiple myeloma cells induces apoptosis through disruption of the

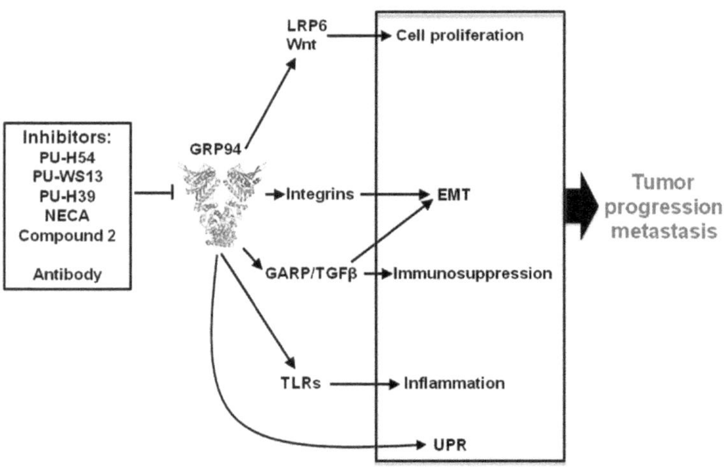

Figure 1 Roles of chaperone GRP94 and its client proteins in oncogenesis. GRP94 mediates the surface expression of the latent TGF-β docking receptor, GARP. GARP is important for TGF-β activation, which plays roles in regulating tumor immunosuppressive environment and metastasis. Furthermore, in macrophages, chronic inflammation coupled with carcinogens trigger gp96 upregulation, which chaperones TLRs and enhances the induction of tumor-associated macrophages (TAMs). TAMs support tumor progression by releasing cytokines. GRP94 chaperones integrin and the Wnt signaling cofactor, LRP6, which are important for tumor cell epithelial–mesenchymal transition and proliferation. (See the color plate.)

Wnt-LRP-survivin pathway, and blocks myeloma growth in a murine xenograft model. These data suggest that GRP94 could serve as a novel therapeutic target for multiple myeloma (Hua et al., 2013). Recent proteomic studies from the Lehner group (Weekes et al., 2012) and our laboratory have further expanded the GRP94-client network. By comparing the difference between plasma membrane proteins isolated from GRP94 WT and mutant/KO cells, Lehner and colleagues (and our unpublished data) showed that, besides the LDL family receptors (LDLR, LRP6, Sorl1, and LRP8), CD180 is a novel GRP94-client protein, which is involved in TLR4 signaling. Binder et al. also found that CD91, a receptor for extracellular GRP94 (Binder, Han, & Srivastava, 2000), is also controlled by GRP94 itself. These findings suggest that GRP94 could self-regulate its expression by controlling levels of its receptor upon stimulation. This property of GRP94 is perhaps utilized by cancer cells to escape from GRP94-peptide complex-induced antitumor immunity.

3.2 Roles of GRP94 in Regulating Tumor-Associated Macrophages

Tumor-associated macrophages (TAMs) belong to the monocytes/macrophage lineage and have been found to promote tumor growth and metastasis by producing cytokines, chemokines, reactive oxygen species, reactive nitrogen species, etc. (DeNardo et al., 2009; Hong, Wu, & Li, 2014; Lin, Nguyen, Russell, & Pollard, 2001; Qian et al., 2011; Reddy et al., 2002; Steidl et al., 2010). GRP94 plays important roles in TAMs' function through folding of both TLRs and integrins (Liu et al., 2010; Randow & Seed, 2001; Yang et al., 2007). Genetic deletion of GRP94 from macrophages reduced colitis and colitis-associated colon tumorigenesis in mice (Morales et al., 2014). Attenuation of azoxymethane/dextran sodium sulfate (AOM/DSS)-induced colon cancer in GRP94-deleted mice correlated with reduced proinflammatory cytokine production (IL17 and IL23) (Morales et al., 2014; Yang et al., 2007). IL17 and IL23 have been shown to be important in cancer biology, substantiated by the finding that exposure of normal colonic murine epithelial cells to a combination of IL17A, IL1β, and TNF-α induce EMT and invasion (Morales et al., 2014). IL17 could promote tumor growth by enhancing angiogenesis (Numasaki et al., 2003). Moreover, elevated IL17 expression has also been shown in patients with ulcerative colitis (Kobayashi et al., 2008). IL23 is overexpressed in a variety of human cancers, including colon cancer (Langowski et al., 2006). IL23 is also elevated in patients with ulcerative colitis and Crohn's disease (Kobayashi et al., 2008), and mutation of the IL23 receptor is highly associated with inflammatory bowel disease (Duerr et al., 2006). Thus, these findings strongly suggest that macrophage-intrinsic GRP94 promotes tumorigenesis via production of IL17 and IL23 in the tumor microenvironment, and provide evidence that TAMs promote tumorigenesis in a GRP94-dependent manner.

TAM-intrinsic GRP94 also plays a role in regulating Wnt activation in colon carcinogenesis. Deletion of GRP94 from TAMs protects the gut epithelium from β-catenin mutation and subsequent activation of the canonical Wnt pathway. GRP94 loss also correlates with increased efficiency of the DNA repair machinery and reduces tumorigenesis in the AOM/DSS-induced colon cancer model (Morales et al., 2014). Morales et al. also observed a reduction of the Wnt receptor Fzd1 and the downstream target p53 in KO mice. This attenuated Wnt pathway activation in KO mice is consistent with decreased tumor burden. Two additional Wnt-activated cytokines, IL1β and TNF-α, are similarly reduced in macrophage-specific GRP94 KO mice (Morales et al., 2014). Together, these observations suggest that TAM-intrinsic GRP94 promotes tumorigenesis by supporting activation of the Wnt signaling pathway.

3.3 Roles of GRP94 in Hepatocellular Carcinogenesis

Several recent studies have indicated that GRP94 plays important roles in hepatic homeostasis, steatosis, and cancer development. Using a liver-specific GRP94 KO mouse model, Rachidi et al. found that GRP94 deletion significantly impairs hepatocyte growth. Interestingly, the percentage of GRP94-null hepatocytes in KO mice decreased from more than 95% in young mice (<6 weeks old) to less than 5% in aged mice (>12 months old), accompanied by exacerbated ongoing steatosis in the regenerated GRP94-positive hepatocytes. In addition, in diethylnitrosamine-induced hepatocarcinoma (HCC), only GRP94-positive hepatocytes developed cancer, suggesting that GRP94 plays critical roles in HCC development. Similarly, development of HCC in aged liver-specific GRP94 KO mice was reported in an independent study (Chen, Ha, Kanel, & Lee, 2014).

The roles of GRP94 in HCC development could be multifactorial. In the liver-specific GRP94 KO mice, regenerated GRP94+ hepatocytes are more prone to carcinogenesis, which is associated with increased production of hepatocyte growth factor from the surrounding tissues (Rachidi et al., 2015). Beyond its role in the unfolded protein response pathway, GRP94 is also an obligate chaperone for several cancer-relevant signaling pathways including those mediated by integrins (Liu & Li, 2008; Wu et al., 2012), Wnt (Liu et al., 2013), and IGF (Barton et al., 2012; Wanderling et al., 2007). GRP94 KO can also affect the microenvironment, which would be expected to modulate immune function in the liver, and in turn regulate HCC development. Further studies are warranted to achieve a more comprehensive understanding of the diverse cell context-dependent mechanisms for GRP94 action.

3.4 Surface GRP94 in Tumor Progression

Even though GRP94 contains an ER retention motif (KDEL), it is also found on the cell surface, where it participates in immune responses. Surface GRP94 was observed to correlate with increased tumor immunogenicity (Robert, Menoret, & Cohen, 1999) and subsequently shown to present peptides in antitumor immune responses (Linderoth, Popowicz, & Sastry, 2000). Zheng et al. demonstrated that surface GRP94 induced DC maturation, leading to inflammation, and an increase of antigen-presenting and costimulatory molecules (Zheng, Dai, Stoilova, & Li, 2001). Interestingly, overexpression of surface GRP94 on tumor cells resulted in tumor regression through T lymphocytes (Zheng et al., 2001), and promoted antigen presentation, which led to increased memory T-cell development (Dai et al., 2003). Tumor-specific surface GRP94 was reported to play a role during activation of DCs and differentiation to Th2 subsets, thus suggesting a potential target for cancer immunotherapies. Indeed, several GRP94-based antitumor therapies have been pursued so far.

For example, immunization with tumor-derived GRP94 induced antitumor immune responses when given early in mice bearing methylcholanthrene-induced fibrosarcomas (Meth A tumors) (Kovalchin, Murthy, Horattas, Guyton, & Chandawarkar, 2001). In addition, DCs pulsed with tumor-derived GRP94 induced antitumor immune responses in T cells and NK cells against multiple myeloma (Qian et al., 2005) and lung cancer (Shinagawa et al., 2008). Furthermore, surface GRP94 is specifically expressed on malignant breast tissues and not on benign tissues (Melendez et al., 2006). Using a GRP94-specific purine-scaffold inhibitor, PU-WS13, Patel et al. found that GRP94 is required for maintaining the architecture of high-density HER2 assemblies at the plasma membrane, and is observed with HER2 at the plasma membrane. Upon GRP94 inhibition, membrane HER2 molecules translocate to early endosomes and plasma membrane-adjacent lysosomes, which in turn blocks downstream signaling. This suggests that surface GRP94 may contribute to the malignant progression of HER2-positive breast cancers because surface GRP94 is required for the folding and dimerization of HER2 (Patel et al., 2013). Indeed, a recent study confirmed that surface GRP94 contributes to the malignant progression of HER2-positive breast cancers by regulating the dimerization of HER2 (Li et al., 2015). In addition, a monoclonal antibody, W9, against surface GRP94, was found to increase the sensitivity of human $BRAF^{V600E}$ melanoma cells to BRAF inhibitors through upregulation of PDGFRα (Sabbatino et al., 2015).

4. STRUCTURAL STUDIES OF GRP94: SIMILAR YET DISTINCT FROM OTHER HSP90 MEMBERS

4.1 Structural Domains

HSP90 is a family of molecular chaperones that are highly conserved from prokaryotes to eukaryotes (Pearl & Prodromou, 2006). It includes cytoplasmic forms of HSP90 encoded by *HSP90AA1* and *HSP90AB1*, an ER-resident paralog GRP94 and TRAP-1 that resides in mitochondria. The structural organization of all HSP90 family members is conserved and consists of four major domains including an N-terminal ATPase domain (NTD), a charged linker region, a middle domain (M domain), and a C-terminal dimerization domain (CTD).

The NTD contains a nucleotide-binding pocket and residues essential for ATP hydrolysis (Prodromou et al., 1997; Stebbins et al., 1997). At the N-terminus of GRP94, the initial amino acids (approx. 70) are different from those of HSP90 both in length and sequence. This difference suggests that HSP90 and GRP94 undergo differential conformational changes in response to ATP binding and NTD dimerization. Other than this variation, the NTD of GRP94 compared to other HSP90s is highly conserved. Moreover, the nucleotide-binding pocket of GRP94 can accommodate conventional inhibitors of HSP90 (geldanamycin (GM), radicicol, and their derivatives), resulting in inhibitory effects on its chaperone function by keeping the protein in a "closed" conformation (Vogen et al., 2002).

A charged linker domain connects the NTD and M domain to mediate conformational changes during ATP hydrolysis (Schulte et al., 1999; Vogen et al., 2002). Deletion of the HSP90 linker domain affects its interaction with cochaperones and compromises its function *in vivo* and *in vitro* (Hainzl, Lapina, Buchner, & Richter, 2009). The charged linker domain in GRP94 also contains calcium-binding sites that allow calcium to induce conformational changes in the NTD and enhance interaction between GRP94 and peptides (Biswas et al., 2007). The calcium-binding capacity of GRP94 is also important for intracellular calcium homeostasis, and GRP94 KO cells are more sensitive to perturbations of calcium concentration. The calcium-binding capability of GRP94 is very relevant to its ER-resident subcellular location, given the importance of this organelle for calcium storage. This appears to be a unique aspect of GRP94 function because calcium-binding activity has not been reported for non-ER-resident HSP90 paralogs.

The M domain is highly conserved between GRP94 and other HSP90s. This domain is essential for the functions of GRP94 and other HSP90s, as it contains the catalytic loop for ATPase activity and interacts with the NTD required for ATP hydrolysis (Dollins et al., 2007; Morra, Potestio, Micheletti, & Colombo, 2012). Further, the C-terminus of the M domain and helices 2, 4, and 5 of the CTD form a constitutive interface for homodimerization. In addition to dimer formation, the CTD domain is essential for client protein binding to both GRP94 and other HSP90s (Chu, Maynard, Chiosis, Nicchitta, & Burlingame, 2006; Wu et al., 2012). At the end of the C-terminus, HSP90 has a conserved tetratricopepetide repeat, which is important for cochaperone interactions, while GRP94 ends with KDEL sequence that is important for its retention in the ER and retrieval should it escape.

4.2 Structural Insights into the Chaperoning Cycle of HSP90 and GRP94

Binding and release of clients are coupled with the ATP hydrolysis cycle of HSP90 and fits well with a "molecular clamp" model (Vaughan et al., 2006). In this model, the ATPase activity is conserved among all HSP90 orthologs and paralogs and is critical for their roles as chaperone proteins. The two-layer α/β-sandwich structure of the N-terminal domain forms an ATPase pocket that binds nucleotides. Surrounding the ATPase pocket area is a "lid" structure consisting of strands and helices. Upon ATP binding, the "lid" structure changes from extending out of the ATPase pocket to closing above the pocket. This conformational change contributes to a transient dimerization in N-terminal domains of the two protomers that takes place in addition to constitutive dimerization of the C-terminus. Thus, HSP90 forms a "closed" dimer conformation upon ATP binding. Assumption of this compact form results in a catalytic Arg residue from the M domain moving into the ATPase pocket and subsequent hydrolysis of ATP to ADP. Upon hydrolysis of ATP to ADP in the ATPase pocket, HSP90 returns to a relaxed open dimer structure (Ali et al., 2006). This model proposes that HSP90 binds its clients in a relaxed conformation, and then releases the client when it is in a tensed structure induced by ATP binding.

Like HSP90, GRP94 also assembles as a "V" shaped homodimer. Nevertheless, Dollins et al. identified several unique structural features of GRP94 (Dollins et al., 2007). Instead of the "closed" dimer structure, ATP-bound GRP94 adopts a twisted "V" structure, in which the N–M domain has a left-handed twist along the axis of the M–C domain. This feature prevents ATP from being hydrolyzed upon binding to GRP94 and probably explains the low ATPase activity of GRP94. However, the structure also showed that simple rotation of a domain can lead to proper alignment of the catalytic residues. Thus, it is possible that a transient catalytic conformation of GRP94 exists which enables its chaperoning function. Currently, it remains unclear whether the formation of a dimer or a higher order complex is essential for the ATPase activity of GRP94 protomers.

5. THE QUESTION OF SPECIFICITY: COCHAPERONE AND CLIENT-BINDING DOMAIN?

Although HSP90 and GRP94 bind to numerous client proteins, each of these chaperones demonstrates client specificity. To date, the factors determining this specificity are not well understood. Thus, more knowledge of the interacting structures between HSP90, client proteins, and accessory factors will be instrumental for understanding the determinants of client specificity.

Studies have revealed that multiple binding sites exist in HSP90 proteins for divergent clients. For example, as early as 1993, Cadepond et al. showed that a couple of negatively charged regions (221–290 and 530–581) and a middle region with a leucine zipper structure (392–419) in HSP90 are important for client binding (Cadepond et al., 1993). In chicken HSP90, AA 381–441 and 601–677 are important for its interaction with the progesterone receptor (Sullivan & Toft, 1993). Pearl's laboratory showed that a conserved hydrophobic patch around Trp300 and an amphipathic loop (327–340) extending out from the M domain may represent substrate-binding sites (Meyer et al., 2003), as deletion of these domains abolished client interaction with HSP90. In crystallographic studies of *E. coli* HtpG protein, Harris et al. found that helix 2 (544–565) of the C-terminal resides in the junction of the HtpG dimer loops out toward the space, and is the most variably positioned region in the C-terminus (Harris, Shiau, & Agard, 2004). Further mutagenesis studies identified the helix 2 region in HSP90 to be important for binding to the glucocorticoid receptor (Fang, Ricketson, Getubig, & Darimont, 2006).

Compared with other HSP90 family proteins, the CBD in GRP94 has been much less studied. Recently, Wu et al. examined whether a C-terminal loop structure (652–678), which corresponds to helix 2 in the CTD of HSP90, could be the CBD in GRP94. This hypothesis was addressed using both genetic and biochemical analyses (Wu et al., 2012). By mutagenesis assay, residues 652–678 of GRP94 were identified as important for client binding and chaperoning of both TLRs and integrins. Importantly, mutations in this loop did not affect the ATPase activity, C-terminal dimerization, and overall stability of the protein. Thus, the CBD domain in the CTD of GRP94 has been identified. However, further structural information for this CBD and other client-interaction regions in GRP94 will be needed to fully understand the specific interactions between client proteins and GRP94.

A major difference between GRP94 and other HSP90s is that, unlike HSP90, GRP94 has an identical structure when bound either with ATP or ADP (Dollins et al., 2007). These crystallographic studies also demonstrate that the ATP-bound dimeric complex adopts a twisted V-shaped conformation that is different from other HSP90 members (Dollins et al., 2007). These differences raise the possibility that the conformational changes of GRP94 necessary for the folding and release of client proteins may not be solely dictated by ATP hydrolysis, but may also be influenced by the client protein or cochaperone interactions. Hitherto, it was thought that GRP94 did not require a cochaperone for client folding, however, recent work has challenged this dogmatic notion (Liu et al., 2010). When TLR function was systemically analyzed in GRP94 KO cells, it was shown that GRP94 is critical for the function of all TLRs except for TLR3. Later, another ER protein called PRAT4A (or CNPY3) was also implicated in promoting the biogenesis of a majority of TLRs except TLR3. Results from biochemical and genetic approaches strongly support the notion that CNPY3 is indeed a TLR-specific cochaperone of GRP94: (i) deletion of either GRP94 or CNPY3 results in indistinguishable posttranslational inactivation of multiple TLRs, (ii) GRP94 and CNPY3 colocalize in the ER compartment, (iii) GRP94 and CNPY3 directly interact with each other in a manner dependent upon adenosine nucleotide, (iv) the interaction between GRP94 and CNPY3 was demonstrated in multiple cell types, (v) disruption of the interaction between GRP94 and CNPY3 by a point mutation in either the GRP94 N domain (E103A) or in CNPY3 (M145K) results in loss of TLR9 function, (vi) GRP94::TLR9 interaction is dependent on the presence of CNPY3, (vii) CNPY3 does not interact efficiently with TLR9 in the absence of GRP94, and (viii) TLR9 cleavage and maturation does not occur in the absence of either GRP94 or CNPY3 (Lammermann et al., 2008). Collectively, these findings explain the complete phenocopy of TLR defects in GRP94 KO and CNPY3-null mice and are consistent with the fact that these two molecules are colocalized in the ER lumen, coexpressed in multiple cell types, and codistributed phylogenetically.

This body of work defining the collaboration of GRP94 with CNPY3 in chaperoning TLRs is a significant advancement in the field of HSP90 biology, because it may enable alternate strategies to specifically inhibit GRP94 function for cancer therapy. However, the mechanistic details of TLR folding by GRP94 and CNPY3 remain unclear, a limitation that has hindered the development of targeted therapeutics against the GRP94–TLR pathway. For example, what are the precise role(s) of the ATPase activity of GRP94 in folding? What is the functional stoichiometry of GRP94 to CNPY3 as a basic functional folding unit? What precisely does CNPY3 do to the TLR folding reaction? Can GRP94–CNPY3 interaction be altered by other ER chaperones during ER stress? Moreover, there are four members in the CNPY family. Can other CNPYs serve as cochaperones for GRP94 to fold other client proteins? Answers to these questions await further biochemical, genetic, and structural studies.

6. TOWARD THE DEVELOPMENT OF GRP94-SPECIFIC INHIBITORS

6.1 Small Molecule Inhibitors

Despite its abundance and ubiquitous expression in most, if not all normal human cells, unique expression patterns of GRP94 have been observed by various groups in multiple cancers. Clinically, GRP94 expression correlates with advanced stage and poor prognosis in a variety of cancers including head and neck cancer (Chiu et al., 2011), gallbladder cancer (Chen et al., 2015), and breast cancer (Li et al., 2015). In some cases, GRP94 is closely linked with promoting the growth and metastasis of certain cancers such as hepatocellular carcinoma (Rachidi et al., 2015), multiple myeloma

(Hua et al., 2013), ovarian cancer, and inflammatory colon carcinomas (Morales et al., 2014). It was initially anticipated that HSP90-targeted drugs might lack specificity and cause damage to normal cells as well as tumor cells. However, the identification of GM, a small-molecule inhibitor of HSP90 changed this mindset and showed that tumor cells are more sensitive to HSP90-targeted drugs than nontransformed cells. Low concentrations of GM selectively induce differentiation, reduce cell proliferation and induce death, via binding specifically to the N-terminal ATPase pocket of HSP90 (Neckers, Schulte, & Mimnaugh, 1999). Since then, multiple inhibitors have been developed for HSP90, and also for other HSP90 (Taldone et al., 2014). Although most published studies have used pan-HSP90 inhibitors to inactivate all HSP90s and the various processes dependent on their functions, there has been great interest in identifying paralog-selective HSP90 inhibitors. Such inhibitors should interfere less with HSP functions in normal cells and thus causeless toxicity, which would allow for the administration of higher drug doses and hopefully greater anticancer activity.

In 2013, Patel et al. identified purine-based ligands that were more than 100-fold selective for GRP94 over HSP90α/β. Despite the high degree of sequence conservation within the ATP-binding pockets of GRP94 and other HSP90s, biochemical and crystallographic studies showed that they do adopt distinct conformations and hydrolyze ATP with notably different rates when bound to nucleotides (Patel et al., 2013). The conformational flexibility of GRP94 allows the identified ligands to "freeze" the protein in a state that unveils a unique pocket in GRP94 due to a five amino acid (QEDGQ) insertion. A study by Chiosis and colleagues utilized a strategy that combined library screening of purine-scaffold compounds and structural studies, to identify PU-H54, PU-WSI3, and PU-H39 as GRP94-selective inhibitory ligands. The authors then noted distinct structural features in the ATP-binding pocket; Phe199 is swung away from the binding pocket revealing a deep hydrophobic cleft. Additionally, the X2-Ar functional group on PU-H54 adopts a backward bend conformation, which allows it to insert into the hydrophobic cleft and be stabilized by GRP94 residue contacts. These distinct structural properties were not observed in the crystal structure of HSP90α-bound paralog-specific inhibitors.

Having confirmed significant biochemical selectivity for their HSP90 paralog inhibitors, the group assessed their activities in whole cells. Using the human breast cancer cell line SKBr3, which expresses high levels of HER2 protein, GRP94 inhibitors induced significant apoptosis with an attendant reduction in cell viability (Patel et al., 2013). Interestingly, no change in HER2 expression was observed in MCF-7 cells in the presence of these GRP94 inhibitors, although both cell lines were sensitive to inhibitors of HSP90α/β. This observation was attributed to distinct requirements for HSP90 paralogs in the chaperoning of HER2 that are dictated by proteome changes in the cell. Initial studies reported that GRP94 associates with newly synthesized HER2 and regulates its trafficking to the plasma membrane (Chavany et al., 1996). Additionally, HSP90 was suggested to maintain the stability of plasma membrane-associated HER2 under steady-state conditions (Xu et al., 2001). However, more recently Patel et al. demonstrated that in high HER2-expressing cells such as SKBr3, GRP94 also binds to HER2 specifically at the plasma membrane and stabilizes the protein. Thus, only membrane but not cytosolic HER2 molecules are substantially reduced in a time-dependent manner upon GRP94 inhibition in SKBr3 cells. By contrast, MCF-7 cells express less HER2 and are not affected by GRP94 inhibitors, but HSP90 inhibitors exhibit profound inhibitory effects in these cells as they target cytosolic HER2.

Hua et al. also confirmed the selective inhibition of GRP94 by PU-WS13 and examined its activity in a xenograft model of multiple myeloma. Targeted inhibition of GRP94 with PU-WS13 induced significant apoptosis and blocked growth of multiple myeloma cells, but not pre-B leukemic cells, demonstrating the dependence of myeloma growth on GRP94 (Hua et al., 2013). In another study, Hong et al. showed that PU-H39 blocked the invasion of pre-B leukemic cells and RAW264.7 cells *in vitro* by interfering with the αI domain (a ligand-binding domain shared by seven integrin α-subunits) of GRP94 (Hong, Liu, Chiosis, Gewirth, & Li, 2013). A different study by Soldano et al. identified another GRP94 inhibitor, 5′-*N*-ethylcarboxamidoadenosine (NECA), which opens up a cavity in GRP94, but not in HSP90, that can be accessed by the 5′-*N*-ethylcarboxamido moiety of NECA (Soldano, Jivan, Nicchitta, & Gewirth, 2003). Other compounds such as compound **2** and Radamide also access the unique 5′-hydrophobic pocket within the ATP-binding site of GRP94 in a similar manner to NECA (Duerfeldt et al., 2012; Immormino et al., 2009).

The usage of HSP90 inhibitors for clinical therapy is still being developed. However, off-target effects and toxicities have been noted, perhaps as a consequence of pan-HSP90 inhibition, as all clinical evaluations revealed simultaneous disruption of other HSP90 isoforms (Biamonte et al., 2010; Kim et al., 2009; Taldone, Gozman, Maharaj, & Chiosis, 2008). Although the more selective GRP94 inhibitors discussed herein should interfere less with other HSP90 isoforms at lower concentrations, the issue of off-target effects cannot be completely ruled out. For example, selective inhibition of GRP94 with low concentrations of compound **2** does not simultaneously target HSP90α/β isoforms, but nevertheless fails to upregulate GRP70 as expected with GRP94 inhibition (McCollum et al., 2008). Though concentrations above 200 μM can induce GRP70, this also destabilizes AKT, a hallmark of cytosolic HSP90 inhibition (Basso et al., 2002). Thus, although compound **C** exhibits higher selectivity for GRP94 against HSP90α/β at lower concentrations, at higher doses, its actions are similar to HSP90 inhibitors indicative of off-target effects. Other inhibitors such as Radamide also exhibit considerable selectivity for GRP94 at low concentrations (Muth et al., 2014), but the nonspecific targeting of other HSP90 isoforms at higher doses cannot be discounted. Although not evaluated for all GRP94 inhibitors, based upon available studies, concentrations greater than 200 μM are expected to nonspecifically target additional HSP90 isoforms. Thus for clinical testing, perhaps two GRP94 inhibitors should be combined and used at lower doses, as opposed to increasing the concentration of one inhibitor.

6.2 Monoclonal Antibodies

In addition to small-molecule inhibition of GRP94, specific monoclonal antibodies against surface GRP94 have also been developed and shown to have potent inhibitory effects on GRP94, and subsequent antitumor effects. Due to the significant toxicity associated with some pan-specific HSP90 inhibitors and even paralog-selective inhibitors, it could be helpful to develop monoclonal antibodies as an alternative strategy to ablate GRP94 functions in cancers. Among these, a monoclonal antibody (W9 mAb) that recognizes an extracellular epitope of GRP94 has been developed (Sabbatino et al., 2015). This antibody binds to a GRP94 epitope, which is selectively expressed on malignant cells but it is not detectable on normal cells. Using this antibody, Sabbatino et al. demonstrated that W9 mAb restores the sensitivity of $BRAF^{V600E}$ melanoma cells to BRAF inhibitors (Sabbatino et al., 2015). Recently, Li et al. reported that another antibody with specificity for membrane GRP94 (GRP94 monoclonal antibody), interfered with GRP94-dependent HER2 dimerization and phosphorylation in breast cancer, suppressed HER2-driven cell growth and induced apoptosis *in vitro* and *in vivo* (Li et al., 2015). Together, these studies have demonstrated context-specific advantages of GRP94-selective inhibition over pan-HSP90 inhibition. However, they also emphasize the importance of further work to identify novel GRP94-selective inhibitors and monoclonal antibodies.

7. TARGETING GRP94 FOR CANCER THERAPY: THE PERSPECTIVE

As a genetic disorder of somatic cells, cancer is a complicated entity with intrinsic molecular alterations that allow tumor cells to escape from normal growth regulation. Additionally, cancer cells are able to overcome extrinsic factors imposed by host immune surveillance mechanisms, to progress. Paradoxically, during the later stages of oncogenesis, tumor cells can also recruit inflammatory cells and soluble mediators to enhance tumor growth. As discussed above, GRP94 can support oncogenesis in a cancer cell-intrinsic manner via the folding of integrins, Wnt coreceptors, and components of other critical pathways. GRP94 can also promote activation of immune suppressive cells via chaperoning of TLRs and activation of TGF-β (through integrin CD103). Therefore, as a critical molecule-linking protein quality control to stress and inflammation, GRP94 is an attractive target for cancer therapy. GRP94-selective inhibitors might be particularly useful for cancers with a high demand for protein folding such as multiple myeloma. They might also be particularly useful against cancers with a dominant component of inflammation such as inflammatory bowel disease-associated colon cancer, and cancers driven by chronic infectious processes including hepatocellular carcinoma, bladder cancer, and mucosa-associated malignancies.

REFERENCES

Ali, M. M., Roe, S. M., Vaughan, C. K., Meyer, P., Panaretou, B., Piper, P. W., et al. (2006). Crystal structure of an Hsp90-nucleotide-p23/Sba1 closed chaperone complex. *Nature*, *440*(7087), 1013–1017.

Barton, E. R., Park, S., James, J. K., Makarewich, C. A., Philippou, A., Eletto, D., et al. (2012). Deletion of muscle GRP94 impairs both muscle and body growth by inhibiting local IGF production. *The FASEB Journal*, *26*(9), 3691–3702. http://dx.doi.org/10.1096/fj.11-203026.

Basso, A. D., Solit, D. B., Chiosis, G., Giri, B., Tsichlis, P., & Rosen, N. (2002). Akt forms an intracellular complex with heat shock protein 90 (Hsp90) and Cdc37 and is destabilized by inhibitors of Hsp90 function. *The Journal of Biological Chemistry*, *277*(42), 39858–39866. http://dx.doi.org/10.1074/jbc.M206322200.

Biamonte, M. A., Van de Water, R., Arndt, J. W., Scannevin, R. H., Perret, D., & Lee, W. C. (2010). Heat shock protein 90: Inhibitors in clinical trials. *Journal of Medicinal Chemistry*, *53*(1), 3–17. http://dx.doi.org/10.1021/jm9004708.

Binder, R. J., Han, D. K., & Srivastava, P. K. (2000). CD91: A receptor for heat shock protein gp96. *Nature Immunology*, *1*(2), 151–155. http://dx.doi.org/10.1038/77835.

Biswas, C., Ostrovsky, O., Makarewich, C. A., Wanderling, S., Gidalevitz, T., & Argon, Y. (2007). The peptide-binding activity of GRP94 is regulated by calcium. *The Biochemical Journal*, *405*(2), 233–241. http://dx.doi.org/10.1042/BJ20061867.

Cadepond, F., Binart, N., Chambraud, B., Jibard, N., Schweizer-Groyer, G., Segard-Maurel, I., et al. (1993). Interaction of glucocorticosteroid receptor and wild-type or mutated 90-kDa heat shock protein coexpressed in baculovirus-infected Sf9 cells. *Proceedings of the National Academy of Sciences of the United States of America*, *90*(22), 10434–10438.

Chavany, C., Mimnaugh, E., Miller, P., Bitton, R., Nguyen, P., Trepel, J., et al. (1996). p185erbB2 binds to GRP94 in vivo. Dissociation of the p185erbB2/GRP94 heterocomplex by benzoquinone ansamycins precedes depletion of p185erbB2. *The Journal of Biological Chemistry*, *271*(9), 4974–4977.

Chen, Y., Chen, C., Ma, C., Sun, S., Zhang, J., & Sun, Y. (2015). Expression of heat-shock protein gp96 in gallbladder cancer and its prognostic clinical significance. *International Journal of Clinical and Experimental Pathology*, *8*(2), 1946–1953.

Chen, W. T., Ha, D., Kanel, G., & Lee, A. S. (2014). Targeted deletion of ER chaperone GRP94 in the liver results in injury, repopulation of GRP94-positive hepatocytes, and spontaneous hepatocellular carcinoma development in aged mice. *Neoplasia*, *16*(8), 617–626. http://dx.doi.org/10.1016/j.neo.2014.07.005.

Chen, B., Piel, W. H., Gui, L., Bruford, E., & Monteiro, A. (2005). The HSP90 family of genes in the human genome: Insights into their divergence and evolution. *Genomics*, *86*(6), 627–637.

Chiu, C. C., Lin, C. Y., Lee, L. Y., Chen, Y. J., Lu, Y. C., Wang, H. M., et al. (2011). Molecular chaperones as a common set of proteins that regulate the invasion phenotype of head and neck cancer. *Clinical Cancer Research*, *17*(14), 4629–4641. http://dx.doi.org/10.1158/1078-0432.CCR-10-2107.

Chu, F., Maynard, J. C., Chiosis, G., Nicchitta, C. V., & Burlingame, A. L. (2006). Identification of novel quaternary domain interactions in the Hsp90 chaperone, GRP94. *Protein Science*, *15*(6), 1260–1269. http://dx.doi.org/10.1110/ps.052065106.

Dai, J., Liu, B., Caudill, M. M., Zheng, H., Qiao, Y., Podack, E. R., et al. (2003). Cell surface expression of heat shock protein gp96 enhances cross-presentation of cellular antigens and the generation of tumor-specific T cell memory. *Cancer Immunity*, *3*, 1.

DeNardo, D. G., Barreto, J. B., Andreu, P., Vasquez, L., Tawfik, D., Kolhatkar, N., et al. (2009). CD4(+) T cells regulate pulmonary metastasis of mammary carcinomas by enhancing protumor properties of macrophages. *Cancer Cell*, *16*(2), 91–102. http://dx.doi.org/10.1016/j.ccr.2009.06.018.

Dollins, D. E., Warren, J. J., Immormino, R. M., & Gewirth, D. T. (2007). Structures of GRP94-nucleotide complexes reveal mechanistic differences between the hsp90 chaperones. *Molecular Cell*, *28*(1), 41–56.

Duerfeldt, A. S., Peterson, L. B., Maynard, J. C., Ng, C. L., Eletto, D., Ostrovsky, O., et al. (2012). Development of a Grp94 inhibitor. *Journal of the American Chemical Society*, *134*(23), 9796–9804. http://dx.doi.org/10.1021/ja303477g.

Duerr, R. H., Taylor, K. D., Brant, S. R., Rioux, J. D., Silverberg, M. S., Daly, M. J., et al. (2006). A genome-wide association study identifies IL23R as an inflammatory bowel disease gene. *Science*, *314*(5804), 1461–1463. http://dx.doi.org/10.1126/science.1135245.

Dunn, G. P., Bruce, A. T., Ikeda, H., Old, L. J., & Schreiber, R. D. (2002). Cancer immunoediting: From immunosurveillance to tumor escape. *Nature Immunology*, *3*(11), 991–998. http://dx.doi.org/10.1038/ni1102-991.

Fang, L., Ricketson, D., Getubig, L., & Darimont, B. (2006). Unliganded and hormone-bound glucocorticoid receptors interact with distinct hydrophobic sites in the Hsp90 C-terminal domain. *Proceedings of the National Academy of Sciences of the United States of America*, *103*(49), 18487–18492.

Hainzl, O., Lapina, M. C., Buchner, J., & Richter, K. (2009). The charged linker region is an important regulator of Hsp90 function. *The Journal of Biological Chemistry*, *284*(34), 22559–22567. http://dx.doi.org/10.1074/jbc.M109.031658.

Harris, S. F., Shiau, A. K., & Agard, D. A. (2004). The crystal structure of the carboxy-terminal dimerization domain of htpG, the Escherichia coli Hsp90, reveals a potential substrate binding site. *Structure*, *12*(6), 1087–1097.

Hong, F., Liu, B., Chiosis, G., Gewirth, D. T., & Li, Z. (2013). alpha7 helix region of alphaI domain is crucial for integrin binding to endoplasmic reticulum chaperone gp96: A potential therapeutic target for cancer metastasis. *The Journal of Biological Chemistry*, *288*(25), 18243–18248. http://dx.doi.org/10.1074/jbc.M113.468850.

Hong, F., Wu, B. X., & Li, Z. (2014). Molecular regulation of macrophages in unleashing cancer-related inflammation. *OncoImmunology*, *3*(1), e27659. http://dx.doi.org/10.4161/onci.27659.

Hua, Y., White-Gilbertson, S., Kellner, J., Rachidi, S., Usmani, S. Z., Chiosis, G., et al. (2013). Molecular chaperone gp96 is a novel therapeutic target of multiple myeloma. *Clinical Cancer Research*, *19*(22), 6242–6251. http://dx.doi.org/10.1158/1078-0432.CCR-13-2083.

Immormino, R. M., Metzger, L. E., 4th, Reardon, P. N., Dollins, D. E., Blagg, B. S., & Gewirth, D. T. (2009). Different poses for ligand and chaperone in inhibitor-bound Hsp90 and GRP94: Implications for paralog-specific drug design. *Journal of Molecular Biology*, *388*(5), 1033–1042. http://dx.doi.org/10.1016/j.jmb.2009.03.071.

Katsuno, Y., Lamouille, S., & Derynck, R. (2013). TGF-beta signaling and epithelial–mesenchymal transition in cancer progression. *Current Opinion in Oncology*, *25*(1), 76–84. http://dx.doi.org/10.1097/CCO.0b013e32835b6371.

Kawai, T., & Akira, S. (2010). The role of pattern-recognition receptors in innate immunity: Update on Toll-like receptors. *Nature Immunology*, *11*(5), 373–384. http://dx.doi.org/10.1038/ni.1863.

Kim, Y. S., Alarcon, S. V., Lee, S., Lee, M. J., Giaccone, G., Neckers, L., et al. (2009). Update on Hsp90 inhibitors in clinical trial. *Current Topics in Medicinal Chemistry*, *9*(15), 1479–1492.

Kobayashi, T., Okamoto, S., Hisamatsu, T., Kamada, N., Chinen, H., Saito, R., et al. (2008). IL23 differentially regulates the Th1/Th17 balance in ulcerative colitis and Crohn's disease. *Gut*, *57*(12), 1682–1689. http://dx.doi.org/10.1136/gut.2007.135053.

Koch, G., Smith, M., Macer, D., Webster, P., & Mortara, R. (1986). Endoplasmic reticulum contains a common, abundant calcium-binding glycoprotein, endoplasmin. *Journal of Cell Science*, *86*, 217–232.

Kovalchin, J. T., Murthy, A. S., Horattas, M. C., Guyton, D. P., & Chandawarkar, R. Y. (2001). Determinants of efficacy of immunotherapy with tumor-derived heat shock protein gp96. *Cancer Immunity*, *1*, 7. pii: 010307.

Kozutsumi, Y., Segal, M., Normington, K., Gething, M. J., & Sambrook, J. (1988). The presence of malfolded proteins in the endoplasmic reticulum signals the induction of glucose-regulated proteins. *Nature*, *332*(6163), 462–464.

Lammermann, T., Bader, B. L., Monkley, S. J., Worbs, T., Wedlich-Soldner, R., Hirsch, K., et al. (2008). Rapid leukocyte migration by integrin-independent flowing and squeezing. *Nature*, *453*(7191), 51–55. http://dx.doi.org/10.1038/nature06887. pii: nature06887.

Langowski, J. L., Zhang, X., Wu, L., Mattson, J. D., Chen, T., Smith, K., et al. (2006). IL-23 promotes tumour incidence and growth. *Nature*, *442*(7101), 461–465. http://dx.doi.org/10.1038/nature04808.

Lee, A. S., Delegeane, A., & Scharff, D. (1981). Highly conserved glucose-regulated protein in hamster and chicken cells: Preliminary characterization of its cDNA clone. *Proceedings of the National Academy of Sciences of the United States of America*, *78*(8), 4922–4925.

Lewis, M. J., Mazzarella, R. A., & Green, M. (1985). Structure and assembly of the endoplasmic reticulum. The synthesis of three major endoplasmic reticulum proteins during lipopolysaccharide-induced differentiation of murine lymphocytes. *The Journal of Biological Chemistry*, *260*(5), 3050–3057.

Li, R., & Emsley, J. (2013). The organizing principle of the platelet glycoprotein Ib–IX–V complex. *Journal of Thrombosis and Haemostasis*, *11*(4), 605–614. http://dx.doi.org/10.1111/jth.12144.

Li, M. O., & Flavell, R. A. (2008). TGF-beta: A master of all T cell trades. *Cell*, *134*(3), 392–404. http://dx.doi.org/10.1016/j.cell.2008.07.025. pii: S0092-8674(08)00945-8.

Li, Y., Lu, J., Cohen, D., & Prochownik, E. V. (2008). Transformation, genomic instability and senescence mediated by platelet/megakaryocyte glycoprotein Ibalpha. *Oncogene*, *27*(11), 1599–1609. http://dx.doi.org/10.1038/sj.onc.1210794.

Li, Z., & Srivastava, P. K. (1993). Tumor rejection antigen gp96/grp94 is an ATPase: Implications for protein folding and antigen presentation. *EMBO Journal*, *12*(8), 3143–3151.

Li, X., Sun, L., Hou, J., Gui, M., Ying, J., Zhao, H., et al. (2015). Cell membrane gp96 facilitates HER2 dimerization and serves as a novel target in breast cancer. *International Journal of Cancer*, *137*(3), 512–524. http://dx.doi.org/10.1002/ijc.29405.

Lin, E. Y., Nguyen, A. V., Russell, R. G., & Pollard, J. W. (2001). Colony-stimulating factor 1 promotes progression of mammary tumors to malignancy. *The Journal of Experimental Medicine*, *193*(6), 727–740.

Linderoth, N. A., Popowicz, A., & Sastry, S. (2000). Identification of the peptide-binding site in the heat shock chaperone/tumor rejection antigen gp96 (Grp94). *The Journal of Biological Chemistry*, *275*(8), 5472–5477.

Liu, B., & Li, Z. (2008). Endoplasmic reticulum HSP90b1 (gp96, grp94) optimizes B-cell function via chaperoning integrin and TLR but not immunoglobulin. *Blood*, *112*(4), 1223–1230. http://dx.doi.org/10.1182/blood-2008-03-143107.

Liu, B., Staron, M., Hong, F., Wu, B. X., Sun, S., Morales, C., et al. (2013). Essential roles of grp94 in gut homeostasis via chaperoning canonical Wnt pathway. *Proceedings of the National Academy of Sciences of the United States of America*, *110*(17), 6877–6882. http://dx.doi.org/10.1073/pnas.1302933110.

Liu, B., Yang, Y., Qiu, Z., Staron, M., Hong, F., Li, Y., et al. (2010). Folding of Toll-like receptors by the HSP90 paralogue gp96 requires a substrate-specific cochaperone. *Nature Communications*, *1*, 79. http://dx.doi.org/10.1038/ncomms1070. pii: ncomms1070.

Luo, B. H., Carman, C. V., & Springer, T. A. (2007). Structural basis of integrin regulation and signaling. *Annual Review of Immunology*, *25*, 619–647. http://dx.doi.org/10.1146/annurev.immunol.25.022106.141618.

Massague, J. (2008). TGFbeta in cancer. *Cell*, *134*(2), 215–230. http://dx.doi.org/10.1016/j.cell.2008.07.001. pii: S0092-8674(08)00878-7.

McCollum, A. K., TenEyck, C. J., Stensgard, B., Morlan, B. W., Ballman, K. V., Jenkins, R. B., et al. (2008). P-glycoprotein-mediated resistance to Hsp90-directed therapy is eclipsed by the heat shock response. *Cancer Research*, *68*(18), 7419–7427. http://dx.doi.org/10.1158/0008-5472.CAN-07-5175.

Medzhitov, R. (2007). TLR-mediated innate immune recognition. *Seminars in Immunology*, *19*(1), 1–2. http://dx.doi.org/10.1016/j.smim.2007.02.001.

Melendez, K., Wallen, E. S., Edwards, B. S., Mobarak, C. D., Bear, D. G., & Moseley, P. L. (2006). Heat shock protein 70 and glycoprotein 96 are differentially expressed on the surface of malignant and nonmalignant breast cells. *Cell Stress & Chaperones*, *11*(4), 334–342.

Meyer, P., Prodromou, C., Hu, B., Vaughan, C., Roe, S. M., Panaretou, B., et al. (2003). Structural and functional analysis of the middle segment of hsp90: Implications for ATP hydrolysis and client protein and cochaperone interactions. *Molecular Cell*, *11*(3), 647–658.

Miranti, C. K., & Brugge, J. S. (2002). Sensing the environment: A historical perspective on integrin signal transduction. *Nature Cell Biology*, *4*(4), E83–E90. http://dx.doi.org/10.1038/ncb0402-e83.

Morales, C., Rachidi, S., Hong, F., Sun, S., Ouyang, X., Wallace, C., et al. (2014). Immune chaperone gp96 drives the contributions of macrophages to inflammatory colon tumorigenesis. *Cancer Research*, *74*(2), 446–459. http://dx.doi.org/10.1158/0008-5472.CAN-13-1677.

Morra, G., Potestio, R., Micheletti, C., & Colombo, G. (2012). Corresponding functional dynamics across the Hsp90 Chaperone family: Insights from a multiscale analysis of MD simulations. *PLoS Computational Biology*, *8*(3), e1002433. http://dx.doi.org/10.1371/journal.pcbi.1002433.

Muth, A., Crowley, V., Khandelwal, A., Mishra, S., Zhao, J., Hall, J., et al. (2014). Development of radamide analogs as Grp94 inhibitors. *Bioorganic & Medicinal Chemistry*, *22*(15), 4083–4098. http://dx.doi.org/10.1016/j.bmc.2014.05.075.

Neckers, L., Schulte, T. W., & Mimnaugh, E. (1999). Geldanamycin as a potential anti-cancer agent: Its molecular target and biochemical activity. *Investigational New Drugs*, *17*(4), 361–373.

Numasaki, M., Fukushi, J., Ono, M., Narula, S. K., Zavodny, P. J., Kudo, T., et al. (2003). Interleukin-17 promotes angiogenesis and tumor growth. *Blood*, *101*(7), 2620–2627. http://dx.doi.org/10.1182/blood-2002-05-1461.

Patel, P. D., Yan, P., Seidler, P. M., Patel, H. J., Sun, W., Yang, C., et al. (2013). Paralog-selective Hsp90 inhibitors define tumor-specific regulation of HER2. *Nature Chemical Biology*, *9*(11), 677–684. http://dx.doi.org/10.1038/nchembio.1335.

Pearl, L. H., & Prodromou, C. (2006). Structure and mechanism of the Hsp90 molecular chaperone machinery. *Annual Review of Biochemistry*, *75*, 271–294.

Prodromou, C., Roe, S. M., O'Brien, R., Ladbury, J. E., Piper, P. W., & Pearl, L. H. (1997). Identification and structural characterization of the ATP/ADP-binding site in the Hsp90 molecular chaperone. *Cell*, *90*(1), 65–75.

Qian, B. Z., Li, J., Zhang, H., Kitamura, T., Zhang, J., Campion, L. R., et al. (2011). CCL2 recruits inflammatory monocytes to facilitate breast-tumour metastasis. *Nature*, *475*(7355), 222–225. http://dx.doi.org/10.1038/nature10138. pii: nature10138.

Qian, J., Wang, S., Yang, J., Xie, J., Lin, P., Freeman, M. E., 3rd, et al. (2005). Targeting heat shock proteins for immunotherapy in multiple myeloma: Generation of myeloma-specific CTLs using dendritic cells pulsed with tumor-derived gp96. *Clinical Cancer Research*, *11*(24 Pt 1), 8808–8815. http://dx.doi.org/10.1158/1078-0432.CCR-05-1553. pii: 11/24/8808.

Qiang, Y. W., & Rudikoff, S. (2004). Wnt signaling in B and T lymphocytes. *Frontiers in Bioscience*, *9*, 1000–1010.

Rachidi, S., Sun, S., Wu, B. X., Jones, E., Drake, R. R., Ogretmen, B., et al. (2015). Endoplasmic reticulum heat shock protein gp96 maintains liver homeostasis and promotes hepatocellular carcinogenesis. *Journal of Hepatology*, *62*(4), 879–888. http://dx.doi.org/10.1016/j.jhep.2014.11.010.

Randow, F., & Seed, B. (2001). Endoplasmic reticulum chaperone gp96 is required for innate immunity but not cell viability. *Nature Cell Biology*, *3*(10), 891–896.

Reddy, R. K., Dubeau, L., Kleiner, H., Parr, T., Nichols, P., Ko, B., et al. (2002). Cancer-inducible transgene expression by the Grp94 promoter: Spontaneous activation in tumors of various origins and cancer-associated macrophages. *Cancer Research*, *62*(24), 7207–7212.

Robert, J., Menoret, A., & Cohen, N. (1999). Cell surface expression of the endoplasmic reticular heat shock protein gp96 is phylogenetically conserved. *Journal of Immunology, 163*(8), 4133–4139.

Sabbatino, F., Favoino, E., Wang, Y., Wang, X., Villani, V., Cai, L., et al. (2015). Grp94-specific monoclonal antibody to counteract BRAF inhibitor resistance in BRAF(V600E) melanoma. *Journal of Translational Medicine, 13*, 2050. http://dx.doi.org/10.1186/1479-5876-13-S1-K12.

Schulte, T. W., Akinaga, S., Murakata, T., Agatsuma, T., Sugimoto, S., Nakano, H., et al. (1999). Interaction of radicicol with members of the heat shock protein 90 family of molecular chaperones. *Molecular Endocrinology, 13*(9), 1435–1448. http://dx.doi.org/10.1210/mend.13.9.0339.

Shinagawa, N., Yamazaki, K., Tamura, Y., Imai, A., Kikuchi, E., Yokouchi, H., et al. (2008). Immunotherapy with dendritic cells pulsed with tumor-derived gp96 against murine lung cancer is effective through immune response of CD8+ cytotoxic T lymphocytes and natural killer cells. *Cancer Immunology, Immunotherapy, 57*(2), 165–174. http://dx.doi.org/10.1007/s00262-007-0359-3.

Soldano, K. L., Jivan, A., Nicchitta, C. V., & Gewirth, D. T. (2003). Structure of the N-terminal domain of GRP94. Basis for ligand specificity and regulation. *The Journal of Biological Chemistry, 278*(48), 48330–48338. http://dx.doi.org/10.1074/jbc.M308661200.

Srivastava, P. K., DeLeo, A. B., & Old, L. J. (1986). Tumor rejection antigens of chemically induced sarcomas of inbred mice. *Proceedings of the National Academy of Sciences of the United States of America, 83*(10), 3407–3411.

Staron, M., Wu, S., Feng, H., Stojanovic, A., Du, X., Bona, R., et al. (2011). Heat shock protein gp96 chaperones platelet glycoprotein Ib–IX–V complex and protects mice from Bernard–Soulier syndrome-like platelet disorder. *Blood, 117*(26), 7136–7144.

Staron, M., Yang, Y., Liu, B., Li, J., Shen, Y., Zuniga-Pflucker, J. C., et al. (2010). gp96, an endoplasmic reticulum master chaperone for integrins and Toll-like receptors, selectively regulates early T and B lymphopoiesis. *Blood, 115*(12), 2380–2390. http://dx.doi.org/10.1182/blood-2009-07-233031.

Stebbins, C. E., Russo, A. A., Schneider, C., Rosen, N., Hartl, F. U., & Pavletich, N. P. (1997). Crystal structure of an Hsp90-geldanamycin complex: Targeting of a protein chaperone by an antitumor agent. *Cell, 89*(2), 239–250.

Steidl, C., Lee, T., Shah, S. P., Farinha, P., Han, G., Nayar, T., et al. (2010). Tumor-associated macrophages and survival in classic Hodgkin's lymphoma. *The New England Journal of Medicine, 362*(10), 875–885. http://dx.doi.org/10.1056/NEJMoa0905680.

Sullivan, W. P., & Toft, D. O. (1993). Mutational analysis of hsp90 binding to the progesterone receptor. *The Journal of Biological Chemistry, 268*(27), 20373–20379.

Swafford, D., & Manicassamy, S. (2015). Wnt signaling in dendritic cells: Its role in regulation of immunity and tolerance. *Discovery Medicine, 19*(105), 303–310.

Taldone, T., Gozman, A., Maharaj, R., & Chiosis, G. (2008). Targeting Hsp90: Small-molecule inhibitors and their clinical development. *Current Opinion in Pharmacology, 8*(4), 370–374. http://dx.doi.org/10.1016/j.coph.2008.06.015.

Taldone, T., Ochiana, S. O., Patel, P. D., & Chiosis, G. (2014). Selective targeting of the stress chaperome as a therapeutic strategy. *Trends in Pharmacological Sciences, 35*(11), 592–603. http://dx.doi.org/10.1016/j.tips.2014.09.001.

Tran, D. Q., Andersson, J., Wang, R., Ramsey, H., Unutmaz, D., & Shevach, E. M. (2009). GARP (LRRC32) is essential for the surface expression of latent TGF-beta on platelets and activated FOXP3+ regulatory T cells. *Proceedings of the National Academy of Sciences of the United States of America, 106*(32), 13445–13450. http://dx.doi.org/10.1073/pnas.0901944106.

Vaughan, C. K., Gohlke, U., Sobott, F., Good, V. M., Ali, M. M., Prodromou, C., et al. (2006). Structure of an Hsp90–Cdc37–Cdk4 complex. *Molecular Cell, 23*(5), 697–707.

Vogen, S., Gidalevitz, T., Biswas, C., Simen, B. B., Stein, E., Gulmen, F., et al. (2002). Radicicol-sensitive peptide binding to the N-terminal portion of GRP94. *The Journal of Biological Chemistry, 277*(43), 40742–40750. http://dx.doi.org/10.1074/jbc.M205323200.

Wanderling, S., Simen, B. B., Ostrovsky, O., Ahmed, N. T., Vogen, S. M., Gidalevitz, T., et al. (2007). GRP94 is essential for mesoderm induction and muscle development because it regulates insulin-like growth factor secretion. *Molecular Biology of the Cell, 18*(10), 3764–3775. http://dx.doi.org/10.1091/mbc.E07-03-0275.

Wang, R., Kozhaya, L., Mercer, F., Khaitan, A., Fujii, H., & Unutmaz, D. (2009). Expression of GARP selectively identifies activated human FOXP3+ regulatory T cells. *Proceedings of the National Academy of Sciences of the United States of America, 106*(32), 13439–13444. http://dx.doi.org/10.1073/pnas.0901965106.

Weekes, M. P., Antrobus, R., Talbot, S., Hor, S., Simecek, N., Smith, D. L., et al. (2012). Proteomic plasma membrane profiling reveals an essential role for gp96 in the cell surface expression of LDLR family members, including the LDL receptor and LRP6. *Journal of Proteome Research, 11*(3), 1475–1484. http://dx.doi.org/10.1021/pr201135e.

Wu, S., Hong, F., Gewirth, D., Guo, B., Liu, B., & Li, Z. (2012). The molecular chaperone gp96/GRP94 interacts with Toll-like receptors and integrins via its C-terminal hydrophobic domain. *The Journal of Biological Chemistry, 287*(9), 6735–6742. http://dx.doi.org/10.1074/jbc.M111.309526.

Xu, W., Mimnaugh, E., Rosser, M. F., Nicchitta, C., Marcu, M., Yarden, Y., et al. (2001). Sensitivity of mature Erbb2 to geldanamycin is conferred by its kinase domain and is mediated by the chaperone protein Hsp90. *The Journal of Biological Chemistry, 276*(5), 3702–3708. http://dx.doi.org/10.1074/jbc.M006864200.

Xue, H. H., & Zhao, D. M. (2012). Regulation of mature T cell responses by the Wnt signaling pathway. *Annals of the New York Academy of Sciences, 1247*, 16–33. http://dx.doi.org/10.1111/j.1749-6632.2011.06302.x.

Yang, Y., Liu, B., Dai, J., Srivastava, P. K., Zammit, D. J., Lefrancois, L., et al. (2007). Heat shock protein gp96 is a master chaperone for toll-like receptors and is important in the innate function of macrophages. *Immunity, 26*(2), 215–226.

Zhang, Y., Ansa-Addo, E., & Li, Z. (2015). GP96: Safeguarding Treg. *Oncotarget, 6*(24), 19936–19937.

Zhang, Y., Wu, B. X., Metelli, A., Thaxton, J. E., Hong, F., Rachidi, S., et al. (2015). GP96 is a GARP chaperone and controls regulatory T cell functions. *The Journal of Clinical Investigation, 125*(2), 859–869. http://dx.doi.org/10.1172/JCI79014.

Zheng, H., Dai, J., Stoilova, D., & Li, Z. (2001). Cell surface targeting of heat shock protein gp96 induces dendritic cell maturation and antitumor immunity. *Journal of Immunology, 167*(12), 6731–6735.

CHAPTER EIGHT

HSP90 and Immune Modulation in Cancer

Michael W. Graner[1]
Department of Neurosurgery, University of Colorado Denver, Aurora, Colorado, USA
[1]Corresponding author: e-mail address: michael.graner@ucdenver.edu

Contents

Abstract

Heat-shock protein 90 (HSP90) is a highly conserved molecular chaperone that plays prominent functional roles in nearly all aspects of cell biology. As a chaperone, it interacts with literally hundreds of "clients," many of which are important drivers, regulators, and promoters of cancer. Thus, HSP90 is a high-value target in the development of anticancer therapeutics. Despite its popularity, our overall knowledge of HSP90 in immune function has lagged behind its well-recognized tumor-supportive roles. The use of inhibitors of HSP90 as chemical biological probes has been invaluable in revealing important roles for the chaperone in multiple aspects of immune function. Given this critical link, we must now consider the question of how immune outcomes may be affected by the HSP90 inhibitors currently in clinical development for the treatment

Advances in Cancer Research, Volume 129
ISSN 0065-230X
http://dx.doi.org/10.1016/bs.acr.2015.10.001

of cancer. This chapter will review some of the immunological aspects of HSP90 function in terms of its intracellular and extracellular roles in antigen presentation, immune effector cell tasks, and regulation of inflammatory processes. This review will further examine the value of HSP90 inhibitors within the context of cancer immunotherapy and will discuss how these drugs might be optimally utilized in combination with immune stimulatory approaches against cancer.

1. INTRODUCTION

Heat-shock protein 90 (HSP90) is one of the most abundant chaperones in the cell, and it is one of the most complex in terms of its protein–protein interactions and overall effects on cellular functions. HSP90 interacts with scores of client proteins, many of which are critical players in tumorigenesis, cancer progression, metastasis, and immune suppression. Thus, the chaperone is a high-value target for pharmacologic inhibitors (Bhat, Tummalapalli, & Rotella, 2014; Jhaveri et al., 2014; Kim, Keum, & Pae, 2013; Sidera & Patsavoudi, 2014). Decades of research in tumor immunology have revealed unique properties of the chaperone in the modulation of immune responses in cancer. This review will now examine some of the roles that have emerged for HSP90 in cancer immunology, including effects in tumor cells, immune cells, and activities in the extracellular environment. As we dissect the literature on this topic, we will highlight the complex and multifunctional roles of HSP90 in modulating cancer immunity.

2. A BRIEF PRIMER ON CANCER IMMUNOLOGY

Given the complexity of the immune system, select aspects and players relevant to cancer immunology are depicted in Fig. 1 to orient the reader to the essential immune cell types impacted by HSP90 functions.

The mammalian immune system consists of innate and adaptive cell types, with antigen-presenting cells (APCs) forming a conduit between the two sides. The innate responder cells are present to respond to pathogens or insults without need for "education" or "training" as is necessary for the adaptive side. Innate response times are often within minutes and are generally "hard-wired" for particular recognition markers. Of the innate effectors, gamma-delta (γδ) T cells and natural killer (NK) cells are directly cytotoxic. γδ T cells are members of the T-lymphocyte family that express an invariant T-cell receptor (TCR) consisting of γ and δ chains that

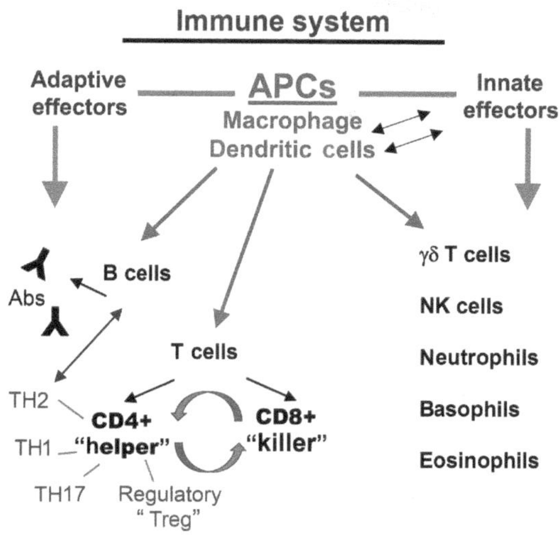

Figure 1 Overview of the cellular immune system. The mammalian immune system consists of innate and adaptive cell types, with antigen-presenting cells (APCs) forming a conduit between the two sides. The innate effectors include gamma-delta (γδ) T cells, natural killer (NK) cells, neutrophils, basophils, and eosinophils. With their phagocytic capacities, macrophage and dendritic cells (DCs) are also in this category (double-headed arrows). The adaptive immune system consists of cells that become specialized to target specific pathogens (or tumor cells). The two main cell types of the adaptive arm are B cells and T cells. These effectors may require "education" by APCs such as DCs. T-cell subclasses include the CD8+ cytotoxic T lymphocytes (CTLs, "killer" T cells), and a variety of CD4+ "helper" T cells (TH1, TH2, TH17), or cells with suppressive/regulatory cell functions (regulatory T cells, Tregs). There is cross talk among many cell types and components, and how HSP90 function affects this complex network is discussed herein. (See the color plate.)

recognizes a variety of damage- or pathogen-associated molecular patterns (DAMPs or PAMPs). NK cells respond to a balance of stimulation through a variety of activating receptors (often C-type lectin family members) and inhibitory receptors (often immunoglobulin-like molecules). Tumor cells often express NK recognition ligands such as major histocompatibility complex (MHC) class I polypeptide-related sequence A or B (MICA, MICB) or UL-16 binding proteins. The phagocytic cell types (neutrophils, but also including macrophage and dendritic cells (DCs)) are attracted to sites of infection or inflammation and can phagocytose pathogens and elaborate microbicidal agents. Basophils and eosinophils release toxins and free radicals. Of the cells of the innate immune system, we will only discuss NK cells, DCs, and macrophages within the context of HSP90.

The adaptive immune system consists of cells that become specialized to target specific pathogens (or tumor cells). This process involves gene rearrangements to produce lymphocyte receptors that solely recognize individual antigenic components of pathogens. This process also allows for the generation of "immunologic memory," subsets of cells that remain available should there be a need to respond to further challenges from the same pathogen. The two main cell types of the adaptive arm are B cells and T cells. B cells display the "B-cell receptor," which is essentially a membrane-bound form of antibody that recognizes antigens. Once activated, B cells mature to "plasma cells" where production and secretion of soluble antibody is at its highest. Antibodies may then bind target molecules, generally surface features of pathogens, which leads to destruction by neutrophils, macrophage, or the complement system. T cells recognize antigen via the "T-cell receptor"; this occurs in the context of presentation of peptides on MHC molecules (see Figs. 3 and 4) and thus requires "education" by APCs such as DCs. T-cell subclasses include the CD8+ cytotoxic T lymphocytes (CTLs, "killer" T cells), and a variety of CD4+ "helper" T cells, largely defined by the nature of their cytokine output (TH1, TH2, TH17), or by their abilities to regulate or suppress activated T-cell functions (regulatory T cells (Tregs)). CD8+ CTLs, after activating interactions with APCs displaying antigenic "pieces" of pathogens, multiply and move into the periphery for destruction of cells displaying the antigen/MHC complex recognized by that CTL's T-cell receptor. CD4+ cells are similarly activated and undergo expansion. As "helper" cells, depending on their cytokine profile, they may aid in CTL activity, may redirect the response, or may regulate and suppress the response. There is cross talk among many aspects of the immune system, where APCs communicate with B and T cells, where T cells communicate with each other, and with B cells, along with interactions and functions inherent in the innate immune response. How HSP90 function supports and shapes this complex network is the subject of the remainder of this review.

3. HSP90 AND ITS ROLES IN IMMUNITY (OR, HOW THIS ALL BEGAN): HSP90 AS AN ANTICANCER VACCINE

From an historical perspective, the initial connections between HSP90 and immune function came about fortuitously, following the "brute force" purification of tumor antigens from murine tumor cells. Almost 30 years ago, Ullrich et al. identified a "tumor-specific transplantation antigen" from chemically induced murine tumors (Meth A) as HSP90, where it appeared to be a cell-surface antigen (Ullrich, Robinson, Law, Willingham, & Appella, 1986). Similarly, Srivastava et al. identified a tumor rejection antigen from Meth A and CMS5 murine tumors (Srivastava, DeLeo, & Old, 1986) that was later recognized as the HSP90 endoplasmic reticulum (ER) paralog, glucose-regulated protein 94 (GRP94/gp96, /HSP90B1) (Maki, Eddy, Byers, Shows, & Srivastava, 1993; Maki, Old, & Srivastava, 1990).

In these experiments, biochemical fractions from cell lysates were assayed by injection into mice, followed by challenge with parental tumor cells. Fractions that provided protection against challenge—i.e., those that generated immune responses against the tumor cells—were further fractionated and tested, until single, purified, proteins were declared the immunizing factor. These purified, tumor-derived proteins were therefore serving as vaccines capable of generating immune responses against the tumor cell line from which the proteins were purified. Curiously, immunization of mice with protein purified from one type of tumor did not cross-protect those mice from tumor challenge with a different tumor type—the definition of immune specificity. This finding was especially intriguing in that the genes for these proteins were not found to be mutated in the cancer cells and appeared to be identical to their counterparts in normal cells (or other cancer cells). This conundrum made the specificity of the vaccines difficult to explain.

Additional work suggested that chaperones such as HSP70 family members could bind peptides (Flynn, Pohl, Flocco, & Rothman, 1991). If such peptide binding was occurring intracellularly, and was stable, purification of the chaperones would also result in isolation of a bound population of peptides—an "antigenic fingerprint" from the cell of origin (Srivastava & Heike, 1991). Assuming tumors produce distinct antigenic profiles, purification of HSPs from a given tumor would theoretically provide a specific antigenic peptide repertoire, and would generate an accordingly specific immune response. Thus, purification of the "same" HSP from different tumor types would lead to tumor-specific vaccination responses, despite the use of the "same" HSP as the vaccine. The true immunogens would be the peptides as part of the chaperone–peptide complexes. It was recognized at that time that 7- to 15-mer peptides were likely the "currency" of the adaptive T-cell response (Livingstone & Fathman, 1987). Transfer of these potential peptide antigens into antigen presentation pathways (demonstrated in the case of HSP70; Udono & Srivastava, 1993) would explain the distinct antigenic properties of what appeared to be otherwise identical proteins purified from different tumor types. The concept of distinct peptide repertoires from different tumors associating with otherwise identical HSPs is illustrated in Fig. 2.

One implication of the peptide-binding capacity of HSP90 is that the chaperone is involved in the process of antigen presentation, particularly during peptide generation. An overview of antigen presentation and areas where chaperones are involved is provided in Fig. 3. Canonically, presentation of intracellularly generated peptides via the "intrinsic" pathway (via major histocompatibility complex class I/MHC I) is shown in the sequences A, B, and C. The "extrinsic" or exogenous pathway (leading to MHC II presentation) is shown in sequences D, E, and F. In A, an ubiquitinated obsolete or unfolded protein from within the cell is guided by chaperones/heat-shock proteins to the proteasome for proteolytic degradation. The resultant peptides are then escorted by HSPs to the transporter(s) associated with antigen processing (TAP 1 and 2) for passage into the ER and loading onto nascent MHC I molecules (B), again with chaperone assistance. The peptide-loaded MHC I molecule enters the secretory transport process via the Golgi apparatus, and eventually appears on the plasma membrane, where it displays the peptide in the context of the MHC

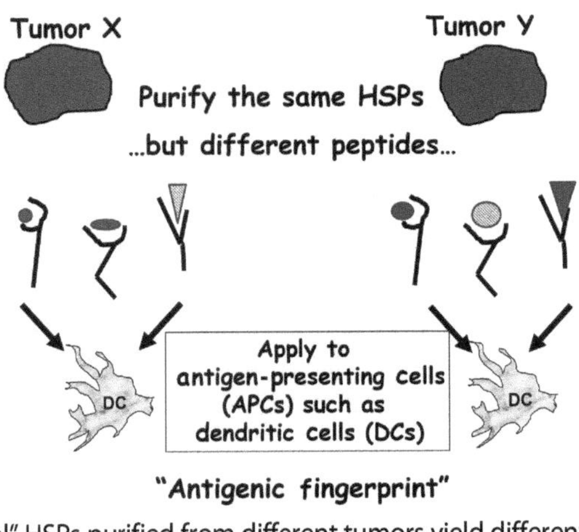

Figure 2 "Identical" HSPs purified from different tumors yield different antigenic peptide populations. Every individual human tumor is considered antigenically distinct, even with identical histological classifications. HSPs chaperone intracellularly generated peptides; thus, purification of apparently identical HSPs, but from different tumor types, harvests distinct peptide populations complexed with the HSPs. In a vaccine scenario, the "identical" HSPs provide tumor-specific "antigenic fingerprints" to the APCs, stimulating unique T cells against each tumor. *Adapted from Oushy and Graner (2015).* (See the color plate.)

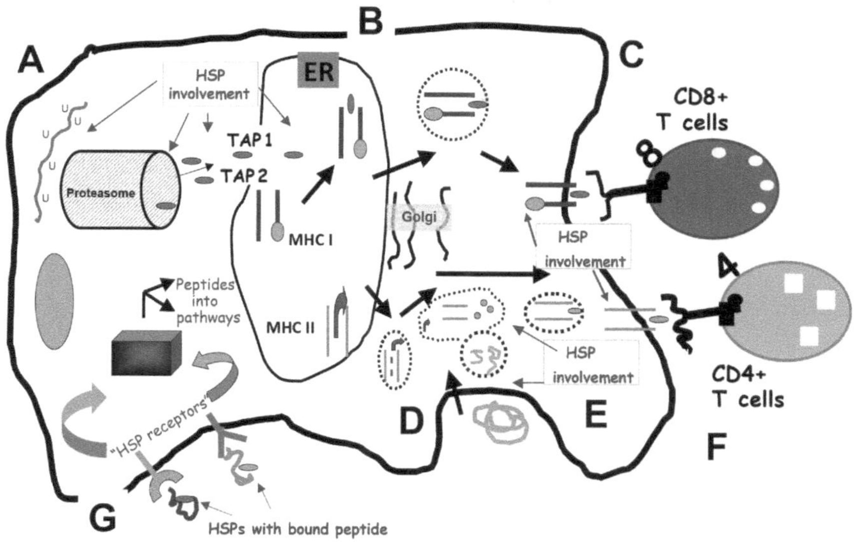

Figure 3 Overview of antigen presentation. Antigen presentation pathways are shown in a highly simplified diagram. Canonically intracellularly generated peptides are ultimately displayed via the "intrinsic" pathway, on major histocompatibility complex class I/MHC I, (sequences A, B, and C). The "extrinsic"/exogenous pathway (leading to MHC II presentation) is shown in sequences D, E, and F. In A, a ubiquitinated protein (red string with "U's") is guided by HSPs to the proteasome (blue cylinder) for degradation Resultant peptides (red ovals) are escorted by HSPs to the transporter(s) associated with antigen processing (TAP 1 and 2) into the endoplasmic reticulum (ER) and loading onto MHC I molecules (B). The peptide-loaded MHC I transports to the cell surface, with MHC/peptide display to CD8+ T cells (blue sphere, C). In the extrinsic pathway (D, E, and F), material from outside the cell (orange globule) enters endocytic vesicles (D). An MHC II molecule forms in the ER with the invariant chain Ii (magenta shape) in place (D). The endocytic vesicle fuses with the exocytic vesicle, with concurrent cleavage of the endocytosed protein (orange pieces) and invariant chain cleavage into CLIP (class II-associated invariant peptide, magenta arrow). Peptides are then loaded onto the MHC II molecule with transport to the cell surface (E) for presentation of MHC II/peptide to CD4+ T cells (green sphere, F). HSPs are involved in numerous steps during endocytosis and processing, including interactions with proteasome subunits and MHC molecules in cell membranes. In (G), exogenous HSPs carrying cellular-derived peptides (e.g., released from dying cells, or from a chaperone-based vaccine) bind and are internalized by specific receptors. The complexed peptides enter into antigen presentation pathways (black box) as described. (See the color plate.)

molecule to CD8+ T cells (C). In the extrinsic pathway, material from outside the cell is endocytosed by potentially numerous mechanisms, ending up in an endocytic vesicle (D). Meanwhile, an MHC II molecule is formed in the ER and packaged for transport with the invariant chain Ii in place (D). The endocytic vesicle fuses with the exocytic vesicle. Here, ongoing peptidase cleavage of the endocytosed protein occurs along with invariant chain cleavage into CLIP (class II-associated invariant peptide) which leads to displacement of CLIP from the peptide-binding pocket. Peptides are then loaded onto the MHC II molecule with continued transport to the cell surface (E) for display and presentation of peptide in the context of MHC II to CD4+ T cells (F). HSPs are involved in numerous steps during endocytosis and processing, including interactions with proteasome subunits and MHC molecules in cell membranes. In (G), exogenous HSPs carrying cellular-derived peptides (e.g., released from dying cells, or perhaps from a chaperone-based vaccine) are bound and internalized by certain receptors (e.g., C-type lectins, scavenger receptors, LOX1, CD91). The bound peptides are displaced and enter into antigen presentation pathways by unknown "black box" mechanisms. Note that these pathways are not mutually exclusive, but overlap and intersect, particularly where the endocytic "exogenous" pathway leads to antigen display on MHC I; this process is called "cross-presentation." While nearly all cells possess the MHC I pathway, generally only specialized, "professional" APCs express the appropriate components needed to utilize the MHC II pathway. Cross-presentation by professional APCs enables simultaneous stimulation of both CD4+ ("helper" T cells) and CD8+ ("killer" T cells) cell types, allowing these cell classes to engage in proximal cross talk and mutual stimulation for a stronger T-cell-based response.

One envisions that HSP90, acting as a chaperone of proteasome-generated peptides (Fig. 3A and B), could be purified from a cell (in this scenario, a tumor cell) after forming complexes with such peptides. In a vaccine scenario, the HSP/peptide complexes are then exposed to an APC, where the complexes are bound and internalized by "HSP receptors" (Fig. 3G), which allows the peptides to enter into both major histocompatibility class I and II (MHC I and MHC II) presentation pathways. This, in turn, can lead to the stimulation of CD8+ CTLs ("killer" T cells) and CD4+ "helper" T cells, respectively, as described above (Fig. 3C and F).

Functionally, tumor-derived HSP90 was found to be immunologically active and capable of antitumor vaccination in additional studies (Graner et al., 2000; Udono & Srivastava, 1994). Such immune responses require that HSP90, involved in multiple sites of antigen processing intracellularly, becomes "externalized" from the cell (in this case, by intentional lysis and subsequent purification), where it can now interact with APCs such as DCs (Palucka & Banchereau, 2013b). As depicted in Fig. 3, this interaction may occur through scavenger receptors such as scavenger receptor class

F member 1 (SCARF1/SREC1), which mediates cross-presentation via MHC I pathways (Murshid, Gong, & Calderwood, 2010) or more standard MHC II mechanisms (Murshid, Gong, & Calderwood, 2014). Notably, these presentation pathways can activate both CD8+ CTLs by MHC I presentation, and CD4+ "helper" T-cell responses by MHC II presentation. However, the process of binding receptors and internalization (eventually leading to antigen display) is alone insufficient to generate a potent T-cell response. Section 4 discusses the innate stimulatory properties of externalized HSPs, including HSP90, which enables a highly activated presentation system capable of promoting effective antitumor responder cells such as T cells.

4. EXTRACELLULAR/CELL SURFACE HSPs AND "DANGER SIGNALS": THE MODE OF CELL DEATH MATTERS TO THE IMMUNE SYSTEM

Although HSP90α/β is predominantly cytosolic, HSP90 can also be cell surface localized, usually under pathologic conditions (Cid, Regidor, Poveda, & Alcazar, 2009; Erkeller-Yuksel, Isenberg, Dhillon, Latchman, & Lydyard, 1992; Ferrarini, Heltai, Zocchi, & Rugarli, 1992; Graner & Bigner, 2006), and also in certain physiological processes such as neuronal migration (Sidera, Samiotaki, Yfanti, Panayotou, & Patsavoudi, 2004). Extracellular release of HSP90 occurs, as well (Hance et al., 2012; Nolan, Franco, Hance, Hayward, & Isaacs, 2015; Suzuki & Kulkarni, 2010), including release via extracellular vesicles (Epple et al., 2012; McCready, Sims, Chan, & Jay, 2010). Thus, HSP90 appears on cell surfaces and/or outside of the cell under both normal and pathologic circumstances, and the context in which this occurs is very important in the resultant immune physiology.

A fundamental tenet for the utility of HSP-based vaccines is that the chaperones provide "danger signals" to the innate (and later, the adaptive) immune systems (Gallucci & Matzinger, 2001; Multhoff, Pockley, Streffer, & Gaipl, 2012) because they represent "danger/damage-associated molecular patterns" (DAMPs) (Dubaniewicz, 2013). There is overlap among the DAMP receptors and the aforementioned HSP receptors (Van Crombruggen, Jacob, Zhang, & Bachert, 2013), implying convergent evolutionary pathways. The basic idea is that if normally intracellularly localized molecules are appearing outside of the cell, a potentially dangerous incident has occurred. From an HSP-based vaccine perspective, this would result in the HSP component of the vaccine serving as its own adjuvant (nonspecific, generalized immune stimulator), while still carrying antigens in the same packet (ultimately providing specific targets). One would envision this complex to be delivered exogenously (e.g., by injection). This scenario exploits a long evolutionary history where immune systems have developed to recognize situations where cell death within the organism needs to provoke an immune response (e.g., pathogen-induced death, tumor cell death).

Conceptually, cells must die in such a manner that chaperones/HSPs (and other molecular danger patterns) are released or displayed in forms that signal to sentinel/scavenging immune cells (e.g., macrophage, DCs) that damage requiring an immune response has occurred. Thus, the release of HSPs—possibly carrying antigenic peptides derived from the pathogen that induced the host cell death—would constitute a form of internal "self-vaccination" to eventually produce adaptive B- and T-cell responses. Furthermore, the HSPs might not have to be completely released extracellularly, but could end up being exposed on the surfaces of dying cells. The danger signal input drives APC expression of costimulatory molecules and release of activating cytokines that yield responder cells (both T cells and NK cells) with more potent activated phenotypes (Dubaniewicz, 2013; Hu et al., 2005; Zeng, Feng, Graner, & Katsanis, 2003). Such a scenario is illustrated in Fig. 4.

As depicted in Fig. 4, extracellular HSPs act as "danger/damage-associated molecular patterns" (DAMPs, danger signals) to drive activating cytokine secretion (e.g., IL-12) and expression of costimulatory molecules (CD80/B7.1 and CD86/B7.2) to provide full stimulation of T cells. The HSPs may be released during cell death or provided as vaccines; in this rendition, those carrying antigenic peptides are bound by receptors on APCs (as mentioned in Fig. 3). Internalization and downstream signaling lead to expression of costimulatory molecules and release of activating cytokines. The display of the peptide antigens on MHC I and II molecules, coupled with the added stimulation, drives potent T-cell responses, leading to more cell death and further release of HSPs.

Initially, it was thought that necrotic cell death, as opposed to apoptosis, was the form of cell death that was immunologically relevant, with HSPs such as HSP90 playing key roles in stimulating sentinel immune cells (i.e., DCs, Basu, Binder, Suto, Anderson, & Srivastava, 2000). Heat shock leading to cell death with necrotic features correlates strongly with increased HSP90 expression (Schueller et al., 2001). However, heat shock of tumor cells dying by an artificially driven "clean" apoptotic mechanism increased surface expression of HSPs 60 and 70 (but not HSP90), and these increases

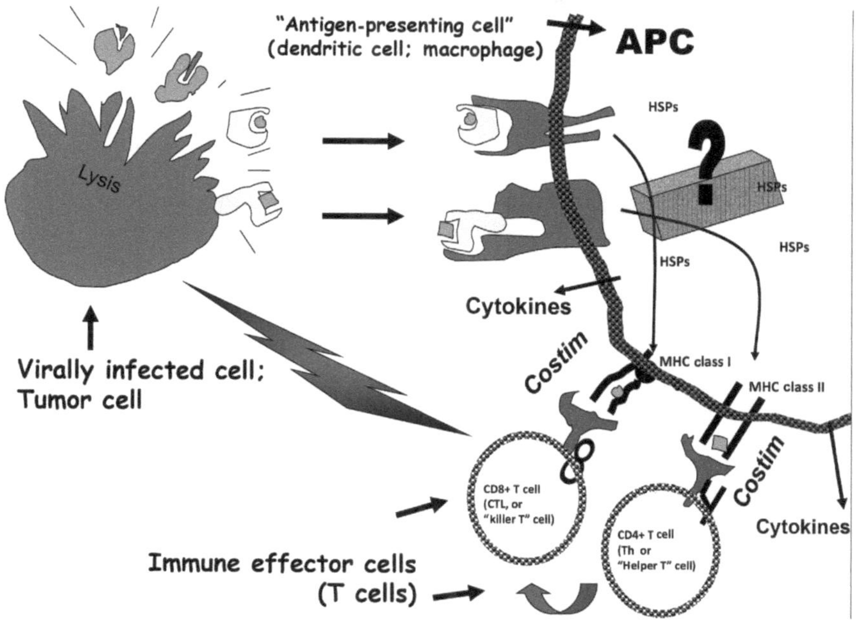

Figure 4 HSPs as "danger signals" prompting APC activation and costimulatory-driven mechanisms for T-cell provocation. Extracellular HSPs act as "danger/damage-associated molecular patterns" (DAMPs, danger signals) promoting cytokine secretion and costimulatory molecules (CD80/B7.1 and CD86/B7.2) expression to provide full stimulation of T cells. The HSPs (green or yellow shapes) may be released during cell death or provided as vaccines. Shown here, HSPs carrying antigenic peptides (small orange shapes) are bound by APCs (see Fig. 3). Internalization and downstream signaling lead to expression of costimulatory molecules and release of activating cytokines. The presentation of peptide antigens on MHC I and II molecules, and accompanying stimulation, drives potent T-cell responses, resulting in more cell death and further release of HSPs. (See the color plate.)

correlated with immune-mediated destruction *in vitro* and *in vivo* (Feng, Zeng, Graner, & Katsanis, 2002; Feng, Zeng, Whitesell, & Katsanis, 2001). This form of immunogenic "stressed apoptosis" could be mimicked by inclusion of exogenous stress proteins (including HSP90; Feng, Zeng, Graner, Likhacheva, & Katsanis, 2003) with tumor cells either when coinjected into mice, or cocultured with APCs, suggesting that even apoptotic tumor cells can be immunogenic if HSPs are exposed during apoptosis.

This effect has since been verified in numerous situations such as chemotherapeutics that drive danger signal expression or relocalization of chaperones during apoptotic cell death (Tesniere et al., 2008), including surface display of HSP90 (Spisek & Dhodapkar, 2007). However, at least in the setting of human primary acute myeloid leukemia, immunogenic apoptosis (with calreticulin-associated cell surface display and release of HSPs 70 and 90) was highly variable across patient tumors, even when different standard antileukemic agents were used (Fredly, Ersvaer, Gjertsen, & Bruserud, 2011). It may be that drug-induced immunogenic demise may be more tumor specific than chemotherapy specific. Bear in mind that in these scenarios the cell surface display or extracellular presence of chaperones/HSPs promotes various levels of innate immune stimulation (e.g., increased expression of cytokines and costimulatory molecules, increased phagocytosis by APCs, etc.). This does not require the presence of HSP–peptide complexes, although these are likely also released and engulfed by APCs and are linked to further adaptive (and thus antigen-specific) immune responses.

Because DCs are considered the most effective APCs, they have received much attention as conduits for cancer immunotherapy, either via *in situ* targeting or by their production *ex vivo* to produce cellular-based vaccines (Palucka & Banchereau, 2013a). In the latter scenario, DCs are cultured and expanded from patient blood cells, and then "pulsed" or exposed to antigen (in any number of formats). The antigen-loaded and -activated DCs are then injected into patients for priming of T cells to generate an anticancer response. As mentioned above, DAMP stimulation of DCs, whether *in situ* or *ex vivo*, is likely to be important in the success of T-cell priming by DCs. Clinically, DC-based, antitumor vaccination using "immunogenically killed" autologous tumor cells (via heat shock and ultraviolet C irradiation) as antigen sources showed some success in a pilot study of patients with indolent non-Hodgkin lymphoma (Di Nicola et al., 2009). Nonresponders had tumor cells that failed to display certain key features, including surface localization of chaperones such as calreticulin and HSP90, and, unlike responders, these patients did not generate serum antibodies against those chaperones (Zappasodi et al., 2010). Thus, HSP90 appears to be a critical requirement for recognition of tumor cell death by the immune system, and this biologic feature is likely exploitable when HSP90 is used in cancer vaccine scenarios.

5. EXTRACELLULAR HSP90 IN/ON EXOSOMES AND THE EFFECTS OF HEAT STRESS ON HSP90 IMMUNOGENICITY

In terms of other forms of extracellular HSP90 (and potentially cancer immunotherapeutics), a version of "extracellular" HSP90 may arise in the form of exosomes. These virus-sized extracellular vesicles are formed in the endosomal pathway as intralumenal vesicles in multivesicular bodies that are released outside the cell (Ung, Madsen, Hellwinkel, Lencioni, & Graner, 2014; Yanez-Mo et al., 2015). We have demonstrated the ability of tumor-derived exosomes harboring significant HSP content to function as vaccines (Graner et al., 2009). Additionally, overexpression of ras-related protein RAB27A (a small GTPase that enables exosome release from cells; Ostrowski et al., 2010) drove increased HSP loading (including HSP90) into exosomes and resulted in enhanced antitumor immunogenicity *in vitro* and *in vivo* (Li, Mu, et al., 2013). Heat stress and chemotherapeutics (irinotecan and carboplatin) drove HSP release (including HSP90), at least some of which is exosome surface-localized, from hepatocellular carcinoma cell lines, and those exosomes promoted NK-cell cytotoxicity *in vitro* (Lv et al., 2012).

While HSP90 is clearly one of the more prevalent proteins found as cargo in exosomes (Roma-Rodrigues, Fernandes, & Baptista, 2014; Yanez-Mo et al., 2015), few publications have identified HSP90 on exosome surfaces (Epple et al., 2012; Lv et al., 2012). One mechanism to explain exosome surface localization of HSP90 could be that HSP90 is found on tumor cell surfaces (Eustace & Jay, 2004; Graner & Bigner, 2006; Graner, Cumming, & Bigner, 2007). This cell surface localization would also result in the localization of HSP90 to exosome surfaces following the process of endosome formation and the intravesicular invagination necessary to form a late endosome/multivesicular body (Ung et al., 2014). The intralumenal vesicles—which are released as exosomes upon fusion of the multivesicular body with the cell membrane—essentially (but selectively) recapitulate the topography of the cell. Of course, there are currently no clear mechanisms explaining HSP90 cell surface localization to begin with.

The immune effects of HSP release during thermal therapies or thermal ablation have not been widely studied (Widenmeyer et al., 2011). One means of upregulating HSPs (including HSP90) and potentiating their extracellular display is through the use of heat shock, or in the clinic, delivery of localized hyperthermia treatments. Using a novel method of electro-hyperthermia in a xenograft colorectal cancer model, treatment resulted in DAMP surface translocation in a temporal sequence (calreticulin, followed by HSP70, then HSP90), setting the stage for testing in immune-competent models (Andocs et al., 2015). Localized tumor heating using magnetite nanoparticles has been postulated to generate immune-stimulating necrotic cell death due to upregulated HSP expression and release (Ito, Honda, & Kobayashi, 2006). In a clinical trial, fever-range temperatures (whole-body hyperthermia) resulted in increased HSP levels (including HSP90) in the blood of colorectal cancer patients pretreated with hyperthermia before surgery (Sulyok et al., 2012), but the immune consequences were not clear. Several types of human anti-HSP90 immune effectors have been reported in cancer patients including antibodies (Luo, Herrera, Soosaipillai, & Diamandis, 2002; Vidal et al., 2004), and even the generation of T cells against HSP90 itself in the form of HSP90-derived peptide antigens displayed by myeloma cells (Li et al., 2014). These findings suggest that extracellular HSP90 may indeed be a target of anticancer immune responses as well as an inducer of such responses.

As mentioned above, HSP90 as a prevalent cargo may be an important factor in the immunogenicity of exosomes. Indeed, exosomes from heat-shocked tumor cells (with increased HSP90 content) demonstrated improved anticancer efficacy in a mouse lymphoma vaccination model (Chen et al., 2006). Similarly, exosomes from heat-shocked cells generated carcinoembryonic antigen (CEA)-specific lymphocytes in a CEA-expressing model (Dai et al., 2005). Local hyperthermia is postulated to lead to exosome release as a potential "self"-vaccination source (Toraya-Brown & Fiering, 2014). Again, it is not clear if increases in HSP expression are accompanied by surface localization of HSP90, as this seems to be rarely assessed.

Other recognized means of inducing immunogenic cell death include forms of photodynamic therapy, but the photosensitizing agent is important (Castano, Mroz, & Hamblin, 2006). Such treatments can induce the surface display of HSPs 60 and 70, as well as GRP94 (Korbelik, Sun, & Cecic, 2005), but so far, only one sensitizer (Rose Bengal Acetate) has been shown to translocate HSP90 to the cell surface (along with calreticulin and protein disulfide isomerase A3 [PDIA3/ERp57/GRP58]) (Panzarini, Inguscio, Fimia, & Dini, 2014). Few reports have evaluated the combination of photodynamic therapy with hyperthermia (He, Ge, & Li, 2012), limiting our understanding of the therapeutic potential of this approach.

6. ...BUT IS EXTERNALIZED HSP90 ALWAYS AN IMMUNE TRIGGER? THE SIGNIFICANCE OF NONIMMUNE ROLES FOR EXTRACELLULAR HSP90

While cell surface and extracellular HSPs are most commonly associated with induction of immune responses (Didelot et al., 2007), it is obvious that surface-localized or extracellularly released HSP90 in situations not involving cell death might also benefit the tumor in some way. One scenario is that noncanonical localization could play a role in modulation of the tumor microenvironment. For instance, extracellular HSP90 appears to chaperone plasminogen and to assist in plasmin generation, leading to increased tumor cell mobility (McCready et al., 2010). Although HSP90 was not demonstrated on the exosomal surface in this instance, such localization has been shown elsewhere (Epple et al., 2012; Lv et al., 2012). Extracellular HSP90 also interacts with matrix metalloproteinase 2 (MMP2), and this complex also promotes tumor migration and invasion (Eustace et al., 2004; Sims, McCready, & Jay, 2011; Yang et al., 2008). Other studies have shown that eHSP90/MMP2 interaction drives tumor angiogenesis (Song et al., 2010). Further, extracellular HSP90 is associated with epithelial-to-mesenchymal transition (EMT) in prostate cancer cells and tumors (Hance et al., 2012), in particular through modulation of enhancer of zeste 2 (EZH2) and the latter's downstream engagement of EMT players and activities (Nolan et al., 2015). Some have argued that most if not all the activity of HSP90 inhibitors seen in clinical trials to date is due to the inadvertent targeting of "'e[xtracellular] Hsp90-dependent' cancer cells" (i.e., extracellular HSP90 is the actual target responsible for anticancer activity because cancer cells rely on it for migration, invasion, and metastasis; Li, Tsen, et al., 2013).

Collectively, the literature suggests that the benefits of extracellular HSP90 in terms of its support of the tumorigenic phenotype, migration, and angiogenesis must outweigh the "risks" of extracellular HSP90 in driving an antitumor immune response. Tumors clearly release HSP90 into the extracellular space and appear to do so with impunity, as tumors grow and metastasize without obvious immune consequences acting against them, at least in the later stages of tumor development and progression. This dichotomy in the effects of extracellular HSP90 may skew away from the chaperone's immunostimulatory role, perhaps due to the overriding effects of tumor-induced immune suppression (Finn, 2012; Khan, Gucalp, & Shapira, 2015).

7. BACK INSIDE THE CELL: HSP90 IN ANTIGEN PRESENTATION: ARE HSP90 INHIBITORS A BOON OR A BANE?

In the literature describing HSP-based vaccines, chaperones/HSPs in the cytosol and ER are hypothesized to be associated with proteasome-derived peptides and thought to function as a "relay line" to pass the peptides from HSP90 to HSP70, through the TAP into the ER, where GRP94 can finally assist in loading of the peptide onto MHC I molecules (Srivastava, Udono, Blachere, & Li, 1994; depicted in Fig. 3). Other studies have implied that GRP94 may not be essential to priming of MHC I with proteasome-derived peptides (Lammert, Arnold, Rammensee, & Schild, 1996), or perhaps may play almost no role at all in MHC priming (Demine & Walden, 2005). However, differential association of model peptides with HSP90 and HSP70 in the cytosol has been shown (Ishii et al., 1999). With increased interest in HSP90 inhibitors as anticancer therapeutics, the effects of these inhibitors on antigen processing, presentation, and overall immune response have become important considerations. Examination of the detailed effects of these agents has, naturally, led to some surprising observations, conflicting information, and more questions.

HSP90 mediates a presentation pathway for only a restricted group of antigenic peptides. The roles of HSP90 and the "immunoproteasome" activator subunits 1 and 2 (PSME1/PA28A; PSME2/PA28B) were assessed in murine tumor cells (EL4 and ovalbumin [OVA]-expressing clone EG.7) following geldanamycin (GA)-mediated inhibition of HSP90, or within the context of acute lipopolysaccharide (LPS) exposure (generating "LPS blasts" as APCs), using PSME1/2 knockout mice. Using various OVA peptides as model antigens, GA treatment of knockout cells abrogated presentation of C-terminal extended OVA peptides (but not of N-terminal extended peptides), while there was no effect in wild-type cells. The same results held for tumor cells (Yamano et al., 2002). These findings are important in that specific peptide sequences may allow for enhanced presentation on MHC molecules, thus enabling a T-cell-based response. Extension of amino acid sequences beyond the peptide region that fits into the MHC binding pocket, either C- or N-terminally, may increase or decrease the eventual display of the peptide. These results were among the first to suggest that HSP90 inhibition may influence processing and presentation of at least specific subsets of antigens. As a caveat, the concentrations of GA used in these studies was relatively high (5–10 μM) compared to amounts used for tumor cell stasis/cytotoxicity studies, which may confound experimental outcomes and consequent interpretation. Still, lending credence to the overall conclusion, additional studies using HSP90 knockdown by silencing (siRNA) also demonstrated a role for HSP90 early in the antigen processing pathway for C-terminal extended OVA peptides (Kunisawa & Shastri, 2006).

HSP90 also promotes antigen display on MHC I of N-terminal extended peptides in both tumor antigen and viral systems. Within this context, HSP90 inhibition did not affect the stability of MHC I molecules, but rather led to generation of peptide-"empty" molecules (which are generally considered unstable; Callahan, Garg, & Srivastava, 2008). These effects were seen with several different HSP90 inhibitors, albeit at high concentrations (10–50 μM). Other studies using 17-AAG and radicicol (at concentrations from 1.5 to 50 μM), or using DCs from heat-shock factor 1 (HSF1)-deficient mice (which reduced levels of HSP90) also demonstrated a role for HSP90 in both cross-presentation of exogenous antigen as well as processing and presentation of endogenous antigens (Ichiyanagi et al., 2010). Furthermore, *HSP90AA1*-knockout mice show impaired ability to translocate antigen into the cytosol of DCs for cross-presentation of exogenous antigen both *in vitro* and importantly, *in vivo* (Imai et al., 2011). These studies implicate HSP90 as an active participant in antigen presentation of both C- and N-terminally extended antigenic peptides leading to MHC I display of such antigens. They also support roles for the chaperone in internalization of antigen into APCs for presentation and generation of T-cell-based immune responses.

As noted above, HSP90 may be involved in other aspects of antigen presentation aside from peptide transfer; the chaperone may stabilize MHC molecules at the cell surface, or it may be involved in transport of potential antigens into the cell. Studies of heat-shock-induced maturation of murine DCs (measured by surface presence of MHC II molecules) revealed a role for HSP90 in the maturation process, possibly by surface MHC II stabilization (Basu & Srivastava, 2003). Altered MHC II expression caused by GA-mediated HSP90 inhibition was also observed using different model antigens and mouse strains, with macrophages as the APCs (Rajagopal, Bal, Mayor, George, & Rath, 2006). Unfortunately, it is unclear whether this was due to an effect on peptide antigen processing or on stabilization of surface MHC by HSP90 (or both). These studies also used relatively high GA concentrations (5–10 μM).

DCs from the aforementioned *HSP90AA1* knockout mice (which have intact HSP90AB1/HSP90β), actually showed *increased* antigen presentation via MHC II (leading to B cell antibody production). This result indicates that HSP90 may facilitate cross-presentation of exogenously derived antigen at the expense of MHC II presentation, in effect, choosing MHC I antigen presentation over MHC II presentation. Such effects could skew the balance between humoral/antibody and cellular immune outputs, suggesting that HSP90 can affect the direction of an immune response of effector cells (toward CTLs, or toward B cell-/antibody-mediated) driven by APCs. However, in a murine autoimmune antibody model of epidermolysis bullosa acquisita, 17-DMAG reduced total B-cell numbers in spleens of treated mice, and reduced serum antibody titer (Tukaj et al., 2014). Bear in mind that the previous report employed specific genetic ablation of *HSP90AA1*, while HSP90 inhibitors presumably target both HSP90 isoforms. Examination of overall human *in vitro* DC function in terms of expression of surface costimulatory molecules, uptake of exogenous antigen, release of activating cytokines, and stimulation of T cells, was dramatically reduced by GA treatment (at only 1 μM concentration; Bae et al., 2007). Cumulatively, these reports suggest a major role for HSP90 in presentation of both MHC I and II antigens, and further suggest that HSP90 inhibition may impair immune responses in treatment scenarios.

As to roles for HSP90 in lymphocytes, Bae et al. (2013) extensively evaluated the effects of GA treatment upon human T cells and NK cells. In T cells, they observed a near-universal reduction of T-cell receptor components, costimulatory receptor expression, proliferation status, and activated cytokine release. A bevy of activating receptors was also reduced on NK cells, as were the cytotoxic capabilities of these cells. Overall, the impression from most of the literature is that HSP90 inhibition impairs antigen presentation by APCs, and that the function of other immune effectors (particularly, T cells, NK cells, and perhaps B cells) is also likely derailed.

8. BEYOND ANTIGEN PRESENTATION: HSP90 IN INFLAMMATORY PROCESSES

The state and type of APC may be important when examining the effects of HSP90 inhibitors on immune cell types. To illustrate these potentially complex context-dependent effects, exposure of human monocyte-derived DCs to therapeutically relevant concentrations of GA (0.1 μM) caused partial activation (measured by MHC II and costimulatory markers) of previously unstimulated (e.g., "immature") DC. However, low-dose GA coadministered with a stimulation cocktail (designed to drive activation) prevented the expected increase in DC activation markers. Stimulated DC migration and allogeneic activation of T cells were also reduced when DCs were treated with GA. Resting T cells themselves were unaffected by GA, but drug treatment inhibited their stimulated proliferation (Trojandt, Reske-Kunz, & Bros, 2014). In a macrophage cell line (J774), HSP90 inhibition by 17-DMAG at relatively low concentrations (0.1 μM) impaired inflammatory responses to the bacterial endotoxin LPS or IFNG with reduction in release of cytokines IL6 and tumor necrosis factor α (TNFA/TNF-α), and the chemical messenger nitric oxide. Expression of I kappa B kinase (IKK, a client of HSP90) was also reduced, apparently leading to sequestration of nuclear factor kappa B (NFκB/NF-κB) in the cytosol and impairment of downstream proinflammatory events, in conjunction with loss of protein kinase B (AKT/PKB) (Shimp et al., 2012). Murine RAW264 macrophage cells treated with the plant flavone luteolin (an anti-inflammatory compound with reported HSP90 inhibitory properties) also demonstrated reduced AKT levels and destabilized Jun protooncogene (JUN/c-Jun), leading to suppressed release of the danger signal high mobility group B-1 protein (HMGB1) following LPS stimulation (Chen et al., 2014). Taken together, these reports convincingly demonstrate that HSP90 inhibition impinges upon multiple APC-related pathways, resulting in biological effects that extend well beyond antigen presentation.

An important role for HSP90 in inflammatory diseases was demonstrated in an extensive study that made use of multiple inflammatory and autoimmune models. In both mice and rats, symptoms were alleviated by administration of a fully synthetic HSP90 inhibitor, EC144 (Yun et al., 2011). This effect involved blockade of auto-antibody responses and suppressed activation of CD4+ T cells. Molecularly, monocyte/macrophage cells harvested from the treated animals had reduced phosphorylation of proteins in the tumor progression locus 2 (TPL2)/MEK1/2/ERK1/2 pathways. Interestingly, NFκB activation seemed unaffected. In several models of inflammatory bowel disease (IBD; Collins et al., 2014), the nonclassical HSP90 inhibitor novobiocin reduced symptoms, pathologic disease characteristics, inflammatory immune infiltrate, CD4+ T-cell activity, and levels of proinflammatory cytokines (particularly TNFA, a major mediator of IBD). T-cell receptor-activated CD4+ T cells exposed to novobiocin also showed reduced NFκB activity. Histopathology also revealed an increase in regulatory T cells (Tregs, detected as increased staining for CD4+/ $CD25^{hi}$/[forkhead box P3] FOXP3+ phenotype, which, as mentioned in Fig. 1, inhibit activity of stimulated CD8+ and other CD4+/TH1 T cells) in the inflamed areas of colon in novobiocin-treated animals. One confounding factor of this study may be the effects of novobiocin's antibacterial activity that is unrelated to its HSP90 inhibitory action.

Histone deacetylase (HDAC) inhibition has been shown to increase both the number and suppressive activity of Tregs (Tao et al., 2007), suggesting that such drugs may be useful in autoimmune inflammatory diseases. Intriguingly, HDAC6 is a known modifier of HSP90 acetylation and function (Kovacs et al., 2005), indicating that effects on HSP90 function may contribute to the immunosuppressive properties of HDAC inhibitors. Indeed, direct inhibition of HSP90 by GA or 17-AAG increases Treg numbers and functions in mice, leading to reduced murine colitis (that is T-cell-driven) as well as suppressed graft rejection (de Zoeten et al., 2011). In yet another report, 17-AAG reduced colonic inflammation and enhanced serum and colon anti-inflammatory IL10 levels in mice with IBD. Interestingly, Treg-mediated suppressive phenomena were coincident with the activation of HSF1 (Collins et al., 2013), which is released from HSP90 following the chaperone's inhibition. Thus, enhanced suppressive T-cell functions are enabled by HSP90 inhibition, reinforcing the global immune-suppressive effects of HSP90 inhibition.

The role of HSP90 in the IKK/NFκB complex deserves another look. This multisubunit super-assembly is one of the main controllers of proinflammatory signaling (Karin & Delhase, 2000). Cell stressors such as heat shock can increase levels of HSPs 70 and 90, leading to the dissociation of the IKK complex and reduced NFκB activity (Salminen, Paimela, Suuronen, & Kaarniranta, 2008). Presumably, this is due to the loss of HSP90/CDC37 association with the IKK signalosome heterocomplex, leading to IKK structural changes that render it detergent insoluble. Inhibitors such as GA cause HSP90 to release HSF1, which then becomes activated and drives expression of HSPs such as HSP27, 70, and 90 (Kim, Kang, & Kim, 1999). Thus, stabilization of IKK (and by inference, functional NFκB) by HSP90 may be impaired by the stress response, which leads to increased HSP90 levels, even as drug treatment attempts to limit HSP90 activity. Of course, during HSP90 inhibitor treatment of tumors, there may be negative consequences from loss of immune inflammatory responses against the tumor, but the inactivation of NFκB in the tumor cells could also prove beneficial.

9. TREATMENT OF TUMORS WITH HSP90 INHIBITORS: A CASE FOR DOSE-DEPENDENT IMMUNE CONSEQUENCES

As described above, the bulk of evidence from cell culture and animal models now indicates (but not unequivocally) that immune responses are reduced by HSP90 inhibition. One might therefore conclude that HSP90 inhibitors as cancer therapeutics would utterly preclude their use in antitumor immune-promoting treatment scenarios, and might even have detrimental effects if antitumor immunity was sufficiently compromised. This could be particularly problematic for cancer patients who are typically in an immune-suppressed state to begin with (Finn, 2012; Khan et al., 2015). Given the complexities of tumor–host interactions, some insights have been gained, but a definitive understanding has yet to emerge.

The effect of HSP90 inhibition upon tumor cell antigen presentation is an important but seldom considered question. If HSP90 client proteins also serve as potential antigens for immune targeting, the increased degradation of those proteins could lead to enhanced presentation of such peptides, particularly in MHC I pathways (see Fig. 3). This would imply greater opportunities for CD8+ T-cell responses. Thus, the potential for a trade-off between general immune suppression systemically versus possible increases in available antigenic peptide display deserves further study. Some examples are mentioned below.

Several combination regimens of HSP90 inhibition and antitumor vaccination have been evaluated (Lin et al., 2007). Mice carrying subcutaneous bladder cancers were treated with intratumoral injections of small quantities of GA (2.5–10 μg). These tumors expressed (human) epidermal growth factor receptor 2 (EGFR2/HER2/neu), which was the target antigen of a plasmid DNA vaccine used for specific immunotherapy of the tumors. In the absence of immunotherapy, GA treatment at the "high" dose (10 μg) reduced tumor growth, but the "low" dose (2.5 μg) resulted in diminished HER2 expression with no decrease in tumor size. Addition of the DNA vaccine to the low GA dose resulted in significant improvement in overall survival, which was dependent upon CD8+ T cytolytic cells, and was augmented by depletion of CD4+ T cells (likely reducing the Treg population and allowing more CD8+ killer activity). The authors proposed that HSP90 inhibition led to greater proteasomal degradation of HER2, leading to enhanced MHC display of antigenic peptides, which were then targeted by the vaccine-generated CD8+ T cells. In support of this explanation, enhanced T and NK lymphocyte infiltration into tumors was noted.

In another example, 17-DMAG was used (oral administration, ~15 mg/kg) to treat mice bearing syngeneic sarcomas expressing ephrin type-A receptor 2 (EPHA2), alone or in combination with a DC-EPHA2 peptide vaccine (Rao et al., 2012). 17-DMAG alone led to increased immune infiltrate (with anti-EPHA2 phenotypes) into the tumor, coincident with degradation of EPHA2. In tandem, myeloid-derived suppressor cells (MDSCs) and Tregs were reduced in the tumor microenvironment. MDSCs are another tumor-driven immune-regulatory cell type that suppresses T-cell and NK-cell responses. Combination treatment with the EPHA2-targeted vaccine eradicated tumors (as did adoptive transfer of anti-EPHA2 CD8+ T cells). Intriguingly, this regimen also benefitted mice bearing EPHA2-negative tumors, because the EPHA2+ vasculature was targeted. Of interest, this report demonstrated a reduction of Tregs (and MDSCs) in the tumor microenvironment. It is not clear whether total Treg numbers were reduced, or whether there might have been alterations in the microenvironment conducive to cytotoxic T-cell responses at the expense of Treg infiltration. As mentioned above, increases in MHC expression with tumor antigens displayed may have been a key factor.

In these cited examples, HSP90 inhibitor doses were relatively low (and delivered locally in one setting), suggesting that low drug exposures may generate a more immune-responsive environment within the tumor compared to the systemic immune effects elicited by higher doses. This suggests that partial HSP90 inhibition achieved with lower drug doses preferentially allows for degradation of client proteins, and enhanced antigen presentation, while preserving components required for an antitumor-responsive environment. There may be a trade-off in terms of increased antigen "production" versus the need for HSP90 to chaperone proteasome-derived peptides (where there is likely task-based redundancy with other HSPs). It is conceivable that the sensitivity to HSP90 inhibition of metastable client chaperoning is greater than that of other aspects of HSP90 function (e.g., stabilization of proteasome and MHC molecules), and that the preferential accumulation demonstrated for inhibitors within tumors spares systemic inhibitory effects of drug upon infiltrating immune cells. It is also possible that some of the degraded HSP90 client proteins include those involved in tumor microenvironment modulation. This is evident for the vascular endothelial growth factor/-receptor (VEGF/VEGF-R) angiogenesis loop (Sanderson et al., 2006), and undoubtedly extends to other microenvironmental factors (Alarcon et al., 2012).

Of the innate immune effectors, NK cells, as noted above (Bae et al., 2013), show reduced activation markers when exposed to HSP90 inhibitors. On the other hand, treatment of tumor cells may actually increase their expression of NK-cell recognition ligands. This scenario was shown for human Hodgkin's lymphoma cells (Boll et al., 2009) treated with the fully synthetic HSP90 inhibitor BIIB021 (CNF2024) (Lundgren et al., 2009). Treated cells increased expression of NK activation ligands (Killer cell lectin-like receptor subfamily K member 1/NKG2-D type II integral membrane protein/CD314/NKG2D) and MHC class I polypeptide-related sequences A and B (MICA and MICB, all mentioned in Fig. 1), which apparently was sufficient for heightened NK-cell targeting. Loss of NFκB activation was also observed irrespective of the IKB mutational state of particular tumor cells. Similar results have been reported in multiple myeloma cell lines using 17-AAG and radicicol as HSP90 inhibitors (Fionda et al., 2009). HSF1 activation upon HSP90 inhibition was considered an important event in the expression of MICA/B. Unfortunately, neither of these studies included *in vivo* treatment of mice with intact NK-cell function. Nonetheless, it is evident that HSP90 inhibitors, perhaps via downregulation of NFκB signaling, or upregulation of HSF1 activity, can increase the level of NK-cell target ligands on tumor cells.

10. CLOSING THOUGHTS

10.1 HSP90 as an Immune Target, as a Drug Target, and at the Intersection of Chemotherapy and Immunotherapy

As discussed near the beginning of this review, HSP90 was first linked to immunity by virtue of its activity as an anticancer vaccine. While that line of research initiated an entire field of chaperone-based cancer immunotherapies, the presumed HSP90–peptide complexes were generally found less effective as anticancer vaccines compared to tumor-derived HSP70, GRP94, or calreticulin (Graner et al., 2000; Udono & Srivastava, 1994). With only a few exceptions since this early work (Corigliano et al., 2013; Kurotaki et al., 2007; Mori et al., 2012), HSP90 has largely been ignored from the perspective of single-protein cancer vaccines. This may be due in part to the fact that its intracellular peptide-binding capacity is relatively low (HSP90 is likely more frequently engaged with whole client proteins or complexed with entire ensembles of other proteins; Schwenkert, Hugel, & Cox, 2014). Others have suggested that the low intrinsic ATPase activity of HSP90 prevents it from efficiently transferring peptide to other molecules (Udono & Srivastava, 1994), but it is unclear if any of these properties are truly responsible for the differential responses that have been reported. Another issue is that purification of HSP90 is considered somewhat more difficult than that of HSP70 or GRP94 (Graner et al., 2000; Menoret & Bell, 2000; Srivastava, 1997). HSP90 is a component of more complex heat shock protein-based vaccines (e.g., chaperone-rich cell lysate; Bu et al., 2015; Graner, Lillehei, & Katsanis, 2014; Graner, Zeng, Feng, &

Katsanis, 2003), but there has not been further development of HSP90 as a vaccine or carrier of antigens. Nevertheless, early studies initiated the concept that HSP90 could participate in antigen processing, a notion reinforced by subsequent use of HSP90 inhibitors.

Over the last decade at least 16 different HSP90-targeting compounds have been tested in clinical trials (http://clinicaltrials.gov) and the development of several continues to be actively pursued. There are also more molecules in preclinical development pipelines, and the role of HSP90 in cancer and other diseases remains of therapeutic interest. The use of HSP90 inhibitors as chemical biological probes of HSP90 function has dramatically contributed to our better understanding of this molecule's activities. In relation to numerous aspects of the immune response, HSP90 inhibitors generally appear a promising approach to achieve therapeutic immune suppression in inflammatory disorders and autoimmune diseases, clinical indications predicted to be quite valuable. Conversely, these same suppressive actions may exert detrimental effects with respect to generation of anticancer immune responses. However, the pleiotropic effects of HSP90 inhibition may actually permit its use at low levels to augment immune responses at the level of the tumor through mechanisms that include enhanced breakdown of tumor antigen targets, increased density of MHC display, NFκB inhibition with downstream effects in the tumor microenvironment, and increased expression of NK-cell targets. Curiously, in the setting of stem cell/bone marrow transplant, inhibition of HSP90 (perhaps combined with AKT/PI3K inhibition) may be valuable in preventing graft versus host disease by suppressing alloreactive T cells and protecting sensitive tissues such as intestinal epithelial cells (Berges et al., 2015; Joly et al., 2015). In this case, the immune-suppressive effects of HSP90 inhibition could actually serve the purpose of a cancer immunotherapy.

In reports where immune responses were improved in cancer models following HSP90 inhibitor treatment, it is worth noting that the doses of inhibitor were relatively low. These data suggest that low drug doses—perhaps in conjunction with preferential drug accumulation in the tumor—might be sufficient to drive molecular changes in the tumors (benefiting immunity) without leading to systemic immune impairment. This area of research appears understudied, and represents a gap in our knowledge of combination therapies. Potential benefits of sustained low-level HSP90 inhibition have been previously highlighted by Whitesell et al. (2014) where it was shown that low nanomolar concentrations of HSP90 inhibitors could prevent evolution of breast tumor cells toward hormone therapy resistance. One might envision similar "resistance" phenomena (i.e., tumor-induced immune suppression) to be downmodulated during low-dose inhibitor treatment, as drug resistance and immune suppression may be linked (Baguley, 2010; Denysenko et al., 2010), particularly in the tumor microenvironment. Because cancer immunotherapies are developing into viable treatment modalities (Datta et al., 2015; de Aquino, Malhotra, Mishra, & Shanker, 2015; Domogala, Madrigal, & Saudemont, 2015; Klinke, 2015; Pico de Coana, Choudhury, & Kiessling, 2015), and HSP90 has a long, albeit complicated history in immunity, the convergence of these therapeutic approaches deserves attention, and may provide substantial hope for improving care for patients with cancer.

REFERENCES

Alarcon, S. V., Mollapour, M., Lee, M. J., Tsutsumi, S., Lee, S., Kim, Y. S., et al. (2012). Tumor-intrinsic and tumor-extrinsic factors impacting hsp90- targeted therapy. *Current Molecular Medicine*, *12*(9), 1125–1141. Retrieved from, http://www.ncbi.nlm.nih.gov/pubmed/22804236.

Andocs, G., Meggyeshazi, N., Balogh, L., Spisak, S., Maros, M. E., Balla, P., et al. (2015). Upregulation of heat shock proteins and the promotion of damage-associated molecular pattern signals in a colorectal cancer model by modulated electrohyperthermia. *Cell Stress & Chaperones*, *20*(1), 37–46. http://dx.doi.org/10.1007/s12192-014-0523-6.

Bae, J., Mitsiades, C., Tai, Y. T., Bertheau, R., Shammas, M., Batchu, R. B., et al. (2007). Phenotypic and functional effects of heat shock protein 90 inhibition on dendritic cell. *Journal of Immunology*, *178*(12), 7730–7737. Retrieved from, http://www.ncbi.nlm.nih.gov/pubmed/17548610.

Bae, J., Munshi, A., Li, C., Samur, M., Prabhala, R., Mitsiades, C., et al. (2013). Heat shock protein 90 is critical for regulation of phenotype and functional activity of human T lymphocytes and NK cells. *Journal of Immunology*, *190*(3), 1360–1371. http://dx.doi.org/10.4049/jimmunol.1200593.

Baguley, B. C. (2010). Multiple drug resistance mechanisms in cancer. *Molecular Biotechnology*, *46*(3), 308–316. http://dx.doi.org/10.1007/s12033-010-9321-2.

Basu, S., Binder, R. J., Suto, R., Anderson, K. M., & Srivastava, P. K. (2000). Necrotic but not apoptotic cell death releases heat shock proteins, which deliver a partial maturation signal to dendritic cells and activate the NF-kappa B pathway. *International Immunology*, *12*(11), 1539–1546. Retrieved from, http://www.ncbi.nlm.nih.gov/pubmed/11058573.

Basu, S., & Srivastava, P. K. (2003). Fever-like temperature induces maturation of dendritic cells through induction of hsp90. *International Immunology*, *15*(9), 1053–1061. Retrieved from, http://www.ncbi.nlm.nih.gov/pubmed/12917257.

Berges, C., Bedke, T., Stuehler, C., Khanna, N., Zehnter, S., Kruhm, M., et al. (2015). Combined PI3K/Akt and Hsp90 targeting synergistically suppresses essential functions of alloreactive T cells and increases Tregs. *Journal of Leukocyte Biology*. http://dx.doi.org/10.1189/jlb.5A0814-413R.

Bhat, R., Tummalapalli, S. R., & Rotella, D. P. (2014). Progress in the discovery and development of heat shock protein 90 (Hsp90) inhibitors. *Journal of Medicinal Chemistry*, *57*(21), 8718–8728. http://dx.doi.org/10.1021/jm500823a.

Boll, B., Eltaib, F., Reiners, K. S., von Tresckow, B., Tawadros, S., Simhadri, V. R., et al. (2009). Heat shock protein 90 inhibitor BIIB021 (CNF2024) depletes NF-kappaB and sensitizes Hodgkin's lymphoma cells for natural killer cell-mediated cytotoxicity. *Clinical Cancer Research*, *15*(16), 5108–5116. http://dx.doi.org/10.1158/1078-0432.CCR-09-0213.

Bu, N., Wu, H., Zhang, G., Zhan, S., Zhang, R., Sun, H., et al. (2015). Exosomes from dendritic cells loaded with chaperone-rich cell lysates elicit a potent T cell immune response against intracranial glioma in mice. *Journal of Molecular Neuroscience*, *56*(3), 631–643. http://dx.doi.org/10.1007/s12031-015-0506-9.

Callahan, M. K., Garg, M., & Srivastava, P. K. (2008). Heat-shock protein 90 associates with N-terminal extended peptides and is required for direct and indirect antigen presentation. *Proceedings of the National Academy of Sciences of the United States of America*, *105*(5), 1662–1667. http://dx.doi.org/10.1073/pnas.0711365105.

Castano, A. P., Mroz, P., & Hamblin, M. R. (2006). Photodynamic therapy and anti-tumour immunity. *Nature Reviews Cancer*, *6*(7), 535–545. http://dx.doi.org/10.1038/nrc1894.

Chen, D., Bi, A., Dong, X., Jiang, Y., Rui, B., Liu, J., et al. (2014). Luteolin exhibits anti-inflammatory effects by blocking the activity of heat shock protein 90 in macrophages. *Biochemical and Biophysical Research Communications*, *443*(1), 326–332. http://dx.doi.org/10.1016/j.bbrc.2013.11.122.

Chen, W., Wang, J., Shao, C., Liu, S., Yu, Y., Wang, Q., et al. (2006). Efficient induction of antitumor T cell immunity by exosomes derived from heat-shocked lymphoma cells. *European Journal of Immunology*, *36*(6), 1598–1607. http://dx.doi.org/10.1002/eji.200535501.

Cid, C., Regidor, I., Poveda, P. D., & Alcazar, A. (2009). Expression of heat shock protein 90 at the cell surface in human neuroblastoma cells. *Cell Stress & Chaperones*, *14*(3), 321–327. http://dx.doi.org/10.1007/s12192-008-0076-7.

Collins, C. B., Aherne, C. M., Yeckes, A., Pound, K., Eltzschig, H. K., Jedlicka, P., et al. (2013). Inhibition of N-terminal ATPase on HSP90 attenuates colitis through enhanced Treg function. *Mucosal Immunology*, *6*(5), 960–971. http://dx.doi.org/10.1038/mi.2012.134.

Collins, C. B., Strassheim, D., Aherne, C. M., Yeckes, A. R., Jedlicka, P., & de Zoeten, E. F. (2014). Targeted inhibition of heat shock protein 90 suppresses tumor necrosis factor-alpha and ameliorates murine intestinal inflammation. *Inflammatory Bowel Diseases*, *20*(4), 685–694. http://dx.doi.org/10.1097/01.MIB.0000442839.28664.75.

Corigliano, M. G., Fenoy, I., Sander, V., Maglioco, A., Goldman, A., & Clemente, M. (2013). Plant heat shock protein 90 as carrier-adjuvant for immunization against a reporter antigen. *Vaccine*, *31*(49), 5872–5878. http://dx.doi.org/10.1016/j.vaccine.2013.09.047.

Dai, S., Wan, T., Wang, B., Zhou, X., Xiu, F., Chen, T., et al. (2005). More efficient induction of HLA-A*0201-restricted and carcinoembryonic antigen (CEA)-specific CTL response by immunization with exosomes prepared from heat-stressed CEA-positive tumor cells. *Clinical Cancer Research*, *11*(20), 7554–7563. http://dx.doi.org/10.1158/1078-0432.CCR-05-0810.

Datta, J., Berk, E., Cintolo, J. A., Xu, S., Roses, R. E., & Czerniecki, B. J. (2015). Rationale for a multimodality strategy to enhance the efficacy of dendritic cell-based cancer immunotherapy. *Frontiers in Immunology*, *6*, 271. http://dx.doi.org/10.3389/fimmu.2015.00271.

de Aquino, M. T., Malhotra, A., Mishra, M. K., & Shanker, A. (2015). Challenges and future perspectives of T cell immunotherapy in cancer. *Immunology Letters*, *166*(2), 117–133. http://dx.doi.org/10.1016/j.imlet.2015.05.018.

de Zoeten, E. F., Wang, L., Butler, K., Beier, U. H., Akimova, T., Sai, H., et al. (2011). Histone deacetylase 6 and heat shock protein 90 control the functions of Foxp3(+) T-regulatory cells. *Molecular and Cellular Biology*, *31*(10), 2066–2078. http://dx.doi.org/10.1128/MCB.05155-11.

Demine, R., & Walden, P. (2005). Testing the role of gp96 as peptide chaperone in antigen processing. *The Journal of Biological Chemistry*, *280*(18), 17573–17578. http://dx.doi.org/10.1074/jbc.M501233200.

Denysenko, T., Gennero, L., Roos, M. A., Melcarne, A., Juenemann, C., Faccani, G., et al. (2010). Glioblastoma cancer stem cells: Heterogeneity, microenvironment and related therapeutic strategies. *Cell Biochemistry and Function*, *28*(5), 343–351. http://dx.doi.org/10.1002/cbf.1666.

Di Nicola, M., Zappasodi, R., Carlo-Stella, C., Mortarini, R., Pupa, S. M., Magni, M., et al. (2009). Vaccination with autologous tumor-loaded dendritic cells induces clinical and immunologic responses in indolent B-cell lymphoma patients with relapsed and measurable disease: A pilot study. *Blood*, *113*(1), 18–27. http://dx.doi.org/10.1182/blood-2008-06-165654.

Didelot, C., Lanneau, D., Brunet, M., Joly, A. L., De Thonel, A., Chiosis, G., et al. (2007). Anti-cancer therapeutic approaches based on intracellular and extracellular heat shock proteins. *Current Medicinal Chemistry*, *14*(27), 2839–2847. Retrieved from, http://www.ncbi.nlm.nih.gov/pubmed/18045130.

Domogala, A., Madrigal, J. A., & Saudemont, A. (2015). Natural killer cell immunotherapy: From bench to bedside. *Frontiers in Immunology*, *6*, 264. http://dx.doi.org/10.3389/fimmu.2015.00264.

Dubaniewicz, A. (2013). Microbial and human heat shock proteins as 'danger signals' in sarcoidosis. *Human Immunology*, *74*(12), 1550–1558. http://dx.doi.org/10.1016/j.humimm.2013.08.275.

Epple, L. M., Griffiths, S. G., Dechkovskaia, A. M., Dusto, N. L., White, J., Ouellette, R. J., et al. (2012). Medulloblastoma exosome proteomics yield functional roles for extracellular vesicles. *PloS One*, 7(7), e42064. http://dx.doi.org/10.1371/journal.pone.0042064. PONE-D-11-08927 [pii].

Erkeller-Yuksel, F. M., Isenberg, D. A., Dhillon, V. B., Latchman, D. S., & Lydyard, P. M. (1992). Surface expression of heat shock protein 90 by blood mononuclear cells from patients with systemic lupus erythematosus. *Journal of Autoimmunity*, *5*(6), 803–814. Retrieved from, http://www.ncbi.nlm.nih.gov/pubmed/1489490.

Eustace, B. K., & Jay, D. G. (2004). Extracellular roles for the molecular chaperone, hsp90. *Cell Cycle*, *3*(9), 1098–1100. Retrieved from, http://www.ncbi.nlm.nih.gov/entrez/query.fcgi?cmd=Retrieve&db=PubMed&dopt=Citation&list_uids=15326368.

Eustace, B. K., Sakurai, T., Stewart, J. K., Yimlamai, D., Unger, C., Zehetmeier, C., et al. (2004). Functional proteomic screens reveal an essential extracellular role for hsp90 alpha in cancer cell invasiveness. *Nature Cell Biology*, *6*(6), 507–514. http://dx.doi.org/10.1038/ncb1131.

Feng, H., Zeng, Y., Graner, M. W., & Katsanis, E. (2002). Stressed apoptotic tumor cells stimulate dendritic cells and induce specific cytotoxic T cells. *Blood*, *100*(12), 4108–4115. http://dx.doi.org/10.1182/blood-2002-05-1389.

Feng, H., Zeng, Y., Graner, M. W., Likhacheva, A., & Katsanis, E. (2003). Exogenous stress proteins enhance the immunogenicity of apoptotic tumor cells and stimulate antitumor immunity. *Blood*, *101*(1), 245–252. http://dx.doi.org/10.1182/blood-2002-05-1580.

Feng, H., Zeng, Y., Whitesell, L., & Katsanis, E. (2001). Stressed apoptotic tumor cells express heat shock proteins and elicit tumor-specific immunity. *Blood*, *97*(11), 3505–3512. Retrieved from, http://www.ncbi.nlm.nih.gov/pubmed/11369644.

Ferrarini, M., Heltai, S., Zocchi, M. R., & Rugarli, C. (1992). Unusual expression and localization of heat-shock proteins in human tumor cells. *International Journal of Cancer*, *51*(4), 613–619. Retrieved from, http://www.ncbi.nlm.nih.gov/pubmed/1601523.

Finn, O. J. (2012). Immuno-oncology: Understanding the function and dysfunction of the immune system in cancer. *Annals of Oncology*, *23*(Suppl. 8), viii6–viii9. http://dx.doi.org/10.1093/annonc/mds256.

Fionda, C., Soriani, A., Malgarini, G., Iannitto, M. L., Santoni, A., & Cippitelli, M. (2009). Heat shock protein-90 inhibitors increase MHC class I-related chain A and B ligand expression on multiple myeloma cells and their ability to trigger NK cell degranulation. *Journal of Immunology*, *183*(7), 4385–4394. http://dx.doi.org/10.4049/jimmunol.0901797.

Flynn, G. C., Pohl, J., Flocco, M. T., & Rothman, J. E. (1991). Peptide-binding specificity of the molecular chaperone BiP. *Nature*, *353*(6346), 726–730. http://dx.doi.org/10.1038/353726a0.

Fredly, H., Ersvaer, E., Gjertsen, B. T., & Bruserud, O. (2011). Immunogenic apoptosis in human acute myeloid leukemia (AML): Primary human AML cells expose calreticulin and release heat shock protein (HSP) 70 and HSP90 during apoptosis. *Oncology Reports*, *25*(6), 1549–1556. http://dx.doi.org/10.3892/or.2011.1229.

Gallucci, S., & Matzinger, P. (2001). Danger signals: SOS to the immune system. *Current Opinion in Immunology*, *13*(1), 114–119. Retrieved from, http://www.ncbi.nlm.nih.gov/pubmed/11154927.

Graner, M. W., Alzate, O., Dechkovskaia, A. M., Keene, J. D., Sampson, J. H., Mitchell, D. A., et al. (2009). Proteomic and immunologic analyses of brain tumor exosomes. *The FASEB Journal*, *23*(5), 1541–1557. http://dx.doi.org/10.1096/fj.08-122184.

Graner, M. W., & Bigner, D. D. (2006). Therapeutic aspects of chaperones/heat-shock proteins in neuro-oncology. *Expert Review of Anticancer Therapy*, *6*(5), 679–695. http://dx.doi.org/10.1586/14737140.6.5.679.

Graner, M. W., Cumming, R. I., & Bigner, D. D. (2007). The heat shock response and chaperones/heat shock proteins in brain tumors: Surface expression, release, and possible immune consequences. *The Journal of Neuroscience*, *27*(42), 11214–11227. http://dx.doi.org/10.1523/JNEUROSCI.3588-07.2007. 27/42/11214 [pii].

Graner, M. W., Lillehei, K. O., & Katsanis, E. (2014). Endoplasmic reticulum chaperones and their roles in the immunogenicity of cancer vaccines. *Frontiers in Oncology*, *4*, 379. http://dx.doi.org/10.3389/fonc.2014.00379.

Graner, M., Raymond, A., Romney, D., He, L., Whitesell, L., & Katsanis, E. (2000). Immunoprotective activities of multiple chaperone proteins isolated from murine B-cell leukemia/lymphoma. *Clinical Cancer Research*, *6*(3), 909–915. Retrieved from, http://www.ncbi.nlm.nih.gov/entrez/query.fcgi?cmd=Retrieve&db=PubMed&dopt=Citation&list_uids=10741715.

Graner, M. W., Zeng, Y., Feng, H., & Katsanis, E. (2003). Tumor-derived chaperone-rich cell lysates are effective therapeutic vaccines against a variety of cancers. *Cancer Immunology, Immunotherapy*, *52*(4), 226–234. http://dx.doi.org/10.1007/s00262-002-0359-2.

Hance, M. W., Dole, K., Gopal, U., Bohonowych, J. E., Jezierska-Drutel, A., Neumann, C. A., et al. (2012). Secreted Hsp90 is a novel regulator of the epithelial to mesenchymal transition (EMT) in prostate cancer. *The Journal of Biological Chemistry*, *287*(45), 37732–37744. http://dx.doi.org/10.1074/jbc.M112.389015.

He, Y., Ge, H., & Li, S. (2012). Haematoporphyrin based photodynamic therapy combined with hyperthermia provided effective therapeutic vaccine effect against colon cancer growth in mice. *International Journal of Medical Sciences*, *9*(8), 627–633. http://dx.doi.org/10.7150/ijms.4865.

Hu, Z., Yang, X. Y., Liu, Y., Morse, M. A., Lyerly, H. K., Clay, T. M., et al. (2005). Release of endogenous danger signals from HIFU-treated tumor cells and their stimulatory effects on APCs. *Biochemical and Biophysical Research Communications*, *335*(1), 124–131. http://dx.doi.org/10.1016/j.bbrc.2005.07.071.

Ichiyanagi, T., Imai, T., Kajiwara, C., Mizukami, S., Nakai, A., Nakayama, T., et al. (2010). Essential role of endogenous heat shock protein 90 of dendritic cells in antigen cross-presentation. *Journal of Immunology, 185*(5), 2693–2700. http://dx.doi.org/10.4049/jimmunol.1000821.

Imai, T., Kato, Y., Kajiwara, C., Mizukami, S., Ishige, I., Ichiyanagi, T., et al. (2011). Heat shock protein 90 (HSP90) contributes to cytosolic translocation of extracellular antigen for cross-presentation by dendritic cells. *Proceedings of the National Academy of Sciences of the United States of America, 108*(39), 16363–16368. http://dx.doi.org/10.1073/pnas.1108372108.

Ishii, T., Udono, H., Yamano, T., Ohta, H., Uenaka, A., Ono, T., et al. (1999). Isolation of MHC class I-restricted tumor antigen peptide and its precursors associated with heat shock proteins hsp70, hsp90, and gp96. *Journal of Immunology, 162*(3), 1303–1309. Retrieved from, http://www.ncbi.nlm.nih.gov/pubmed/9973383.

Ito, A., Honda, H., & Kobayashi, T. (2006). Cancer immunotherapy based on intracellular hyperthermia using magnetite nanoparticles: A novel concept of "heat-controlled necrosis" with heat shock protein expression. *Cancer Immunology, Immunotherapy, 55*(3), 320–328. http://dx.doi.org/10.1007/s00262-005-0049-y.

Jhaveri, K., Ochiana, S. O., Dunphy, M. P., Gerecitano, J. F., Corben, A. D., Peter, R. I., et al. (2014). Heat shock protein 90 inhibitors in the treatment of cancer: Current status and future directions. *Expert Opinion on Investigational Drugs, 23*(5), 611–628. http://dx.doi.org/10.1517/13543784.2014.902442.

Joly, A. L., Deepti, A., Seignez, A., Goloudina, A., Hebrard, S., Schmitt, E., et al. (2015). The HSP90 inhibitor, 17AAG, protects the intestinal stem cell niche and inhibits graft versus host disease development. *Oncogene*. http://dx.doi.org/10.1038/onc.2015.242.

Karin, M., & Delhase, M. (2000). The I kappa B kinase (IKK) and NF-kappa B: Key elements of proinflammatory signalling. *Seminars in Immunology, 12*(1), 85–98. http://dx.doi.org/10.1006/smim.2000.0210.

Khan, H., Gucalp, R., & Shapira, I. (2015). Evolving concepts: Immunity in oncology from targets to treatments. *Journal of Oncology, 2015*, 847383. http://dx.doi.org/10.1155/2015/847383.

Kim, H. R., Kang, H. S., & Kim, H. D. (1999). Geldanamycin induces heat shock protein expression through activation of HSF1 in K562 erythroleukemic cells. *IUBMB Life, 48*(4), 429–433. http://dx.doi.org/10.1080/713803536.

Kim, T., Keum, G., & Pae, A. N. (2013). Discovery and development of heat shock protein 90 inhibitors as anticancer agents: A review of patented potent geldanamycin derivatives. *Expert Opinion on Therapeutic Patents, 23*(8), 919–943. http://dx.doi.org/10.1517/13543776.2013.780597.

Klinke, D. J., 2nd. (2015). Enhancing the discovery and development of immunotherapies for cancer using quantitative and systems pharmacology: Interleukin-12 as a case study. *Journal for Immunotherapy of Cancer, 3*, 27. http://dx.doi.org/10.1186/s40425-015-0069-x.

Korbelik, M., Sun, J., & Cecic, I. (2005). Photodynamic therapy-induced cell surface expression and release of heat shock proteins: Relevance for tumor response. *Cancer Research, 65*(3), 1018–1026. Retrieved from, http://www.ncbi.nlm.nih.gov/pubmed/15705903.

Kovacs, J. J., Murphy, P. J., Gaillard, S., Zhao, X., Wu, J. T., Nicchitta, C. V., et al. (2005). HDAC6 regulates Hsp90 acetylation and chaperone-dependent activation of glucocorticoid receptor. *Molecular Cell, 18*(5), 601–607. http://dx.doi.org/10.1016/j.molcel.2005.04.021.

Kunisawa, J., & Shastri, N. (2006). Hsp90alpha chaperones large C-terminally extended proteolytic intermediates in the MHC class I antigen processing pathway. *Immunity, 24*(5), 523–534. http://dx.doi.org/10.1016/j.immuni.2006.03.015.

Kurotaki, T., Tamura, Y., Ueda, G., Oura, J., Kutomi, G., Hirohashi, Y., et al. (2007). Efficient cross-presentation by heat shock protein 90-peptide complex-loaded dendritic cells via an endosomal pathway. *Journal of Immunology, 179*(3), 1803–1813. Retrieved from, http://www.ncbi.nlm.nih.gov/pubmed/17641047.

Lammert, E., Arnold, D., Rammensee, H. G., & Schild, H. (1996). Expression levels of stress protein gp96 are not limiting for major histocompatibility complex class I-restricted antigen presentation. *European Journal of Immunology, 26*(4), 875–879. http://dx.doi.org/10.1002/eji.1830260423.

Li, W., Mu, D., Tian, F., Hu, Y., Jiang, T., Han, Y., et al. (2013). Exosomes derived from Rab27a-overexpressing tumor cells elicit efficient induction of antitumor immunity. *Molecular Medicine Reports, 8*(6), 1876–1882. http://dx.doi.org/10.3892/mmr.2013.1738.

Li, R., Qian, J., Zhang, W., Fu, W., Du, J., Jiang, H., et al. (2014). Human heat shock protein-specific cytotoxic T lymphocytes display potent antitumour immunity in multiple myeloma. *British Journal of Haematology, 166*(5), 690–701. http://dx.doi.org/10.1111/bjh.12943.

Li, W., Tsen, F., Sahu, D., Bhatia, A., Chen, M., Multhoff, G., et al. (2013). Extracellular Hsp90 (eHsp90) as the actual target in clinical trials: Intentionally or unintentionally. *International Review of Cell and Molecular Biology, 303*, 203–235. http://dx.doi.org/10.1016/B978-0-12-407697-6.00005-2.

Lin, C. C., Tu, C. F., Yen, M. C., Chen, M. C., Hsieh, W. J., Chang, W. C., et al. (2007). Inhibitor of heat-shock protein 90 enhances the antitumor effect of DNA vaccine targeting clients of heat-shock protein. *Molecular Therapy, 15*(2), 404–410. http://dx.doi.org/10.1038/sj.mt.6300014.

Livingstone, A. M., & Fathman, C. G. (1987). The structure of T-cell epitopes. *Annual Review of Immunology, 5*, 477–501. http://dx.doi.org/10.1146/annurev.iy.05.040187.002401.

Lundgren, K., Zhang, H., Brekken, J., Huser, N., Powell, R. E., Timple, N., et al. (2009). BIIB021, an orally available, fully synthetic small-molecule inhibitor of the heat shock protein Hsp90. *Molecular Cancer Therapeutics, 8*(4), 921–929. http://dx.doi.org/10.1158/1535-7163.MCT-08-0758.

Luo, L. Y., Herrera, I., Soosaipillai, A., & Diamandis, E. P. (2002). Identification of heat shock protein 90 and other proteins as tumour antigens by serological screening of an ovarian carcinoma expression library. *British Journal of Cancer, 87*(3), 339–343. http://dx.doi.org/10.1038/sj.bjc.6600439.

Lv, L. H., Wan, Y. L., Lin, Y., Zhang, W., Yang, M., Li, G. L., et al. (2012). Anticancer drugs cause release of exosomes with heat shock proteins from human hepatocellular carcinoma cells that elicit effective natural killer cell antitumor responses in vitro. *The Journal of Biological Chemistry, 287*(19), 15874–15885. http://dx.doi.org/10.1074/jbc.M112.340588.

Maki, R. G., Eddy, R. L., Jr., Byers, M., Shows, T. B., & Srivastava, P. K. (1993). Mapping of the genes for human endoplasmic reticular heat shock protein gp96/grp94. *Somatic Cell and Molecular Genetics, 19*(1), 73–81. Retrieved from, http://www.ncbi.nlm.nih.gov/pubmed/8460400.

Maki, R. G., Old, L. J., & Srivastava, P. K. (1990). Human homologue of murine tumor rejection antigen gp96: 5'-regulatory and coding regions and relationship to stress-induced proteins. *Proceedings of the National Academy of Sciences of the United States of America, 87*(15), 5658–5662. Retrieved from, http://www.ncbi.nlm.nih.gov/pubmed/2377606.

McCready, J., Sims, J. D., Chan, D., & Jay, D. G. (2010). Secretion of extracellular hsp90alpha via exosomes increases cancer cell motility: A role for plasminogen activation. *BMC Cancer, 10*, 294. http://dx.doi.org/10.1186/1471-2407-10-294.

Menoret, A., & Bell, G. (2000). Purification of multiple heat shock proteins from a single tumor sample. *Journal of Immunological Methods, 237*(1–2), 119–130. Retrieved from, http://www.ncbi.nlm.nih.gov/pubmed/10725457.

Mori, T., Nishizawa, S., Hirohashi, Y., Torigoe, T., Tamura, Y., Takahashi, A., et al. (2012). Efficiency of G2/M-related tumor-associated antigen-targeting cancer immunotherapy depends on antigen expression in the cancer stem-like population. *Experimental and Molecular Pathology*, *92*(1), 27–32. http://dx.doi.org/10.1016/j.yexmp.2011.09.016.

Multhoff, G., Pockley, A. G., Streffer, C., & Gaipl, U. S. (2012). Dual role of heat shock proteins (HSPs) in anti-tumor immunity. *Current Molecular Medicine*, *12*(9), 1174–1182. Retrieved from, http://www.ncbi.nlm.nih.gov/pubmed/22804240.

Murshid, A., Gong, J., & Calderwood, S. K. (2010). Heat shock protein 90 mediates efficient antigen cross presentation through the scavenger receptor expressed by endothelial cells-I. *Journal of Immunology*, *185*(5), 2903–2917. http://dx.doi.org/10.4049/jimmunol.0903635.

Murshid, A., Gong, J., & Calderwood, S. K. (2014). Hsp90-peptide complexes stimulate antigen presentation through the class II pathway after binding scavenger receptor SREC-I. *Immunobiology*, *219*(12), 924–931. http://dx.doi.org/10.1016/j.imbio.2014.08.001.

Nolan, K. D., Franco, O. E., Hance, M. W., Hayward, S. W., & Isaacs, J. S. (2015). Tumor-secreted Hsp90 subverts polycomb function to drive prostate tumor growth and invasion. *The Journal of Biological Chemistry*, *290*(13), 8271–8282. http://dx.doi.org/10.1074/jbc.M115.637496.

Ostrowski, M., Carmo, N. B., Krumeich, S., Fanget, I., Raposo, G., Savina, A., et al. (2010). Rab27a and Rab27b control different steps of the exosome secretion pathway. *Nature Cell Biology*, *12*(1), 19–30. http://dx.doi.org/10.1038/ncb2000. sup pp 11–13.

Oushy, S., & Graner, M. W. (2015). Immunology and immunotherapy in brain tumors—Immune failure and potential counteractions. In T. Lichtor (Ed.), *Molecular considerations and evolving surgical management issues in the treatment of patients with a brain tumor*. Rijeka, Croatia: InTech.

Palucka, K., & Banchereau, J. (2013a). Dendritic-cell-based therapeutic cancer vaccines. *Immunity*, *39*(1), 38–48. http://dx.doi.org/10.1016/j.immuni.2013.07.004.

Palucka, K., & Banchereau, J. (2013b). Human dendritic cell subsets in vaccination. *Current Opinion in Immunology*, *25*(3), 396–402. http://dx.doi.org/10.1016/j.coi.2013.05.001.

Panzarini, E., Inguscio, V., Fimia, G. M., & Dini, L. (2014). Rose Bengal acetate photodynamic therapy (RBAc-PDT) induces exposure and release of damage-associated molecular patterns (DAMPs) in human HeLa cells. *PloS One*, *9*(8), e105778. http://dx.doi.org/10.1371/journal.pone.0105778.

Pico de Coana, Y., Choudhury, A., & Kiessling, R. (2015). Checkpoint blockade for cancer therapy: Revitalizing a suppressed immune system. *Trends in Molecular Medicine*, *21*(8), 482–491. http://dx.doi.org/10.1016/j.molmed.2015.05.005.

Rajagopal, D., Bal, V., Mayor, S., George, A., & Rath, S. (2006). A role for the Hsp90 molecular chaperone family in antigen presentation to T lymphocytes via major histocompatibility complex class II molecules. *European Journal of Immunology*, *36*(4), 828–841. http://dx.doi.org/10.1002/eji.200535326.

Rao, A., Taylor, J. L., Chi-Sabins, N., Kawabe, M., Gooding, W. E., & Storkus, W. J. (2012). Combination therapy with HSP90 inhibitor 17-DMAG reconditions the tumor microenvironment to improve recruitment of therapeutic T cells. *Cancer Research*, *72*(13), 3196–3206. http://dx.doi.org/10.1158/0008-5472.CAN-12-0538.

Roma-Rodrigues, C., Fernandes, A. R., & Baptista, P. V. (2014). Exosome in tumour microenvironment: Overview of the crosstalk between normal and cancer cells. *BioMed Research International*, *2014*, 179486. http://dx.doi.org/10.1155/2014/179486.

Salminen, A., Paimela, T., Suuronen, T., & Kaarniranta, K. (2008). Innate immunity meets with cellular stress at the IKK complex: Regulation of the IKK complex by HSP70 and HSP90. *Immunology Letters*, *117*(1), 9–15. http://dx.doi.org/10.1016/j.imlet.2007.12.017.

Sanderson, S., Valenti, M., Gowan, S., Patterson, L., Ahmad, Z., Workman, P., et al. (2006). Benzoquinone ansamycin heat shock protein 90 inhibitors modulate multiple functions required for tumor angiogenesis. *Molecular Cancer Therapeutics*, *5*(3), 522–532. http://dx.doi.org/10.1158/1535-7163.MCT-05-0439.

Schueller, G., Paolini, P., Friedl, J., Stift, A., Dubsky, P., Bachleitner-Hofmann, T., et al. (2001). Heat treatment of hepatocellular carcinoma cells: Increased levels of heat shock proteins 70 and 90 correlate with cellular necrosis. *Anticancer Research*, *21*(1A), 295–300. Retrieved from, http://www.ncbi.nlm.nih.gov/pubmed/11299751.

Schwenkert, S., Hugel, T., & Cox, M. B. (2014). The Hsp90 ensemble: Coordinated Hsp90-cochaperone complexes regulate diverse cellular processes. *Nature Structural & Molecular Biology*, *21*(12), 1017–1021. http://dx.doi.org/10.1038/nsmb.2927.

Shimp, S. K., 3rd., Parson, C. D., Regna, N. L., Thomas, A. N., Chafin, C. B., Reilly, C. M., et al. (2012). HSP90 inhibition by 17-DMAG reduces inflammation in J774 macrophages through suppression of Akt and nuclear factor-kappaB pathways. *Inflammation Research*, *61*(5), 521–533. http://dx.doi.org/10.1007/s00011-012-0442-x.

Sidera, K., & Patsavoudi, E. (2014). HSP90 inhibitors: Current development and potential in cancer therapy. *Recent Patents on Anti-Cancer Drug Discovery*, *9*(1), 1–20. Retrieved from, http://www.ncbi.nlm.nih.gov/pubmed/23312026.

Sidera, K., Samiotaki, M., Yfanti, E., Panayotou, G., & Patsavoudi, E. (2004). Involvement of cell surface HSP90 in cell migration reveals a novel role in the developing nervous system. *The Journal of Biological Chemistry*, *279*(44), 45379–45388. http://dx.doi.org/10.1074/jbc.M405486200.

Sims, J. D., McCready, J., & Jay, D. G. (2011). Extracellular heat shock protein (Hsp)70 and Hsp90alpha assist in matrix metalloproteinase-2 activation and breast cancer cell migration and invasion. *PloS One*, *6*(4), e18848. http://dx.doi.org/10.1371/journal.pone.0018848.

Song, X., Wang, X., Zhuo, W., Shi, H., Feng, D., Sun, Y., et al. (2010). The regulatory mechanism of extracellular Hsp90{alpha} on matrix metalloproteinase-2 processing and tumor angiogenesis. *The Journal of Biological Chemistry*, *285*(51), 40039–40049. http://dx.doi.org/10.1074/jbc.M110.181941.

Spisek, R., & Dhodapkar, M. V. (2007). Towards a better way to die with chemotherapy: Role of heat shock protein exposure on dying tumor cells. *Cell Cycle*, *6*(16), 1962–1965. Retrieved from, http://www.ncbi.nlm.nih.gov/pubmed/17721082.

Srivastava, P. K. (1997). Purification of heat shock protein-peptide complexes for use in vaccination against cancers and intracellular pathogens. *Methods*, *12*(2), 165–171. http://dx.doi.org/10.1006/meth.1997.0464.

Srivastava, P. K., DeLeo, A. B., & Old, L. J. (1986). Tumor rejection antigens of chemically induced sarcomas of inbred mice. *Proceedings of the National Academy of Sciences of the United States of America*, *83*(10), 3407–3411. Retrieved from, http://www.ncbi.nlm.nih.gov/pubmed/3458189.

Srivastava, P. K., & Heike, M. (1991). Tumor-specific immunogenicity of stress-induced proteins: Convergence of two evolutionary pathways of antigen presentation? *Seminars in Immunology*, *3*(1), 57–64. Retrieved from, http://www.ncbi.nlm.nih.gov/entrez/query.fcgi?cmd=Retrieve&db=PubMed&dopt=Citation&list_uids=1893123.

Srivastava, P. K., Udono, H., Blachere, N. E., & Li, Z. (1994). Heat shock proteins transfer peptides during antigen processing and CTL priming. *Immunogenetics*, *39*(2), 93–98. Retrieved from, http://www.ncbi.nlm.nih.gov/pubmed/8276462.

Sulyok, I., Fleischmann, E., Stift, A., Roth, G., Lebherz-Eichinger, D., Kasper, D., et al. (2012). Effect of preoperative fever-range whole-body hyperthermia on immunological markers in patients undergoing colorectal cancer surgery. *British Journal of Anaesthesia*, *109*(5), 754–761. http://dx.doi.org/10.1093/bja/aes248.

Suzuki, S., & Kulkarni, A. B. (2010). Extracellular heat shock protein HSP90beta secreted by MG63 osteosarcoma cells inhibits activation of latent TGF-beta1. *Biochemical and Biophysical Research Communications, 398*(3), 525–531. http://dx.doi.org/10.1016/j.bbrc.2010.06.112.

Tao, R., de Zoeten, E. F., Ozkaynak, E., Chen, C., Wang, L., Porrett, P. M., et al. (2007). Deacetylase inhibition promotes the generation and function of regulatory T cells. *Nature Medicine, 13*(11), 1299–1307. http://dx.doi.org/10.1038/nm1652.

Tesniere, A., Panaretakis, T., Kepp, O., Apetoh, L., Ghiringhelli, F., Zitvogel, L., et al. (2008). Molecular characteristics of immunogenic cancer cell death. *Cell Death and Differentiation, 15*(1), 3–12. http://dx.doi.org/10.1038/sj.cdd.4402269.

Toraya-Brown, S., & Fiering, S. (2014). Local tumour hyperthermia as immunotherapy for metastatic cancer. *International Journal of Hyperthermia, 30*(8), 531–539. http://dx.doi.org/10.3109/02656736.2014.968640.

Trojandt, S., Reske-Kunz, A. B., & Bros, M. (2014). Geldanamycin-mediated inhibition of heat shock protein 90 partially activates dendritic cells, but interferes with their full maturation, accompanied by impaired upregulation of RelB. *Journal of Experimental & Clinical Cancer Research, 33*, 16. http://dx.doi.org/10.1186/1756-9966-33-16.

Tukaj, S., Tiburzy, B., Manz, R., de Castro Marques, A., Orosz, A., Ludwig, R. J., et al. (2014). Immunomodulatory effects of heat shock protein 90 inhibition on humoral immune responses. *Experimental Dermatology, 23*(8), 585–590. http://dx.doi.org/10.1111/exd.12476.

Udono, H., & Srivastava, P. K. (1993). Heat shock protein 70-associated peptides elicit specific cancer immunity. *The Journal of Experimental Medicine, 178*(4), 1391–1396. Retrieved from, http://www.ncbi.nlm.nih.gov/pubmed/8376942.

Udono, H., & Srivastava, P. K. (1994). Comparison of tumor-specific immunogenicities of stress-induced proteins gp96, hsp90, and hsp70. *Journal of Immunology, 152*(11), 5398–5403. Retrieved from, http://www.ncbi.nlm.nih.gov/pubmed/8189059.

Ullrich, S. J., Robinson, E. A., Law, L. W., Willingham, M., & Appella, E. (1986). A mouse tumor-specific transplantation antigen is a heat shock-related protein. *Proceedings of the National Academy of Sciences of the United States of America, 83*(10), 3121–3125. Retrieved from, http://www.ncbi.nlm.nih.gov/pubmed/3458168.

Ung, T. H., Madsen, H. J., Hellwinkel, J. E., Lencioni, A. M., & Graner, M. W. (2014). Exosome proteomics reveals transcriptional regulator proteins with potential to mediate downstream pathways. *Cancer Science, 105*(11), 1384–1392. http://dx.doi.org/10.1111/cas.12534.

Van Crombruggen, K., Jacob, F., Zhang, N., & Bachert, C. (2013). Damage-associated molecular patterns and their receptors in upper airway pathologies. *Cellular and Molecular Life Sciences, 70*(22), 4307–4321. http://dx.doi.org/10.1007/s00018-013-1356-7.

Vidal, C. I., Mintz, P. J., Lu, K., Ellis, L. M., Manenti, L., Giavazzi, R., et al. (2004). An HSP90-mimic peptide revealed by fingerprinting the pool of antibodies from ovarian cancer patients. *Oncogene, 23*(55), 8859–8867. http://dx.doi.org/10.1038/sj.onc.1208082.

Whitesell, L., Santagata, S., Mendillo, M. L., Lin, N. U., Proia, D. A., & Lindquist, S. (2014). HSP90 empowers evolution of resistance to hormonal therapy in human breast cancer models. *Proceedings of the National Academy of Sciences of the United States of America, 111*(51), 18297–18302. http://dx.doi.org/10.1073/pnas.1421323111.

Widenmeyer, M., Shebzukhov, Y., Haen, S. P., Schmidt, D., Clasen, S., Boss, A., et al. (2011). Analysis of tumor antigen-specific T cells and antibodies in cancer patients treated with radiofrequency ablation. *International Journal of Cancer, 128*(11), 2653–2662. http://dx.doi.org/10.1002/ijc.25601.

Yamano, T., Murata, S., Shimbara, N., Tanaka, N., Chiba, T., Tanaka, K., et al. (2002). Two distinct pathways mediated by PA28 and hsp90 in major histocompatibility complex class

I antigen processing. *The Journal of Experimental Medicine*, *196*(2), 185–196. Retrieved from, http://www.ncbi.nlm.nih.gov/pubmed/12119343.

Yanez-Mo, M., Siljander, P. R., Andreu, Z., Zavec, A. B., Borras, F. E., Buzas, E. I., et al. (2015). Biological properties of extracellular vesicles and their physiological functions. *Journal of Extracellular Vesicles*, *4*, 27066. http://dx.doi.org/10.3402/jev.v4.27066.

Yang, Y., Rao, R., Shen, J., Tang, Y., Fiskus, W., Nechtman, J., et al. (2008). Role of acetylation and extracellular location of heat shock protein 90alpha in tumor cell invasion. *Cancer Research*, *68*(12), 4833–4842. http://dx.doi.org/10.1158/0008-5472.CAN-08-0644.

Yun, T. J., Harning, E. K., Giza, K., Rabah, D., Li, P., Arndt, J. W., et al. (2011). EC144, a synthetic inhibitor of heat shock protein 90, blocks innate and adaptive immune responses in models of inflammation and autoimmunity. *Journal of Immunology*, *186*(1), 563–575. http://dx.doi.org/10.4049/jimmunol.1000222.

Zappasodi, R., Pupa, S. M., Ghedini, G. C., Bongarzone, I., Magni, M., Cabras, A. D., et al. (2010). Improved clinical outcome in indolent B-cell lymphoma patients vaccinated with autologous tumor cells experiencing immunogenic death. *Cancer Research*, *70*(22), 9062–9072. http://dx.doi.org/10.1158/0008-5472.CAN-10-1825.

Zeng, Y., Feng, H., Graner, M. W., & Katsanis, E. (2003). Tumor-derived, chaperone-rich cell lysate activates dendritic cells and elicits potent antitumor immunity. *Blood*, *101*(11), 4485–4491. http://dx.doi.org/10.1182/blood-2002-10-3108. 2002-10-3108 [pii].

CHAPTER NINE

Hsp90: A Global Regulator of the Genotype-to-Phenotype Map in Cancers

Daniel Jarosz[*,†,1]
[*]Chemical & Systems Biology, Stanford University School of Medicine, Stanford, California, USA
[†]Developmental Biology, Stanford University School of Medicine, Stanford, California, USA
[1]Corresponding author: e-mail address: jarosz@stanford.edu

Contents

Abstract

Cancer cells have the unusual capacity to limit the cost of the mutation load that they harbor and simultaneously harness its evolutionary potential. This property fuels drug resistance, a key failure mode in oncogene-directed therapy. However, the factors that regulate this capacity might also provide an Achilles' heel that could be exploited therapeutically. Recently, insight has come from a seemingly distant field: protein folding. It is now clear that protein homeostasis broadly supports malignancy and fuels the rapid evolution of drug resistance. Among protein homeostatic mechanisms that influence cancer biology, the essential ATP-driven molecular chaperone heat-shock protein 90 (Hsp90) is especially important. Hsp90 catalyzes folding of many proteins that regulate growth and development. These "client" kinases, transcription factors, and ubiquitin ligases often play critical roles in human disease, especially cancer. Studies in a wide range of systems—from single-celled organisms to human tumor samples—suggest that Hsp90 can broadly reshape the map between genotype and phenotype, acting as a "capacitor" and "potentiator" of genetic variation. Indeed, it has likely done so to such a degree that it has left an impress on diverse genome sequences. Hsp90 can constitute as much as 5% of total protein in transformed cells

Advances in Cancer Research, Volume 129
ISSN 0065-230X
http://dx.doi.org/10.1016/bs.acr.2015.11.001

and increased levels of heat-shock activation correlate with poor prognosis in breast cancer. These findings and others have motivated a flurry of interest in Hsp90 inhibitors as cancer therapeutics, which have met with rather limited success as single agents, but may eventually prove invaluable in limiting the emergence of resistance to other chemotherapeutics, both genotoxic and molecularly targeted. Here, we provide an overview of Hsp90 function, review its relationship to genetic variation and the evolution of new traits, and discuss the importance of these findings for cancer biology and future efforts to drug this pathway.

1. INTRODUCTION

Biological systems can remain robust for long periods of time, but they can also diversify rapidly. Although such evolutionary versatility is critical for the long-term survival of species, it can also have devastating consequences for human health. It is well known that cancer cells and many microbial pathogens can thrive despite accumulating large numbers of mutations (Fares, Barrio, Sabater-Munoz, & Moya, 2002; Maisnier-Patin et al., 2005; Tarafa et al., 2008). Furthermore, these cells can harness their mutation reservoir to drive the rapid evolution of new traits. Molecular insight into this problem has recently emerged from the realization that most biological polymers must fold to function, and most mutations are destabilizing (Bloom, Wilke, Arnold, & Adami, 2004). For example, protein chaperones stabilize numerous mutations, including many that activate protooncogenes (Falsone, Leptihn, Osterauer, Haslbeck, & Buchner, 2004; Whitesell, Mimnaugh, De Costa, Myers, & Neckers, 1994). Thus, as discussed in GRP94/gp96 in Cancer (Wu et al.) and Impact of post-translational modifications of anticancer activity of Hsp90 inhibitors (Woodford et al.), several classes of chaperone inhibitors (especially those targeting Hsp90) are being developed both preclinically and clinically as potential cancer chemotherapeutics (Taldone, Ochiana, Patel, & Chiosis, 2014; Trepel, Mollapour, Giaccone, & Neckers, 2010). However, it is unknown how this strategy might influence the evolution of new traits (e.g., drug resistance) in highly mutated cells.

Several lines of evidence suggest that one of the greatest therapeutic impacts of chaperone inhibition may be the ability to tune the evolution of drug resistance. Cancer cells and some bacterial pathogens have the unusual capacity to limit the cost of the enormous mutation load that they harbor (Fares et al., 2002; Maisnier-Patin et al., 2005). However, they can simultaneously harness its evolutionary potential (Oliver, Canton, Campo, Baquero, & Blazquez, 2000). This property fuels the acquisition of target-related drug resistance, a key failure mode in oncogene-directed therapy. However, factors that regulate this capacity also constitute a novel type of nononcogene addiction that could, at least in principle, be exploited therapeutically. Hsp90 constitutes up to 5% of total protein in transformed cells, and increased levels of heat-shock system activation correlate with poor prognosis in many cancer types, as has recently been highlighted for breast cancer (Mendillo et al., 2012; Santagata et al., 2011). These findings and others have motivated a flurry of interest in Hsp90 inhibitors as cancer therapeutics: over two dozens have been in clinical trials over the past decade (Trepel et al., 2010). Preclinical studies suggest that the combinatorial disruption of multiple oncogenic pathways by Hsp90 inhibitors could prove a unique, particularly strong benefit of this therapeutic strategy, but something which remains to be demonstrated in patients (Drysdale, Brough, Massey, Jensen, & Schoepfer, 2006; Sawai et al., 2008; Sos et al., 2009; Wang, Trepel, Neckers, & Giaccone, 2010).

However, several obstacles prevent the full realization of potential therapeutic benefit from perturbing Hsp90 as a global regulator of the genotype-to-phenotype map. First, Hsp90 inhibition is a blunt tool and could in principle affect virtually every regulatory pathway in the cell. Deeper knowledge about how Hsp90 rewires signaling pathways will allow the development of more targeted therapeutic strategies. Second, many functionally important mutations in cancer do not occur in proteins, but rather in noncoding regulatory sequences. Third, Hsp90 can also influence epigenetic variation and the production of new genetic variation through its role in supporting elements of the genome maintenance machinery. Finally, although Hsp90 inhibition may be able to slow the acquisition of drug resistance, it is unlikely to stop it altogether because other mechanisms with overlapping abilities may be important in promoting the profound evolvability that characterizes malignancy. This chapter reviews these topics and provides insight into the basic design principles that govern the Hsp90 chaperone's influence on evolution. The contribution of genetic and epigenetic variation to the process is described, and the most prominent issues remaining to be resolved are discussed in the context of how we can gain the ability to exploit the full potential of this system therapeutically.

2. PROTEIN FOLDING, ENVIRONMENTAL STRESS, AND PHENOTYPIC VARIABILITY

Although it has long been appreciated that biological systems are remarkably robust to perturbations (Barkai & Leibler, 1997; Masel &

Trotter, 2010; Waddington, 1942), molecular explanations for the origin of this behavior have been scarce. The best understood example comes from studies in a seemingly distant field—protein folding. Surprisingly, most proteins are strikingly unstable, with native folds typically maintained with a free energy of unfolding (ΔG) of about -5 to -15 kcal/mol (Fersht, 1998). As a frame of reference, this roughly corresponds to the free energy of ATP hydrolysis (Berg, Tymoczko, & Stryer, 2006). The intrinsic tolerance of protein structures to mutation varies greatly. However, most point mutations alter protein stability by 0.5–5 kcal/mol (Depristo, Weinreich, & Hartl, 2005). There is some evidence that combined mutations can have disproportionately large effects in some circumstances (Bershtein, Segal, Bekerman, Tokuriki, & Tawfik, 2006; Serrano, Day, & Fersht, 1993). Since decreased stability increases the fraction of unfolded protein molecules, protein function can be negatively affected by mutations that decrease stability but do not otherwise affect a protein's function. Many mutations associated with human disease, and in particular cancer, are predicted to affect protein stability (Sahni et al., 2015; Wang & Moult, 2001).

A major mechanism by which cells contend with the potentially deleterious effects of misfolded proteins relies on molecular chaperones. Chaperones assist in *de novo* folding, refolding, stabilization, and degradation of other proteins. They are ancient, universally conserved, and often essential for survival (Lindquist & Craig, 1988; Taipale, Jarosz, & Lindquist, 2010; Young, Agashe, Siegers, & Hartl, 2004). By virtue of enabling the folding and therefore function (Borkovich, Farrelly, Finkelstein, Taulien, & Lindquist, 1989) of proteins that would otherwise misfold and aggregate, chaperones have a central role in protecting organisms from environmental perturbations and endogenous stresses as well. Findings in a wide variety of organisms have led to a compelling hypothesis: that the protein folding machinery can buffer genetic variation in addition to environmental perturbations in a wide variety of organisms (Burga, Casanueva, & Lehner, 2011; Jarosz & Lindquist, 2010; Queitsch, Sangster, & Lindquist, 2002; Rohner et al., 2013; Rutherford & Lindquist, 1998; Yeyati, Bancewicz, Maule, & van Heyningen, 2007).

As discussed in Hsp90 as a 'chaperone' of the Epigenome: Insights and Opportunities for Cancer Therapy (Issaacs) and Unusual Suspects in the Twilight Zone between the Hsp90 Interactome and Carcinogenesis (Vartholomaiou), Hsp90 is essential and extremely abundant. It constitutes as much as 1–2% of the total protein in the eukaryotic cytosol under basal conditions (Borkovich et al., 1989). It has a well-characterized role in regulating and stabilizing a wide variety of substrate proteins, commonly referred to as Hsp90 "clients" (Taipale et al., 2010). As an example, Hsp90 associates with steroid hormone receptors and enables them to adopt conformations that are competent for ligand binding (Picard et al., 1990; Smith & Toft, 2008). Hsp90 is also required for the maturation of many kinases, E3 ubiquitin ligases, and some transcription factors (Citri et al., 2006; Taipale et al., 2012). Genome- and proteome-wide experiments in yeast have revealed that Hsp90 physically or genetically interacts with a large and diverse set of clients involved in processes ranging from signal transduction to chromatin remodeling, protein trafficking, and the cell cycle (McClellan et al., 2007; Millson et al., 2005; Zhao & Houry, 2005). Whether as a result of Hsp90's abundance, its stress-regulated activity, its select "clientele" of substrate proteins, all of the above, or some other yet-defined property, studies in multiple organisms from yeast to vertebrates have revealed that this protein exerts a strong influence on the revelation of new phenotypes.

3. A "HUB OF HUBS" THAT BROADLY REGULATES INHERITANCE

Intriguing results from a wide range of eukaryotic model organisms have illustrated that Hsp90 can sculpt phenotypes by several mechanisms. In *Drosophila melanogaster*, impairment of Hsp90 function pharmacologically, by mutation, or with temperature reveals remarkable morphological traits that affect diverse adult structures (Rutherford & Lindquist, 1998). These phenotypes are dependent on the genetic background. That is, the traits have a heritable basis, indicating that Hsp90 buffers the potential morphological and phenotypic effects of genetic variation. These observations also suggest that Hsp90 can prevent the manifestation of phenotypic diversity, thereby contributing to canalization (Milton, Ulane, & Rutherford, 2006; Salathia et al., 2007; Sangster, Lindquist, & Queitsch, 2004). Interestingly, traits affected by Hsp90 inhibition are among the most canalized in *D. melanogaster* (Milton et al., 2006), consistent with a central role for Hsp90 in ensuring uniformity of phenotype.

Similar increased phenotypic diversity has also been observed in evolutionarily distant eukaryotes such as the mustard plant *Arabidopsis thaliana* (Queitsch et al., 2002), the budding yeast *Saccharomyces cerevisiae* (Jarosz & Lindquist, 2010), the zebrafish *Danio rerio* (Yeyati et al., 2007), the Mexican blind cavefish *Astyanax mexicanus* (Rohner et al., 2013), and the nematode

Caenorhabditis elegans (Burga et al., 2011). In *A. thaliana*, an inbreeding organism with considerably less heterozygosity than *Drosophila*, reduced Hsp90 function reveals phenotypic diversity in many features of the plant, including cotyledons, hypocotyls, root morphology, rosettes, and pigmentation (Queitsch et al., 2002; Sangster et al., 2007). A modest temperature shift from 22 to 27 °C induces a similar response, indicating that subtle environmental changes are sufficient to elicit dramatic Hsp90-dependent alterations in phenotype. In zebrafish, Hsp90 inhibition similarly affects diverse phenotypes (Yeyati et al., 2007). Hsp90 inhibition did not alter the somitogenesis defect of a *yot* mutant, but it did reveal anophthalmia, microphthalmia, and nanophthalmia (AMN) in a small percentage of organisms examined (Yeyati et al., 2007). Moreover, Hsp90 inhibition greatly exacerbated the eye defects of the *pax6b* mutant *sunrise*, which causes a decrease in lens size relative to retinal size. In contrast, the severity of the *sufu* mutant *dreumes*, which causes a similar eye defect to *sunrise*, was reduced by inhibiting Hsp90 (Yeyati et al., 2007). These observations are broadly similar to the effects of Hsp90 inhibition in *Drosophila* and *Arabidopsis*. Hsp90 clearly contributes to canalization of development, and even its modest inhibition can dramatically alter the effects of underlying mutations.

4. Hsp90 AS A CAPACITOR AND POTENTIATOR OF GENETIC VARIATION

The evolution of new traits requires not only access to adaptive phenotypes but also underlying genetic mechanisms to ensure their heritability. Protein chaperones such as Hsp90 can promote phenotypic robustness by masking or "buffering" destabilizing mutations in other proteins (Burga et al., 2011; Cowen & Lindquist, 2005; Jarosz & Lindquist, 2010; Queitsch et al., 2002; Rutherford & Lindquist, 1998). This allows populations to accumulate standing genetic variation without a corresponding increase in phenotypic variance (Sangster et al., 2004). Yet Hsp90's position as a central hub in the environmental stress response (Bergman & Siegal, 2003) endows this system with an off switch. When chaperone capacity is exceeded under stressful conditions, widespread changes in protein homeostasis reveal novel morphologies and fitness consequences (Burga et al., 2011; Cowen & Lindquist, 2005; Jarosz, Lancaster, Brown, & Lindquist, 2014; Jarosz & Lindquist, 2010; Queitsch et al., 2002; Rutherford & Lindquist, 1998). Thus, Hsp90 has been described as an "evolutionary capacitor" in that it stores phenotypic variance under normal conditions and releases it in times of stress (Jarosz, Taipale, & Lindquist, 2010).

Hsp90 also facilitates the immediate phenotypic manifestation of mutations that destabilize protein structure. Because many proteins are only marginally stable, folding represents a major constraint on the evolution of new function (Bloom, Labthavikul, Otey, & Arnold, 2006). Empirically, Hsp90 stabilizes gain-of-function mutations in both pathogenic fungi (Cowen, 2009; Cowen & Lindquist, 2005; Shapiro et al., 2009) and tumors (Neckers, 2007; Trepel et al., 2010; Tsutsumi, Beebe, & Neckers, 2009; Whitesell et al., 1994; Xu & Neckers, 2007; Xu et al., 2007). In *C. albicans*, resistance to antifungals relies in large part on Hsp90's stabilization of calcineurin and protein kinase C (Vincent, Lancaster, Scherz-Shouval, Whitesell, & Lindquist, 2013). In human tumor cells, Hsp90 directly binds to and stabilizes numerous oncogene families including SRC-family kinases, receptor tyrosine kinases, steroid hormone receptors, and many more (Falsone et al., 2004; Taipale et al., 2010; Whitesell et al., 1994; Xu & Lindquist, 1993; Xu, Singer, & Lindquist, 1999). Seeking to exploit this dependence, Hsp90 inhibitors are undergoing clinical trials for a variety of cancers including breast, lung, prostate, and others (Trepel et al., 2010).

Although a genetic basis for Hsp90-buffered phenotypic variation has been established in several cases (Queitsch et al., 2002; Rutherford & Lindquist, 1998; Yeyati et al., 2007), the full complement of genetic loci involved in this phenomenon has yet to be elucidated. In *A. thaliana*, quantitative trait locus (QTL) analysis was employed by Sangster et al. to identify chromosomal regions that contribute to Hsp90-dependent phenotypes (Sangster et al., 2007, 2008). Genotyped control and Hsp90-reduced recombinant inbred lines (Salathia & Queitsch, 2007) were used to map numerous quantitative life-history traits. Hsp90-dependent buffering of cryptic genetic variation appears to be widespread in *A. thaliana*. Of particular interest is a QTL on chromosome 2 revealed by Hsp90 inhibition that is associated with hypocotyl length in the dark (Sangster et al., 2008), a trait that is a clear example of adaptive phenotypic plasticity. This observation suggests that the effects of Hsp90 on phenotypic plasticity in plants also have a genetic basis.

The extraordinary genetic power of budding yeast permitted Jarosz and Lindquist to broadly identify genetic loci, the phenotypic manifestations of which are influenced by Hsp90 (Jarosz & Lindquist, 2010). They identified this genetic variation by examining over 100 growth traits in genotyped meiotic progeny from a cross (Brem, Yvert, Clinton, & Kruglyak, 2002; Perlstein, Ruderfer, Roberts, Schreiber, & Kruglyak, 2007) between a wild yeast strain (RM11-1a—from a California vineyard) and a laboratory yeast strain (BY4716) with linkage analysis. As a frame of reference, these strains are only slightly more polymorphic than two human beings. Using an exhaustive suite of reciprocal allele replacements, they confirmed the causal variants for four loci (Mec1, Pdr8, Nsf1, and Ndi1). Some of these causal polymorphisms occurred in Hsp90 clients. Others occurred in nonclients, and even in noncoding regulatory sequences (Jarosz & Lindquist, 2010).

Broadly speaking, Hsp90 acted to reshape the map between genotype and phenotype in these experiments. Forty-four robust associations between genotype and phenotype disappeared with modest Hsp90 inhibition; sixty-three previously absent associations became apparent (out of over 500 total associations mapped). Many traits involved multiple Hsp90-contingent alleles. Fully one-third involved two and nearly 15% involved three or more. Thus, reducing Hsp90 function produced genetically complex traits in a single step, providing access to phenotypes that would take much longer to achieve via mutation alone. Most remarkably, reducing Hsp90 function improved the correlation between genotype and phenotype in sequenced yeast strains ($P < 10^{-7}$), establishing that Hsp90 has likely played an important role in shaping genome evolution (Jarosz & Lindquist, 2010). Thus, across large tracts of evolution, impairment of Hsp90 function reveals traits associated with cryptic genetic variation (Gibson & Dworkin, 2004) that are normally phenotypically silent. What might all of this mean for the process of malignant progression in cancers? Mechanistic investigation of this question stands as a goalpost for future studies. However, it is tempting to speculate that many so-called passenger mutations may indeed constitute a vast reservoir of cryptic genetic variation inherent to many malignancies. This Hsp90-buffered genetic variation would then accumulate in a relatively silent fashion until chaperone function becomes overwhelmed by the burden of mutation or environmental insults, whether endogenous or therapeutic.

5. GENETIC ASSIMILATION OF ENVIRONMENTALLY INDUCED TRAITS

Waddington made a second seminal observation related to the acquisition of new phenotypes under environmental stress: that morphological characteristics revealed by environmental stress can become genetically assimilated, or "hard-wired" in the genome, through repeated selection

(Waddington, 1942, 1953, 1956, 1959). In *D. melanogaster*, Rutherford and Lindquist observed that selection for traits revealed by Hsp90 inhibition results in their enrichment and ultimately in Hsp90 independence (Rutherford & Lindquist, 1998). During selection for a deformed eye morphology initially revealed by *Hsp83* heterozygosity, an increasing fraction of the progeny exhibited the trait with each generation, until it was eventually expressed in 80–100% of the population. In contrast, no such response was observed in a control experiment in which unaffected lines from the same crosses were selected. The near complete fixation of this trait by selection is remarkable given that Mendelian segregation prohibits the presence of *Hsp83* mutation in such a large proportion of the population.

Indeed, none of the 50 flies examined at the conclusion of selection retained the initial *Hsp83* mutation (Rutherford & Lindquist, 1998). The deformed eye morphology lost its connection to Hsp90 function during the course of selection. Provocatively, backcrosses to a parental strain restore the ability of Hsp90 to buffer the deformed eye phenotype. Thus, selection fundamentally alters the reaction norm for the deformed eye trait (Rutherford & Lindquist, 1998). Just as in Waddington's classical foundational experiments (Waddington, 1942, 1953, 1956, 1959), the combinatorial effects of environmentally sensitive polymorphisms would appear to underlie genetic assimilation.

Similar efforts to assimilate the AMN trait revealed by Hsp90 inhibition in *D. rerio* were largely unsuccessful, although they are complicated by embryonic lethality of the trait during inbreeding (Yeyati et al., 2007). AMN likely involves at least two loci and its remarkably complex inheritance patterns demonstrate that Hsp90 can act in a myriad of ways even in the context of a single trait. Nonetheless, AMN-affected progeny were observed in 2–13% of the offspring from out-breeding even without Hsp90 inhibition, compared to an absence of AMN-affected fish in a control cross.

Hsp90 has also been implicated in the acquisition of cave traits during the adaptation of Mexican cavefish. In multiple instances, geographic isolation of *Astyanax mexicanus* in caves led to rapid eye loss in a manner that is generally agreed to have been adaptive. How this could occur in a handful of generations remained enigmatic. Rohner et al. (2013) asked whether Hsp90 might have fueled this rapid evolutionary process, by experimentally inhibiting Hsp90 during the development of river populations of *Astyanax*. They found that inhibition decanalized the trait to produce some fish with small eye size. More strikingly, the progeny of small eye-size fish retained this trait even in the absence of Hsp90 inhibition. Thus, in a single generation, the trait could be "hard-wired" by selection. Finally, the authors attempted to discern whether features of the cave environment might have engaged the Hsp90 response. Indeed, the altered osmotic environment of the cave water induced a transcriptional response that was nearly identical to chemical inhibition of Hsp90. Thus, even in vertebrates selection can act to increase the fraction of the population expressing a trait that was initially revealed by Hsp90 inhibition.

Perhaps with more obvious relevance to cancer and resistance to therapeutic interventions, genetic assimilation can also occur in the context of Hsp90-driven antifungal drug resistance. In *S. cerevisiae*, rapid selection results in the growth of variants that express Hsp90-dependent resistance to the commonly used antimycotic fluconazole, while gradual selection produces resistance that is stable to Hsp90 inhibition (Cowen & Lindquist, 2005). Genetic assimilation of resistance was investigated in a series of *C. albicans* clinical isolates from an HIV-infected patient undergoing fluconazole therapy over a 2-year period; this collection represents a single strain evolving under selective pressure within a human host (White, 1997). Hsp90's effect on the azole-resistance trait arises largely from its ability to chaperone calcineurin (Cowen, Carpenter, Matangkasombut, Fink, & Lindquist, 2006). The fluconazole resistance of isolates from early time points during treatment was abolished by impairing Hsp90 or calcineurin function. In contrast, later isolates possessed more robust fluconazole resistance that was not affected by the Hsp90 inhibitor geldanamycin, the calcineurin inhibitor cyclosporin A, or thermal stress. This finding demonstrates that continued selective pressure within a human host results in the genetic assimilation of an Hsp90-dependent trait (fluconazole resistance) in *C. albicans* just as repeated inbreeding decoupled the relationship between Hsp90 activity and the deformed eye trait in *D. melanogaster*.

The mechanism(s) by which genetic assimilation operates has been the subject of considerable debate. Waddington argued that continued selection enriched the population in environmentally responsive polymorphisms to the extent that the trait was no longer conditionally expressed. Consistent with this theory, Rutherford and Lindquist observed a complex genetic basis for the traits selected under Hsp90 inhibition in *D. melanogaster* (Rutherford & Lindquist, 1998). Sangster et al. (Salathia et al., 2007; Sangster et al., 2008) and Yeyati et al. (Yeyati et al., 2007) similarly observed a complex genetic basis for Hsp90 responsive traits in *A. thaliana* and *D. rerio*, respectively. Likewise, the eye traits that Rohner et al. examined in Mexican cavefish have a known complex genetic basis (Rohner et al., 2013). While the phenotypic effects of polymorphisms can be individually buffered by Hsp90 function, collectively they may surpass the ability of Hsp90 to canalize a given trait. Although results in *D. melanogaster* are broadly consistent with such a model, the precise mechanisms of genetic assimilation remain to be experimentally elucidated, and some have argued for a contribution of epigenetic mechanisms (Gangaraju et al., 2011; Lu, Wang, & Ruden, 2012). Indeed, in advanced cancers, both genetic and epigenetic mechanisms are likely to be at work during malignant progression and the acquisition of high-grade drug resistance.

6. Hsp90 AND PARTIAL PENETRANCE

The same mutations that cause devastating consequences in some individuals can have no effect in others. The reasons for such "partial penetrance" are typically unknown, although environmental differences and genetic or epigenetic variation are often invoked as explanations. Working in *C. elegans*, Burga et al. (2011) examined whether Hsp90 might influence these processes. Using an incompletely penetrant null mutation in the T-box transcription factor gene *tbx-9*, the authors tested whether Hsp90 activity would influence the developmental phenotype of the mutation. The authors used a transcriptional reporter for Hsp90 (*daf-21* in worms) and were able to observe animal-to-animal variation in the *tbx-9* mutants. Strikingly, these differences correlated strongly with the phenotypic penetrance of the mutant allele. That is, animals with lower Hsp90 function were more likely to demonstrate an abnormal phenotype resulting from the *tbx-9* mutation that they harbored. Notably, this mechanism may also be engaged by environmental stress. Also working in *C. elegans*, Casanueva, Burga, and Lehner (2012) discovered that stimulation of a stress response (e.g., heat shock) can decrease the penetrance of several mutations. However, this robustness comes at a penalty in terms of reproductive fitness. Animals that did not develop mutant phenotypes were significantly less fertile than those that did. Broadly speaking, these studies suggest that interindividual variation in buffering systems can exert a strong influence on the outcome of inherited mutations. In cancers, such Hsp90-dependent effects would be expected to dramatically increase the phenotypic heterogeneity available within tumors for selective pressures to act upon and fuel tumor evolution. Implications for the use of Hsp90 inhibitors in the treatment of cancer are profound but remain largely unexplored.

7. Hsp90 AS A POTENTIATOR OF EPIGENETIC VARIATION

Other work has suggested that the relationship between Hsp90 and phenotypic diversity may be partly epigenetic (Gangaraju et al., 2011; Ruden, Garfinkel, Sollars, & Lu, 2003; Sawarkar, Sievers, & Paro, 2012; Sollars et al., 2003; Tariq, Nussbaumer, Chen, Beisel, & Paro, 2009). Sollars and colleagues identified mutations that promote ectopic outgrowth in a *Krüppel* mutant strain of flies (Sollars et al., 2003), and among these were five mutations in *Hsp83*. Pharmacological inhibition of Hsp90 also induced ectopic outgrowth in the mutant strain. In selection experiments, however, further Hsp90 inhibition was not required for maintenance or enrichment of the ectopic outgrowth trait. Curiously, treatment with histone deacetylase inhibitors did reduce the fraction of flies showing ectopic outgrowth (Sollars et al., 2003), suggesting that enrichment of this trait may be the result of epigenetic mechanisms. Interpretation of this result is complicated by the nonselective nature of HDAC inhibitors (Carey & La Thangue, 2006), and the observation that the histone deacetylase HDAC6 associates with Hsp90 and regulates its function *in vivo* (Bali et al., 2005; Boyault, Sadoul, Pabion, & Khochbin, 2007; Kovacs et al., 2005; Scroggins et al., 2007). Recent investigations in yeast have also revealed a potential link between Hsp90 and epigenetic gene regulation. Hsp90 cofactors in budding yeast include Tah1 and Pih1, which physically and functionally associate with components of the chromatin remodeling complexes Ino80 and SWR-C and the TIP60 histone acetyltransferase complex (Zhao et al., 2005, 2008; Zhao & Houry, 2005).

Other work in *Drosophila* has placed Hsp90 on chromatin (Sawarkar & Paro, 2013; Sawarkar et al., 2012). Most Hsp90 is cytoplasmic, but a small percentage (~2–3% in *Drosophila*) can be found in the nucleus. Sawarkar et al. (2012) mapped chromatin-binding sites of Hsp90 at high resolution across the *Drosophila* genome (see also chapters "Hsp90 in Cancer: Transcriptional Roles in the Nucleus" by Calderwood and Neckers, and "Hsp90 as a 'Chaperone' of the Epigenome: Insights and Opportunities for Cancer Therapy" by Isaacs, in this volume). They found that far from being randomly distributed, the protein localizes near promoters of both coding and noncoding genes, including many microRNAs (miRNAs). Using a variety of computational and biochemical analyses, the authors found that Hsp90 exerted much of its influence on transcription through stabilization of the negative elongation factor complex. Inhibition of Hsp90 led to upregulation of target genes. Indeed, Hsp90 was required for maximal activation of paused genes both in *Drosophila* and in mammalian cells in response to environmental stimuli. The overall effects were modest, but the breadth of this mechanism suggests that it may unify the effects of Hsp90 on epigenetic variation with its effects on genetic variation.

Another central, cancer-relevant epigenetic pathway that is highly dependent on Hsp90 involves Trithorax. Tariq et al. (2009) discovered a direct interaction between Hsp90 and Trithorax in *Drosophila*. Trithorax is a member of the TrxG chromatin proteins that, together with the members of the Polycomb group, modulate epigenetic signals to control the developmental fate of cells. Hsp90 cooperates with Trithorax at chromatin to maintain the active expression state of targets including *Hox* genes, which play key roles in development and have also been implicated in cancer progression (Shah & Sukumar, 2010). Pharmacological inhibition of Hsp90 leads to degradation of Trithorax and downregulation of homeotic gene expression. A similar effect was observed with the human ortholog, mixed lineage leukemia protein (MLL).

Extracellular Hsp90 (eHsp90) has also been linked to the induction of an epigenetic program mediated by (see also the chapter "Hsp90 as a 'Chaperone' of the Epigenome: Insights and Opportunities for Cancer Therapy" by Isaacs, in this volume). Briefly, Nolan, Franco, Hance, Hayward, and Isaacs (2015) investigated the relationship between eHsp90 and activation of the epithelial-to-mesenchymal transition (EMT). EMT activation is a major risk factor for cancer progression. The authors found that eHsp90 can robustly initiate the EMT in prostate cancer cells by modulating EZH2 expression and activity. EZH2 is a methyltransferase of the polycomb repressor signaling complex. eHsp90 initiates sustained activation of MEK/ERK, which is critical to facilitate EZH2 transcriptional upregulation. This pathway engages many EMT-related processes including *Snail* and *Twist* expression, tumor cell motility, and anoikis resistance (anchorage-independent survival and growth). eHsp90 was sufficient to induce tumor growth, suppress E-cadherin, and initiate localized invasion, all events that are highly dependent upon EZH2 function. These examples implicate eHsp90 in a highly conserved epigenetic process that promotes a critical developmental transition in cancer.

8. Hsp90 AS A DRIVER OF NEW GENETIC AND EPIGENETIC VARIATION IN RESPONSE TO STRESS

Most studies addressing the influence of Hsp90 on the phenotypic manifestation of genetic or epigenetic variation have not considered whether new variation might be elicited by reduction of chaperone activity. This was a reasonable assumption at the time, given the lack of known mechanistic underpinnings for such an effect, the frequency with which the variants arose, and the extraordinary reproducibility of the traits produced in model organisms. Although a vast body of literature from cancer research, development, and even *in vitro* reconstitution supports a strong influence of Hsp90 on preexisting variation, it now appears that Hsp90 also influences the generation of *de novo* genetic and epigenetic variation. These mechanisms, too, expand Hsp90's function as a hub of hubs through which environmental stimuli are integrated to produce phenotypic diversity.

At the most basic level, Hsp90 can regulate the rate at which mutations are produced in transformed mammalian cells. Sekimoto et al. (2010) examined the relationship between activity of the Y-family DNA polymerase pol eta and Hsp90. Pol eta is a member of the Y-family of DNA polymerases and catalyzes the preferential error-free bypass of UV-induced lesions in the genome. The authors found that Hsp90 promotes the induction of nuclear pol eta foci upon UV irradiation. This occurs because Hsp90 controls pol eta's ability to bind to the monoubiquitinated form of the processivity clamp, PCNA. Inhibition of Hsp90 promotes UV-induced cell toxicity and mutagenesis in a pol eta-dependent fashion. Although the magnitude of the mutagenic effects is modest, these observations establish that Hsp90 has the power to influence the fidelity of DNA replication in transformed cells.

Profound inhibition of Hsp90 can also fuel other types of mutagenesis. Chen, Bradford, Seidel, and Li (2012) found that Hsp90 inhibition could increase chromosomal instability in budding yeast to produce a population with increased karyotypic diversity. Subsequent challenge with diverse stressors led to the emergence of drug-resistant clones with specific chromosome gains. These findings are the first to implicate Hsp90 in the acquisition of aneuploidy and provide strong evidence that this mechanism of mutagenesis can fuel the emergence of adaptive traits under stress in eukaryotic cells. This relationship is not surprising given the established role of Hsp90 in the function of the mitotic spindle (Aligue, Akhavan-Niak, & Russell, 1994), and the efficacy of combining Hsp90 and Wee1 inhibitors in chemotherapy (Haarberg et al., 2013; Iwai et al., 2012).

Hsp90 participates in the maturation of many additional proteins that regulate mutagenesis. Key among these is Piwi, which enables piRNA-mediated silencing of transposons in animals. Specchia and colleagues (Piacentini et al., 2014; Specchia et al., 2010) examined the influence of

Hsp90 mutations on the emergence of phenotypic diversity in *Drosophila* and noted an increased frequency of flies with a *Stellate* phenotype. Because this phenotype can arise from transposition, the authors investigated whether there was a general increase in transposition in the Hsp90 mutant flies. Indeed there was, and these events occurred at a fairly high frequency. Transposition events drove the emergence of many different kinds of new phenotypes in this study, albeit at a somewhat lower frequency than in prior work, suggesting that this Piwi-based mechanism may be an important contributor to Hsp90-dependent phenotypic diversification, at least in *Drosophila*.

The engagement of Hsp90 in the Piwi machinery is especially robust. Studies from Xiol et al. (2012) revealed that the cochaperone Fkbp6 is a critical factor in delivering piRNAs to the Piwi protein Miwi2. Mice that lacked Fkbp6 had increased LINE1 retrotransposon insertions and reduced DNA methylation. Both arose from reduced nuclear accumulation of Miwi2. Turning to an insect cell culture model, the authors discovered that inhibition of Hsp90 leads to accumulation of short antisense RNAs. They identified these as by-products of piRNA amplifications that accumulate only in nuage-localized Piwi proteins. The authors proposed that the chaperone machinery normally ejects these inhibitory RNAs, allowing turnover of Piwi complexes for their participation in piRNA amplification.

Piwi may also function in Hsp90-mediated suppression of epigenetic variation. Gangaraju et al. (2011) reported that the *Drosophila* piRNA pathway, but not siRNA or miRNA, was required for decanalization of eye outgrowth caused by Hsp90 inhibition in *Kruppel* mutants. They isolated a stable protein complex consisting of Hsp90, Piwi, and the Hsp90/Hsp70-organizing protein (HOP), and demonstrated the involvement of this complex in canalization of eye outgrowth. Collectively, these observations suggest that epigenetic silencing of existing genetic variation and suppression of *de novo* mutagenesis, karyotypic diversity, and transposition are all important mechanisms through which Hsp90 couples environmental stimuli to the generation of phenotypic diversity. Although less well understood, the relationship between Hsp90 function and Piwi may have particular relevance to cancer. Aberrant expression of Piwi proteins often correlates with increased invasion, downregulation of p53, and repressed apoptotic cell death. Although the precise mechanisms driving many of these relationships remain to be established, expression of Piwi proteins is clearly correlated with poor outcomes in several tumor types (Suzuki, Honda, & Kirino, 2012).

9. Hsp90 AS A REGULATOR OF POSTTRANSCRIPTIONAL GENE SILENCING

RNA-induced silencing complexes (RISCs) play a central role in posttranscriptional gene regulation. Work in multiple organisms has established a central role for Hsp90 in supporting this biology. In plants, Iki et al. (2010) showed that Hsp90 was integrated into a complex with the plant argonaute protein AGO1. This complex incorporates small interfering RNA (siRNA) and miRNA duplexes to form RISCs that sequester the single-stranded siRNA guide strand and miRNA strand. In these experiments, the siRNA duplex was specifically incorporated into Hsp90-bound AGO1; subsequent removal of the passenger strand was triggered by hydrolysis of Hsp90-bound ATP. Removal of the siRNA passenger strand then required the ribonuclease activity of AGO1. Thus, Hsp90 plays an integral role in RISC formation, extending beyond a simple client/chaperone relationship. Contemporaneous work in *Drosophila* by Miyoshi, Takeuchi, Siomi, and Siomi (2010) came to very similar conclusions, establishing the evolutionary conservation of this relationship between Hsp90 and RISC. Indeed, this may provide yet another mechanism through which Hsp90 can influence canalization. miRNAs can provide developmental robustness by suppressing the impact of genetic variation between individuals (Cassidy et al., 2013; Posadas & Carthew, 2014). In cancer, miRNAs play myriad roles and have been implicated in the development of tumor heterogeneity, providing an additional link between Hsp90 function and the acquisition of phenotypic diversity (Li et al., 2014).

In addition to the epigenetic and mutational consequences of Hsp90 involvement in Piwi biology described above, Hsp90 may also exert effects on posttranscriptional gene silencing via Piwi (Ichiyanagi et al., 2014). Analogous to the results from Xiol et al., referenced above, Ichiyanagi and colleagues found that knockout of Hsp90α in mouse germ cells led to a large reduction in the expression of primary and secondary piRNAs and mislocalization of Miwi2, a Piwi homolog. However, whereas mutation of Fkbp6 led to reduced piRNAs of 28–32 nt in length, the Hsp90α mutation reduced piRNAs of 24–32 nts, suggesting that chaperone function may alter Piwi activity at multiple levels. DNA methylation and mRNA levels of L1 retrotransposon were unchanged in Hsp90α mutant testes. However, L1 protein levels were increased, suggesting that the mechanisms involved proceed through a posttranscriptional form of regulation. More generally, the relationship between Hsp90 activity and Piwi links the activity of this chaperone to the function of long-lived RNAs. Although these RNAs are not generally heritable over a very large number of generations, they do have the power to drive transgenerational epigenetic inheritance; Piwi itself has been proposed as a diagnostic and prognostic marker for many types of cancer—and for cancer progression (Weick & Miska, 2014). Moreover, L1 retrotransposons have been hotly debated as a potential driver of tumorigenesis and the evolution of malignancy (Doucet-O'Hare et al., 2015; Rodic & Burns, 2013; Rodic et al., 2014, 2015).

10. BROADER IMPLICATIONS FOR EVOLUTION AND DISEASE

Hsp90 is one of the most highly connected proteins in the cell. Its function is naturally coupled to the activities of its diverse repertoire of clientele—kinases, E3 ubiquitin ligases, and transcription factors that are central to virtually every cellular and developmental pathway. By controlling and coordinating the activity of these client proteins, Hsp90 has the power to broadly impact phenotypic diversification and tumor heterogeneity—fueling malignancy and the acquisition of drug resistance. In principle, this diversification can occur by a variety of mechanisms, genetic, epigenetic, and posttranscriptional. The environmental regulation of Hsp90's function serves to couple these relationships to diverse stresses, providing a means through which external stimuli can create heritable phenotypic change.

In this chapter, we have surveyed many of the ways that this interface between environmental stressors, both internal and external, and the phenotypic diversification enabled by Hsp90 drives evolutionary change—often with profound effects on cancer and other human diseases. Much additional work is needed to reveal the precise molecular mechanisms that permit Hsp90 to remodel the genotype-to-phenotype map in cancers and uncover the most important types of genetic and epigenetic variation on which these mechanisms can act. Such an understanding will be of enormous value because, by revealing the mechanisms that highly mutated cells employ to survive and evolve new traits, we are likely to identify new Achilles' heels that are ripe for therapeutic intervention. Indeed, the insights gained from this work will hopefully allow us to move away from the blunt and to date rather disappointing tool of global, maximally tolerated Hsp90 inhibition to the development of precise, tumor-specific, and more effective therapeutic strategies.

REFERENCES

Aligue, R., Akhavan-Niak, H., & Russell, P. (1994). A role for Hsp90 in cell cycle control: Wee1 tyrosine kinase activity requires interaction with Hsp90. *The EMBO Journal, 13*, 6099–6106.

Bali, P., Pranpat, M., Bradner, J., Balasis, M., Fiskus, W., Guo, F., et al. (2005). Inhibition of histone deacetylase 6 acetylates and disrupts the chaperone function of heat shock protein 90: A novel basis for antileukemia activity of histone deacetylase inhibitors. *The Journal of Biological Chemistry, 280*, 26729–26734.

Barkai, N., & Leibler, S. (1997). Robustness in simple biochemical networks. *Nature, 387*, 913–917.

Berg, J. M., Tymoczko, J. L., & Stryer, L. (2006). Biochemistry. (6th ed.): W.H. Freeman, New York.

Bergman, A., & Siegal, M. L. (2003). Evolutionary capacitance as a general feature of complex gene networks. *Nature, 424*, 549–552.

Bershtein, S., Segal, M., Bekerman, R., Tokuriki, N., & Tawfik, D. (2006). Robustness-epistasis link shapes the fitness landscape of a randomly drifting protein. *Nature, 444*, 929–932.

Bloom, J. D., Labthavikul, S. T., Otey, C. R., & Arnold, F. H. (2006). Protein stability promotes evolvability. *Proceedings of the National Academy of Sciences of the United States of America, 103*, 5869–5874.

Bloom, J. D., Wilke, C. O., Arnold, F. H., & Adami, C. (2004). Stability and the evolvability of function in a model protein. *Biophysical Journal, 86*, 2758–2764.

Borkovich, K. A., Farrelly, F. W., Finkelstein, D. B., Taulien, J., & Lindquist, S. (1989). hsp82 is an essential protein that is required in higher concentrations for growth of cells at higher temperatures. *Molecular and Cellular Biology, 9*, 3919–3930.

Boyault, C., Sadoul, K., Pabion, M., & Khochbin, S. (2007). HDAC6, at the crossroads between cytoskeleton and cell signaling by acetylation and ubiquitination. *Oncogene, 26*, 5468–5476.

Brem, R. B., Yvert, G., Clinton, R., & Kruglyak, L. (2002). Genetic dissection of transcriptional regulation in budding yeast. *Science, 296*, 752–755.

Burga, A., Casanueva, M. O., & Lehner, B. (2011). Predicting mutation outcome from early stochastic variation in genetic interaction partners. *Nature, 480*, 250–253.

Carey, N., & La Thangue, N. B. (2006). Histone deacetylase inhibitors: Gathering pace. *Current Opinion in Pharmacology, 6*, 369–375.

Casanueva, M. O., Burga, A., & Lehner, B. (2012). Fitness trade-offs and environmentally induced mutation buffering in isogenic C. Elegans. *Science, 335*, 82–85.

Cassidy, J. J., Jha, A. R., Posadas, D. M., Giri, R., Venken, K. J., Ji, J., et al. (2013). miR-9a minimizes the phenotypic impact of genomic diversity by buffering a transcription factor. *Cell, 155*, 1556–1567.

Chen, G., Bradford, W. D., Seidel, C. W., & Li, R. (2012). Hsp90 stress potentiates rapid cellular adaptation through induction of aneuploidy. *Nature, 482*, 246–250.

Citri, A., Harari, D., Shohat, G., Ramakrishnan, P., Gan, J., Lavi, S., et al. (2006). Hsp90 recognizes a common surface on client kinases. *The Journal of Biological Chemistry, 281*, 14361–14369 (Supplementary Data).

Cowen, L. E. (2009). Hsp90 orchestrates stress response signaling governing fungal drug resistance. *PLoS Pathogens, 5*, e1000471.

Cowen, L. E., Carpenter, A. E., Matangkasombut, O., Fink, G. R., & Lindquist, S. (2006). Genetic architecture of Hsp90-dependent drug resistance. *Eukaryotic Cell, 5*, 2184–2188.

Cowen, L. E., & Lindquist, S. (2005). Hsp90 potentiates the rapid evolution of new traits: Drug resistance in diverse fungi. *Science, 309*, 2185–2189.

Depristo, M. A., Weinreich, D. M., & Hartl, D. L. (2005). Missense meanderings in sequence space: A biophysical view of protein evolution. *Nature Reviews. Genetics*, *6*, 678–687.

Doucet-O'Hare, T. T., Rodic, N., Sharma, R., Darbari, I., Abril, G., Choi, J. A., et al. (2015). LINE-1 expression and retrotransposition in Barrett's esophagus and esophageal carcinoma. *Proceedings of the National Academy of Sciences of the United States of America*, *112*, E4894–E4900.

Drysdale, M. J., Brough, P. A., Massey, A., Jensen, M. R., & Schoepfer, J. (2006). Targeting Hsp90 for the treatment of cancer. *Current Opinion in Drug Discovery & Development*, *9*, 483–495.

Falsone, S. F., Leptihn, S., Osterauer, A., Haslbeck, M., & Buchner, J. (2004). Oncogenic mutations reduce the stability of SRC kinase. *Journal of Molecular Biology*, *344*, 281–291.

Fares, M. A., Barrio, E., Sabater-Munoz, B., & Moya, A. (2002). The evolution of the heat-shock protein GroEL from Buchnera, the primary endosymbiont of aphids, is governed by positive selection. *Molecular Biology and Evolution*, *19*, 1162–1170.

Fersht, A. R. (1998). Structure and mechanism in protein science: A guide to enzyme catalysis and protein folding. (1st ed.): W.H. Freeman, New York.

Gangaraju, V. K., Yin, H., Weiner, M. M., Wang, J., Huang, X. A., & Lin, H. (2011). Drosophila Piwi functions in Hsp90-mediated suppression of phenotypic variation. *Nature Genetics*, *43*, 153–158.

Gibson, G., & Dworkin, I. (2004). Uncovering cryptic genetic variation. *Nature Reviews. Genetics*, *5*, 681–690.

Haarberg, H. E., Paraiso, K. H., Wood, E., Rebecca, V. W., Sondak, V. K., Koomen, J. M., et al. (2013). Inhibition of Wee1, AKT, and CDK4 underlies the efficacy of the HSP90 inhibitor XL888 in an in vivo model of NRAS-mutant melanoma. *Molecular Cancer Therapeutics*, *12*, 901–912.

Ichiyanagi, T., Ichiyanagi, K., Ogawa, A., Kuramochi-Miyagawa, S., Nakano, T., Chuma, S., et al. (2014). HSP90alpha plays an important role in piRNA biogenesis and retrotransposon repression in mouse. *Nucleic Acids Research*, *42*, 11903–11911.

Iki, T., Yoshikawa, M., Nishikiori, M., Jaudal, M. C., Matsumoto-Yokoyama, E., Mitsuhara, I., et al. (2010). In vitro assembly of plant RNA-induced silencing complexes facilitated by molecular chaperone HSP90. *Molecular Cell*, *39*, 282–291.

Iwai, A., Bourboulia, D., Mollapour, M., Jensen-Taubman, S., Lee, S., Donnelly, A. C., et al. (2012). Combined inhibition of Wee1 and Hsp90 activates intrinsic apoptosis in cancer cells. *Cell Cycle*, *11*, 3649–3655.

Jarosz, D. F., Lancaster, A. K., Brown, J. C. S., & Lindquist, S. (2014). An evolutionarily conserved epigenetic element converts wild fungi from metabolic specialists to generalists. *Cell*, *158*, 1072–1082.

Jarosz, D. F., & Lindquist, S. (2010). Hsp90 and environmental stress transform the adaptive value of natural genetic variation. *Science*, *330*, 1820–1824.

Jarosz, D. F., Taipale, M., & Lindquist, S. (2010). Protein homeostasis and the phenotypic manifestation of genetic diversity: Principles and mechanisms. *Annual Review of Genetics*, *44*, 189–216.

Kovacs, J. J., Murphy, P. J., Gaillard, S., Zhao, X., Wu, J. T., Nicchitta, C. V., et al. (2005). HDAC6 regulates Hsp90 acetylation and chaperone-dependent activation of glucocorticoid receptor. *Molecular Cell*, *18*, 601–607.

Li, Y., Li, W., Ying, Z., Tian, H., Zhu, X., Li, J., et al. (2014). Metastatic heterogeneity of breast cancer cells is associated with expression of a heterogeneous TGFbeta-activating miR424-503 gene cluster. *Cancer Research*, *74*, 6107–6118.

Lindquist, S. L., & Craig, E. A. (1988). The heat-shock proteins. *Annual Review of Genetics*, *22*, 631–677.

Lu, X., Wang, L., & Ruden, D. M. (2012). Hsp90 inhibitors and the reduction of anti-cancer drug resistance by non-genetic and genetic mechanisms. *Pharmaceuticals*, *5*, 890–898.

Maisnier-Patin, S., Roth, J. R., Fredriksson, A., Nystrom, T., Berg, O. G., & Andersson, D. I. (2005). Genomic buffering mitigates the effects of deleterious mutations in bacteria. *Nature Genetics*, *37*, 1376–1379.

Masel, J., & Trotter, M. V. (2010). Robustness and evolvability. *Trends in Genetics*, *26*, 406–414.

McClellan, A. J., Xia, Y., Deutschbauer, A. M., Davis, R. W., Gerstein, M., & Frydman, J. (2007). Diverse cellular functions of the hsp90 molecular chaperone uncovered using systems approaches. *Cell*, *131*, 121–135.

Mendillo, M. L., Santagata, S., Koeva, M., Bell, G. W., Hu, R., Tamimi, R. M., et al. (2012). HSF1 drives a transcriptional program distinct from heat shock to support highly malignant human cancers. *Cell*, *150*, 549–562.

Millson, S. H., Truman, A. W., King, V., Prodromou, C., Pearl, L. H., & Piper, P. W. (2005). A two-hybrid screen of the yeast proteome for Hsp90 interactors uncovers a novel Hsp90 chaperone requirement in the activity of a stress-activated mitogen-activated protein kinase, Slt2p (Mpk1p). *Eukaryotic Cell*, *4*, 849–860.

Milton, C. C., Ulane, C. M., & Rutherford, S. (2006). Control of canalization and evolvability by Hsp90. *PloS One*, *1*, e75.

Miyoshi, T., Takeuchi, A., Siomi, H., & Siomi, M. C. (2010). A direct role for Hsp90 in pre-RISC formation in Drosophila. *Nature Structural & Molecular Biology*, *17*, 1024–1026.

Neckers, L. (2007). Heat shock protein 90: The cancer chaperone. *Journal of Biosciences*, *32*, 517–530.

Nolan, K. D., Franco, O. E., Hance, M. W., Hayward, S. W., & Isaacs, J. S. (2015). Tumor-secreted Hsp90 subverts polycomb function to drive prostate tumor growth and invasion. *The Journal of Biological Chemistry*, *290*, 8271–8282.

Oliver, A., Canton, R., Campo, P., Baquero, F., & Blazquez, J. (2000). High frequency of hypermutable Pseudomonas aeruginosa in cystic fibrosis lung infection. *Science*, *288*, 1251–1254.

Perlstein, E. O., Ruderfer, D. M., Roberts, D. C., Schreiber, S. L., & Kruglyak, L. (2007). Genetic basis of individual differences in the response to small-molecule drugs in yeast. *Nature Genetics*, *39*, 496–502.

Piacentini, L., Fanti, L., Specchia, V., Bozzetti, M. P., Berloco, M., Palumbo, G., et al. (2014). Transposons, environmental changes, and heritable induced phenotypic variability. *Chromosoma*, *123*, 345–354.

Picard, D., Khursheed, B., Garabedian, M. J., Fortin, M. G., Lindquist, S. L., & Yamamoto, K. R. (1990). Reduced levels of hsp90 compromise steroid receptor action in vivo. *Nature*, *348*, 166–168.

Posadas, D. M., & Carthew, R. W. (2014). MicroRNAs and their roles in developmental canalization. *Current Opinion in Genetics & Development*, *27*, 1–6.

Queitsch, C., Sangster, T. A., & Lindquist, S. L. (2002). Hsp90 as a capacitor of phenotypic variation. *Nature*, *417*, 618–624.

Rodic, N., & Burns, K. H. (2013). Long interspersed element-1 (LINE-1): Passenger or driver in human neoplasms? *PLoS Genetics*, *9*, e1003402.

Rodic, N., Sharma, R., Sharma, R., Zampella, J., Dai, L., Taylor, M. S., et al. (2014). Long interspersed element-1 protein expression is a hallmark of many human cancers. *The American Journal of Pathology*, *184*, 1280–1286.

Rodic, N., Steranka, J. P., Makohon-Moore, A., Moyer, A., Shen, P., Sharma, R., et al. (2015). Retrotransposon insertions in the clonal evolution of pancreatic ductal adenocarcinoma. *Nature Medicine*, *21*, 1060–1064.

Rohner, N., Jarosz, D. F., Kowalko, J. E., Yoshizawa, M., Jeffery, W. R., Borowsky, R. L., et al. (2013). Cryptic variation in morphological evolution: HSP90 as a capacitor for loss of eyes in cavefish. *Science*, *342*, 1372–1375.

Ruden, D. M., Garfinkel, M. D., Sollars, V. E., & Lu, X. (2003). Waddington's widget: Hsp90 and the inheritance of acquired characters. *Seminars in Cell & Developmental Biology*, *14*, 301–310.

Rutherford, S. L., & Lindquist, S. (1998). Hsp90 as a capacitor for morphological evolution. *Nature*, *396*, 336–342.

Sahni, N., Yi, S., Taipale, M., Fuxman Bass, J. I., Coulombe-Huntington, J., Yang, F., et al. (2015). Widespread macromolecular interaction perturbations in human genetic disorders. *Cell*, *161*, 647–660.

Salathia, N., Lee, H. N., Sangster, T. A., Morneau, K., Landry, C. R., Schellenberg, K., et al. (2007). Indel arrays: An affordable alternative for genotyping. *The Plant Journal*, *51*, 727–737.

Salathia, N., & Queitsch, C. (2007). Molecular mechanisms of canalization: Hsp90 and beyond. *Journal of Biosciences*, *32*, 457–463.

Sangster, T. A., Bahrami, A., Wilczek, A., Watanabe, E., Schellenberg, K., McLellan, C., et al. (2007). Phenotypic diversity and altered environmental plasticity in Arabidopsis thaliana with reduced Hsp90 levels. *PloS One*, *2*, e648.

Sangster, T. A., Lindquist, S. L., & Queitsch, C. (2004). Under cover: Causes, effects and implications of Hsp90-mediated genetic capacitance. *Bioessays*, *26*, 348–362.

Sangster, T. A., Salathia, N., Lee, H. N., Watanabe, E., Schellenberg, K., Morneau, K., et al. (2008). HSP90-buffered genetic variation is common in Arabidopsis thaliana. *Proceedings of the National Academy of Sciences of the United States of America*, *105*, 2969–2974.

Santagata, S., Hu, R., Lin, N. U., Mendillo, M. L., Collins, L. C., Hankinson, S. E., et al. (2011). High levels of nuclear heat-shock factor 1 (HSF1) are associated with poor prognosis in breast cancer. *Proceedings of the National Academy of Sciences of the United States of America*, *108*, 18378–18383.

Sawai, A., Chandarlapaty, S., Greulich, H., Gonen, M., Ye, Q., Arteaga, C. L., et al. (2008). Inhibition of Hsp90 down-regulates mutant epidermal growth factor receptor (EGFR) expression and sensitizes EGFR mutant tumors to paclitaxel. *Cancer Research*, *68*, 589–596.

Sawarkar, R., & Paro, R. (2013). Hsp90@chromatin.nucleus: An emerging hub of a networker. *Trends in Cell Biology*, *23*, 193–201.

Sawarkar, R., Sievers, C., & Paro, R. (2012). Hsp90 globally targets paused RNA polymerase to regulate gene expression in response to environmental stimuli. *Cell*, *149*, 807–818.

Scroggins, B. T., Robzyk, K., Wang, D., Marcu, M. G., Tsutsumi, S., Beebe, K., et al. (2007). An acetylation site in the middle domain of Hsp90 regulates chaperone function. *Molecular Cell*, *25*, 151–159.

Sekimoto, T., Oda, T., Pozo, F. M., Murakumo, Y., Masutani, C., Hanaoka, F., et al. (2010). The molecular chaperone Hsp90 regulates accumulation of DNA polymerase eta at replication stalling sites in UV-irradiated cells. *Molecular Cell*, *37*, 79–89.

Serrano, L., Day, A. G., & Fersht, A. R. (1993). Step-wise mutation of barnase to binase. A procedure for engineering increased stability of proteins and an experimental analysis of the evolution of protein stability. *Journal of Molecular Biology*, *233*, 305–312.

Shah, N., & Sukumar, S. (2010). The Hox genes and their roles in oncogenesis. *Nature Reviews. Cancer*, *10*, 361–371.

Shapiro, R. S., Uppuluri, P., Zaas, A. K., Collins, C., Senn, H., Perfect, J. R., et al. (2009). Hsp90 orchestrates temperature-dependent Candida albicans morphogenesis via Ras1-PKA signaling. *Current Biology: CB*, *19*, 621–629.

Smith, D. F., & Toft, D. O. (2008). The intersection of steroid receptors with molecular chaperones: Observations and questions. *Molecular Endocrinology*, *22*, 2229–2240.
Sollars, V., Lu, X., Xiao, L., Wang, X., Garfinkel, M. D., & Ruden, D. M. (2003). Evidence for an epigenetic mechanism by which Hsp90 acts as a capacitor for morphological evolution. *Nature Genetics*, *33*, 70–74.
Sos, M. L., Michel, K., Zander, T., Weiss, J., Frommolt, P., Peifer, M., et al. (2009). Predicting drug susceptibility of non-small cell lung cancers based on genetic lesions. *The Journal of Clinical Investigation*, *119*, 1727–1740.
Specchia, V., Piacentini, L., Tritto, P., Fanti, L., D'Alessandro, R., Palumbo, G., et al. (2010). Hsp90 prevents phenotypic variation by suppressing the mutagenic activity of transposons. *Nature*, *463*, 662–665.
Suzuki, R., Honda, S., & Kirino, Y. (2012). PIWI expression and function in cancer. *Frontiers in Genetics*, *3*, 204.
Taipale, M., Jarosz, D. F., & Lindquist, S. (2010). HSP90 at the hub of protein homeostasis: Emerging mechanistic insights. *Nature Reviews. Molecular Cell Biology*, *11*, 515–528.
Taipale, M., Krykbaeva, I., Koeva, M., Kayatekin, C., Westover, K. D., Karras, G. I., et al. (2012). Quantitative analysis of HSP90-client interactions reveals principles of substrate recognition. *Cell*, *150*, 987–1001.
Taldone, T., Ochiana, S. O., Patel, P. D., & Chiosis, G. (2014). Selective targeting of the stress chaperome as a therapeutic strategy. *Trends in Pharmacological Sciences*, *35*, 592–603.
Tarafa, G., Tuck, D., Ladner, D., Topazian, M., Brand, R., Deters, C., et al. (2008). Mutational load distribution analysis yields metrics reflecting genetic instability during pancreatic carcinogenesis. *Proceedings of the National Academy of Sciences of the United States of America*, *105*, 4306–4311.
Tariq, M., Nussbaumer, U., Chen, Y., Beisel, C., & Paro, R. (2009). Trithorax requires Hsp90 for maintenance of active chromatin at sites of gene expression. *Proceedings of the National Academy of Sciences of the United States of America*, *106*, 1157–1162.
Trepel, J., Mollapour, M., Giaccone, G., & Neckers, L. (2010). Targeting the dynamic HSP90 complex in cancer. *Nature Reviews. Cancer*, *10*, 537–549.
Tsutsumi, S., Beebe, K., & Neckers, L. (2009). Impact of heat-shock protein 90 on cancer metastasis. *Future Oncology*, *5*, 679–688.
Vincent, B. M., Lancaster, A. K., Scherz-Shouval, R., Whitesell, L., & Lindquist, S. (2013). Fitness trade-offs restrict the evolution of resistance to amphotericin B. *PLoS Biology*, *11*, e1001692.
Waddington, C. H. (1942). Canalization of development and the inheritance of acquired characters. *Nature*, *150*, 563–565.
Waddington, C. H. (1953). Genetic assimilation of an acquired character. *Evolution*, *7*, 118–126.
Waddington, C. H. (1956). Genetic assimilation of the bithorax phenotype. *Evolution*, *10*, 1–13.
Waddington, C. H. (1959). Canalization of development and genetic assimilation of acquired characters. *Nature*, *183*, 1654–1655.
Wang, Z., & Moult, J. (2001). SNPs, protein structure, and disease. *Human Mutation*, *17*, 263–270.
Wang, Y., Trepel, J. B., Neckers, L. M., & Giaccone, G. (2010). STA-9090, a small-molecule Hsp90 inhibitor for the potential treatment of cancer. *Current Opinion in Investigational Drugs*, *11*, 1466–1476.
Weick, E. M., & Miska, E. A. (2014). PiRNAs: From biogenesis to function. *Development*, *141*, 3458–3471.
White, T. C. (1997). Increased mRNA levels of ERG16, CDR, and MDR1 correlate with increases in azole resistance in Candida albicans isolates from a patient infected with human immunodeficiency virus. *Antimicrobial Agents and Chemotherapy*, *41*, 1482–1487.

Whitesell, L., Mimnaugh, E. G., De Costa, B., Myers, C. E., & Neckers, L. M. (1994). Inhibition of heat shock protein HSP90-pp60v-src heteroprotein complex formation by benzoquinone ansamycins: Essential role for stress proteins in oncogenic transformation. *Proceedings of the National Academy of Sciences of the United States of America, 91*, 8324–8328.

Xiol, J., Cora, E., Koglgruber, R., Chuma, S., Subramanian, S., Hosokawa, M., et al. (2012). A role for Fkbp6 and the chaperone machinery in piRNA amplification and transposon silencing. *Molecular Cell, 47*, 970–979.

Xu, Y., & Lindquist, S. L. (1993). Heat-shock protein hsp90 governs the activity of pp60v-src kinase. *Proceedings of the National Academy of Sciences of the United States of America, 90*, 7074–7078.

Xu, W., & Neckers, L. (2007). Targeting the molecular chaperone heat shock protein 90 provides a multifaceted effect on diverse cell signaling pathways of cancer cells. *Clinical Cancer Research: An Official Journal of the American Association for Cancer Research, 13*, 1625–1629.

Xu, Y., Singer, M. A., & Lindquist, S. L. (1999). Maturation of the tyrosine kinase c-src as a kinase and as a substrate depends on the molecular chaperone Hsp90. *Proceedings of the National Academy of Sciences of the United States of America, 96*, 109–114.

Xu, W., Soga, S., Beebe, K., Lee, M. J., Kim, Y. S., Trepel, J., et al. (2007). Sensitivity of epidermal growth factor receptor and ErbB2 exon 20 insertion mutants to Hsp90 inhibition. *British Journal of Cancer, 97*, 741–744.

Yeyati, P. L., Bancewicz, R. M., Maule, J., & van Heyningen, V. (2007). Hsp90 selectively modulates phenotype in vertebrate development. *PLoS Genetics, 3*, e43.

Young, J. C., Agashe, V. R., Siegers, K., & Hartl, F. U. (2004). Pathways of chaperone-mediated protein folding in the cytosol. *Nature Reviews. Molecular Cell Biology, 5*, 781–791.

Zhao, R., Davey, M., Hsu, Y. C., Kaplanek, P., Tong, A., Parsons, A. B., et al. (2005). Navigating the chaperone network: An integrative map of physical and genetic interactions mediated by the hsp90 chaperone. *Cell, 120*, 715–727.

Zhao, R., & Houry, W. A. (2005). Hsp90: A chaperone for protein folding and gene regulation. *Biochemistry and Cell Biology, 83*, 703–710.

Zhao, R., Kakihara, Y., Gribun, A., Huen, J., Yang, G., Khanna, M., et al. (2008). Molecular chaperone Hsp90 stabilizes Pih1/Nop17 to maintain R2TP complex activity that regulates snoRNA accumulation. *The Journal of Cell Biology, 180*, 563–578.

INDEX

Note: Page numbers followed by "*f*" indicate figures, and "*t*" indicate tables.

F

G

H

Q

R

S

T

U

V

W

X

Y

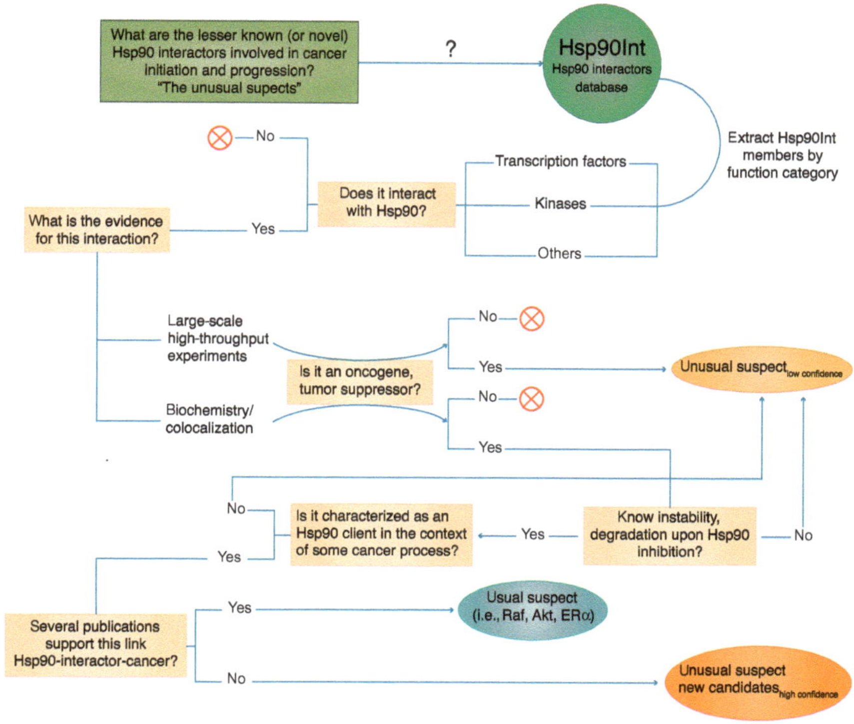

Figure 1, Evangelia Vartholomaiou *et al.* (See Page 3 of this volume.)

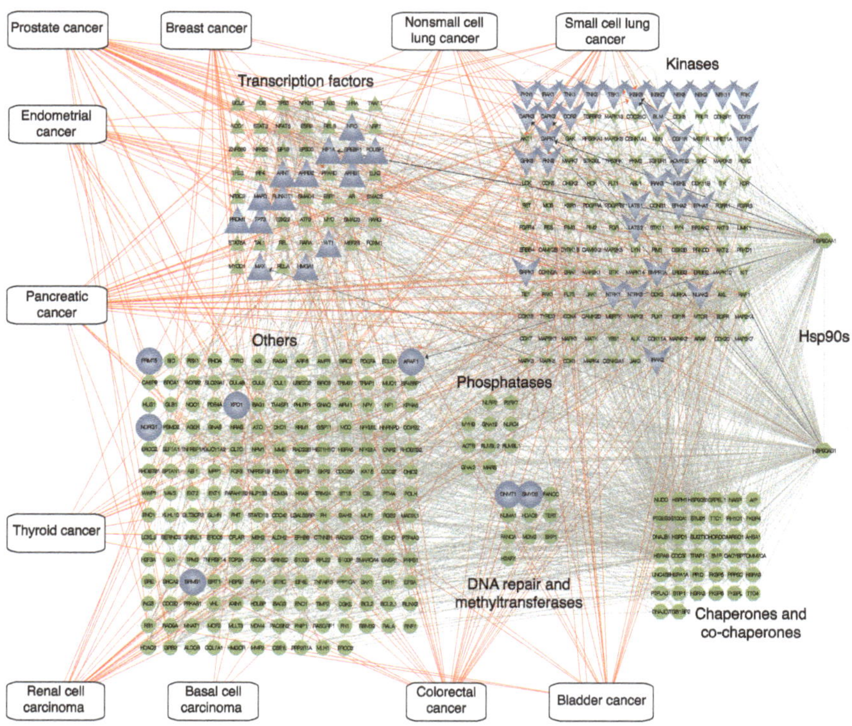

Figure 2, Evangelia Vartholomaiou *et al.* (See Page 5 of this volume.)

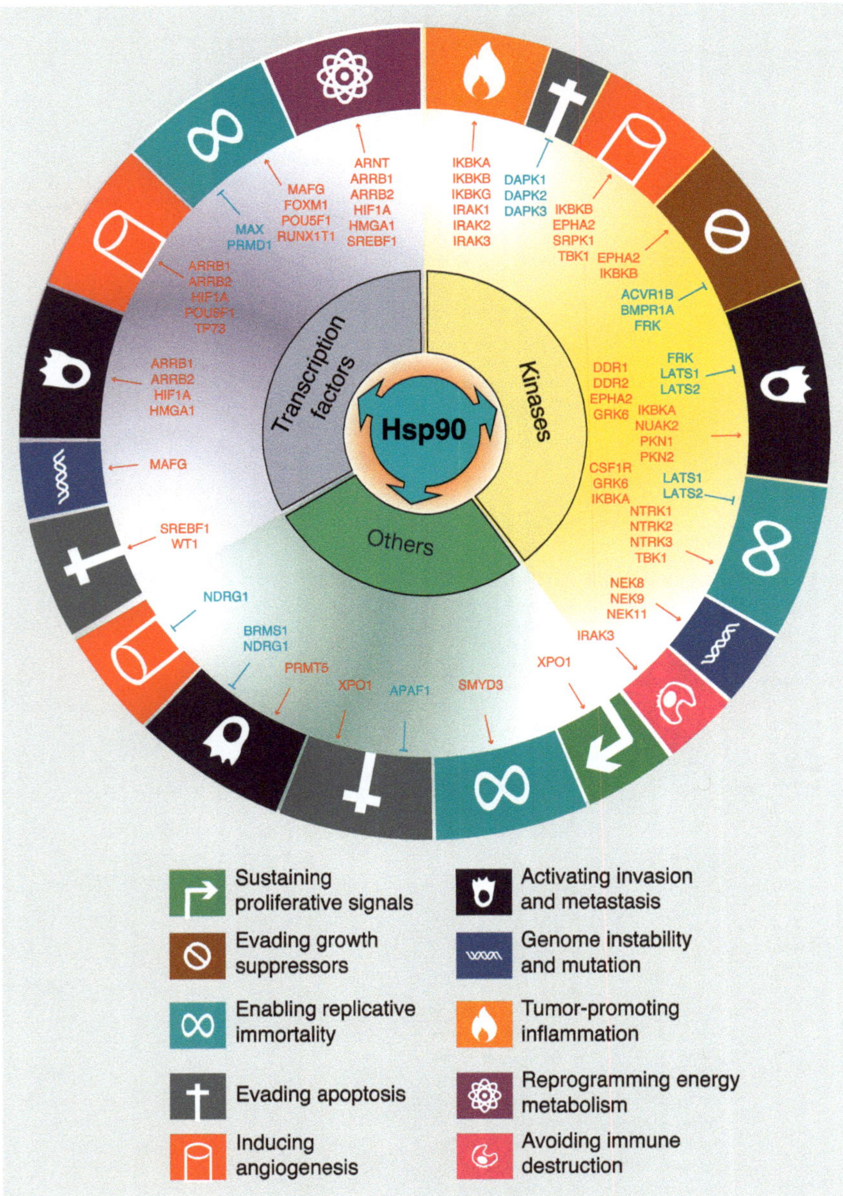

Figure 3, Evangelia Vartholomaiou *et al.* (See Page 21 of this volume.)

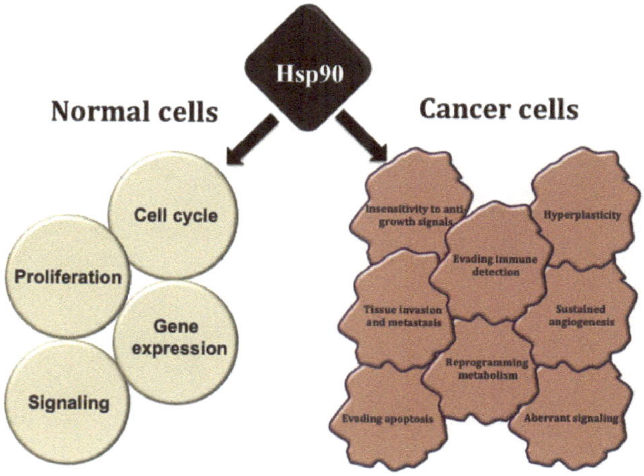

Figure 1, Mark R. Woodford *et al.* (See Page 32 of this volume.)

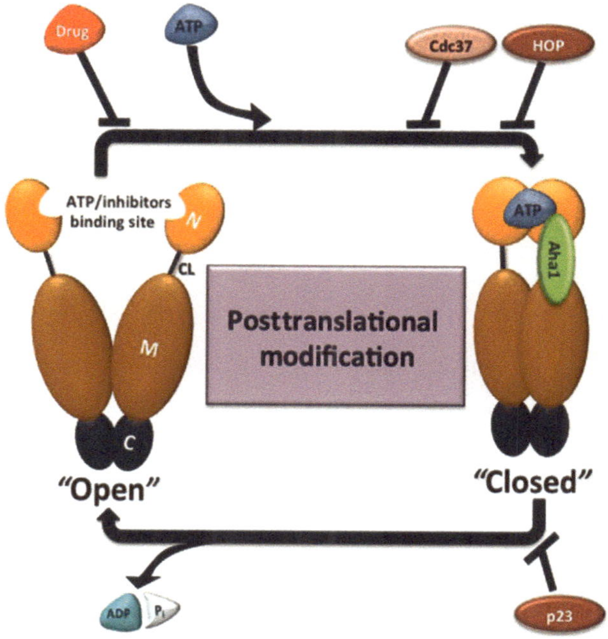

Figure 2, Mark R. Woodford *et al.* (See Page 33 of this volume.)

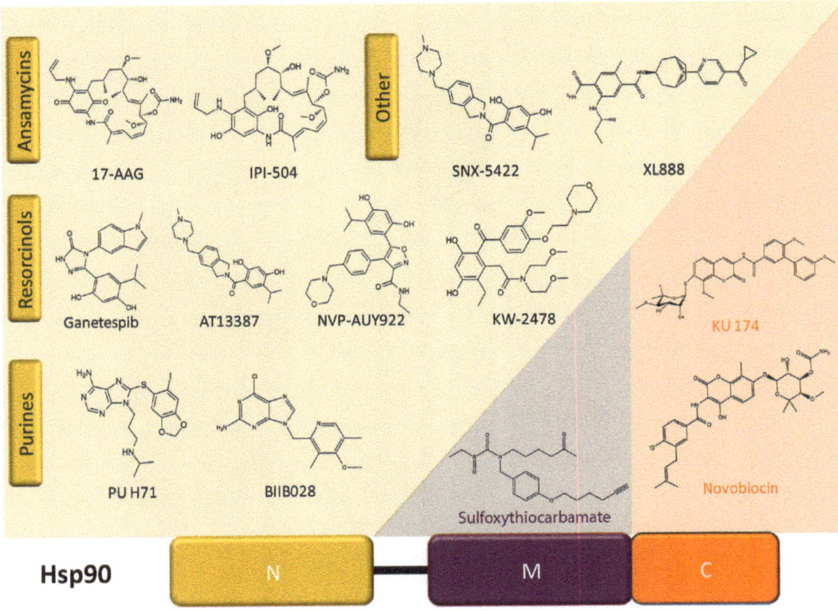

Figure 3, Mark R. Woodford *et al.* (See Page 35 of this volume.)

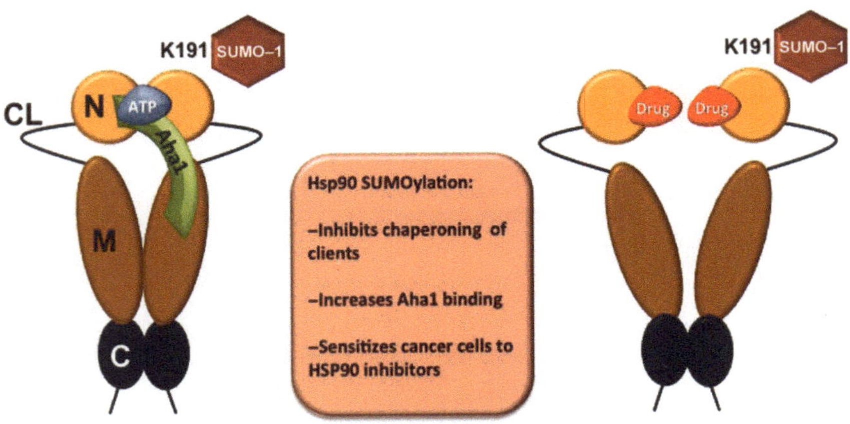

Figure 4, Mark R. Woodford *et al.* (See Page 45 of this volume.)

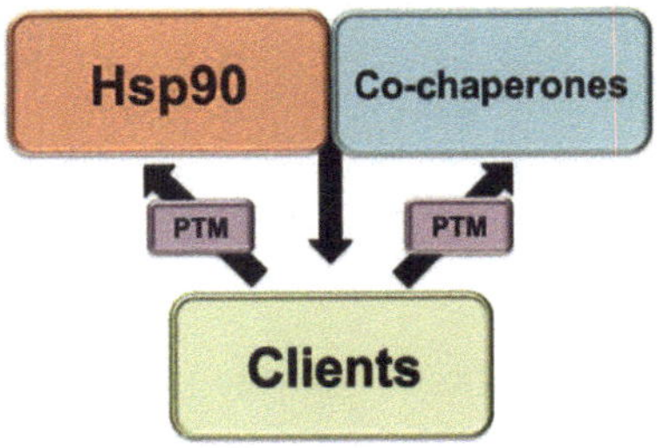

Figure 5, Mark R. Woodford *et al.* (See Page 47 of this volume.)

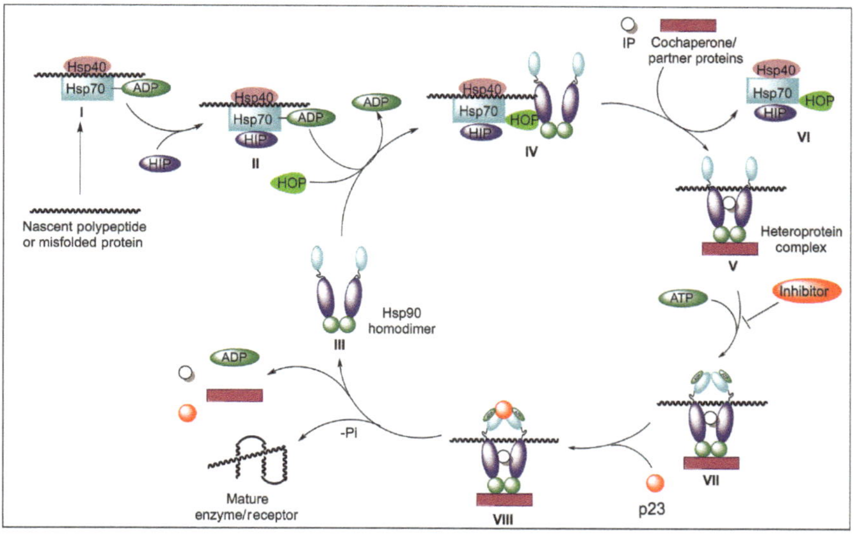

Figure 1, Gaurav Garg *et al.* (See Page 55 of this volume.)

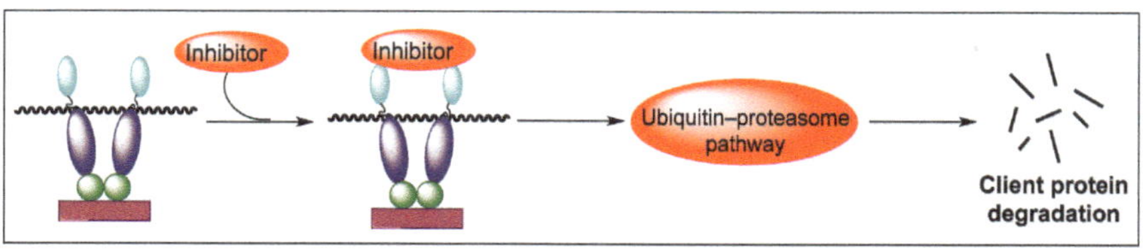

Figure 2, Gaurav Garg *et al.* (See Page 58 of this volume.)

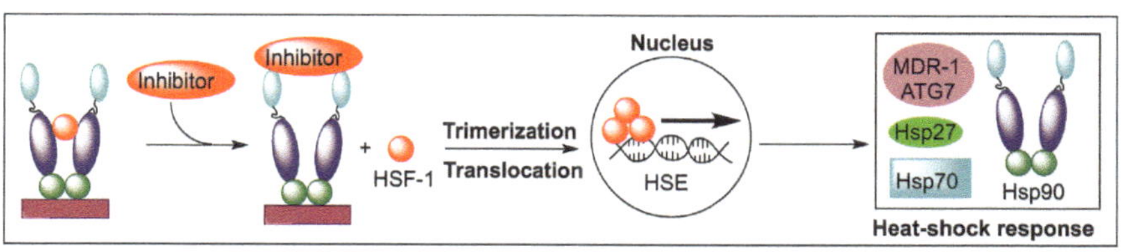

Figure 4, Gaurav Garg *et al.* (See Page 60 of this volume.)

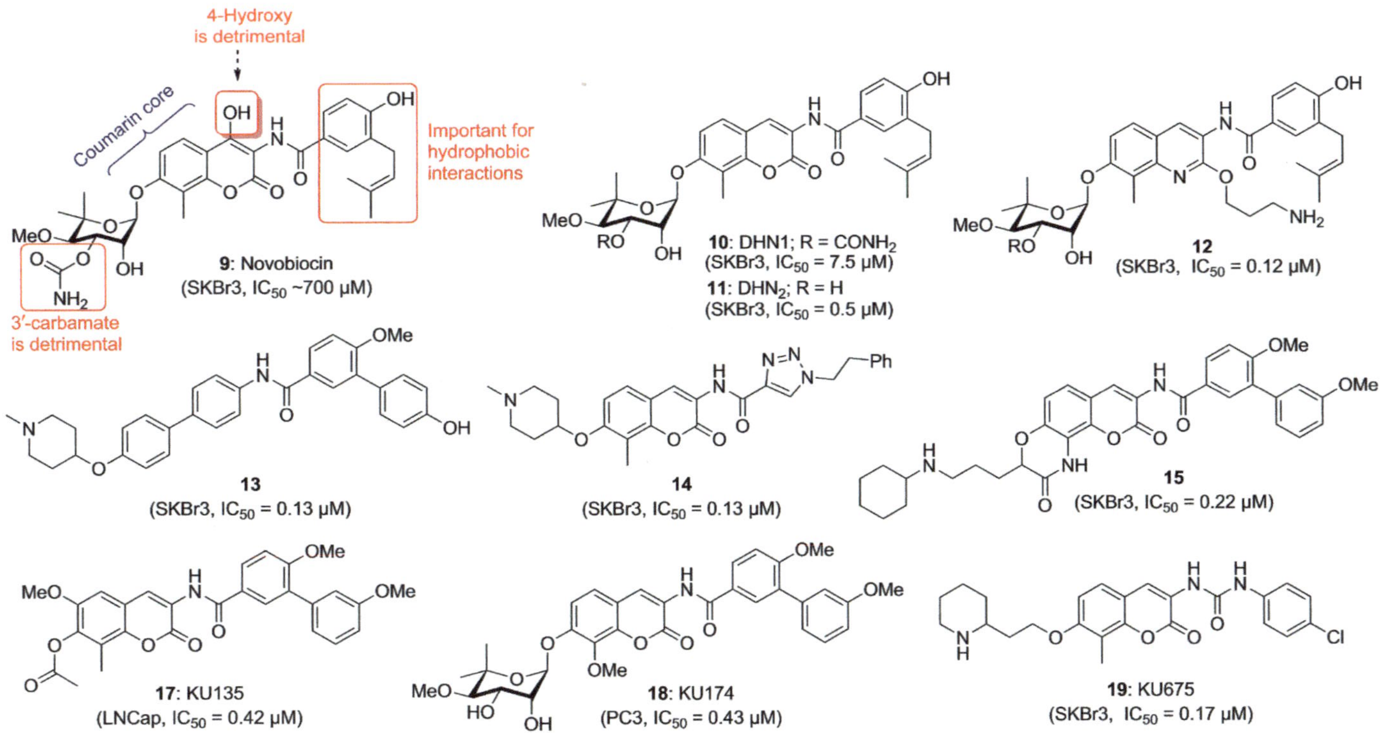

Figure 5, Gaurav Garg *et al.* (See Page 62 of this volume.)

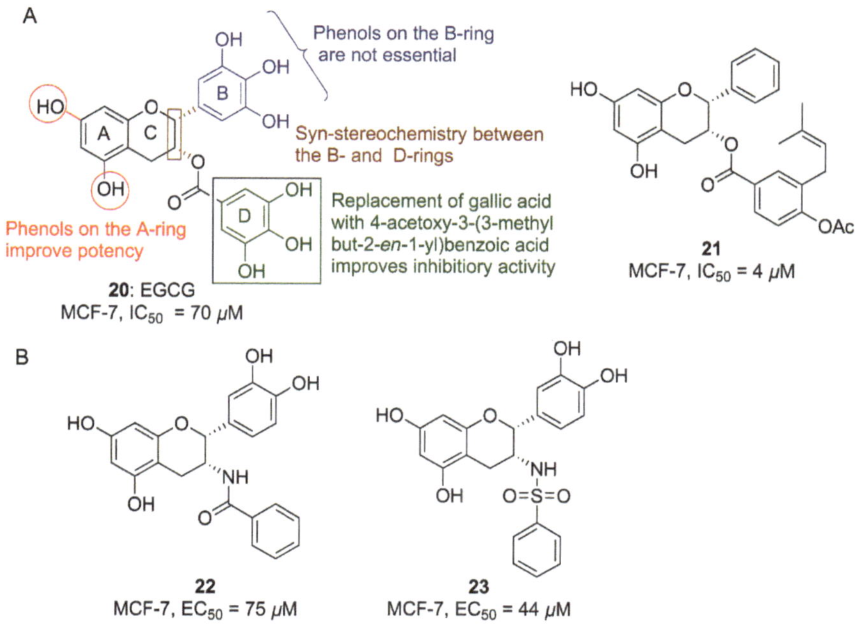

Figure 6, Gaurav Garg *et al.* (See Page 64 of this volume.)

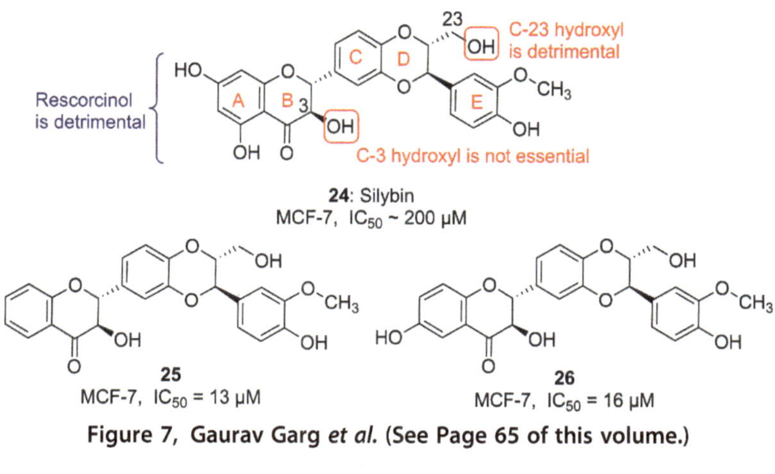

Figure 7, Gaurav Garg *et al.* (See Page 65 of this volume.)

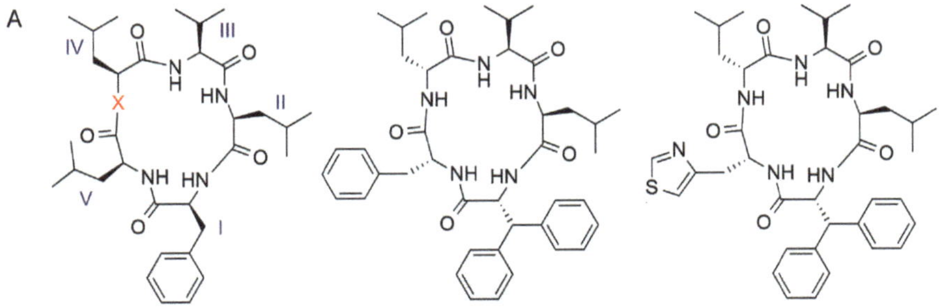

30: Sansalvamide A, X = O
31: San A-amide, X = NH

32: SM145
(HCT116, IC_{50} = 7.4 μM)

33: SM253
(HCT116, IC_{50} = 5.0 μM)

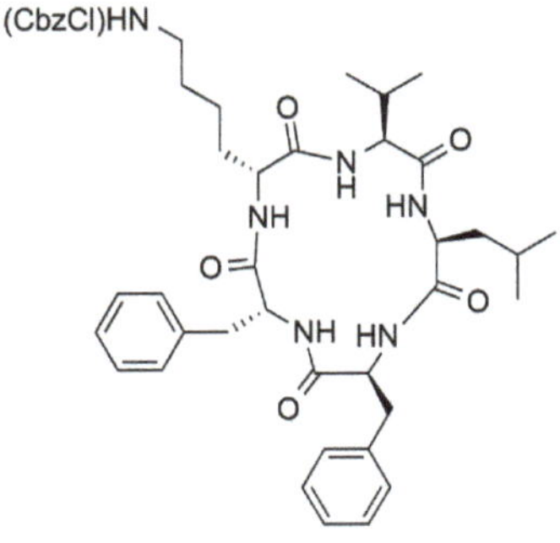

34
(HCT116, IC_{50} = 5.0 μM)

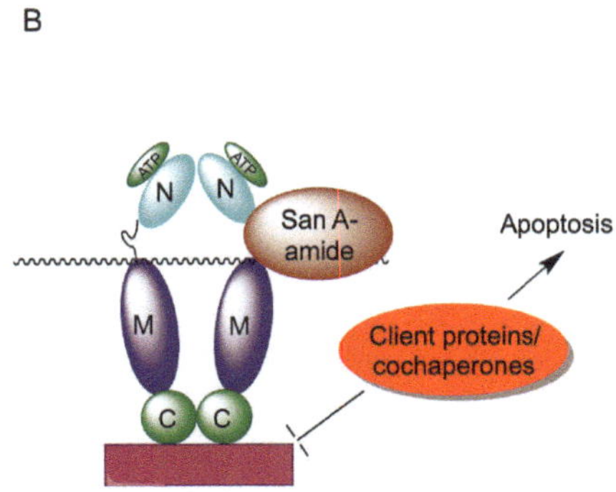

Figure 9, Gaurav Garg *et al.* (See Page 68 of this volume.)

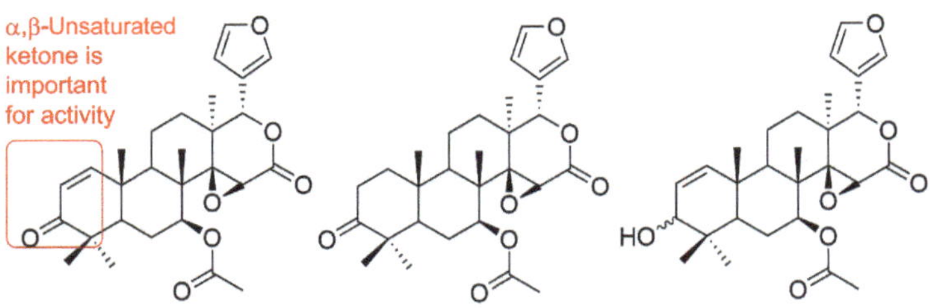

43: Gedunin
(MCF-7, IC_{50} = 8.84 μM)

44
(MCF-7, IC_{50} >100 μM)

45
(MCF-7, IC_{50} = 54.3 μM)

Figure 12, Gaurav Garg *et al.* (See Page 73 of this volume.)

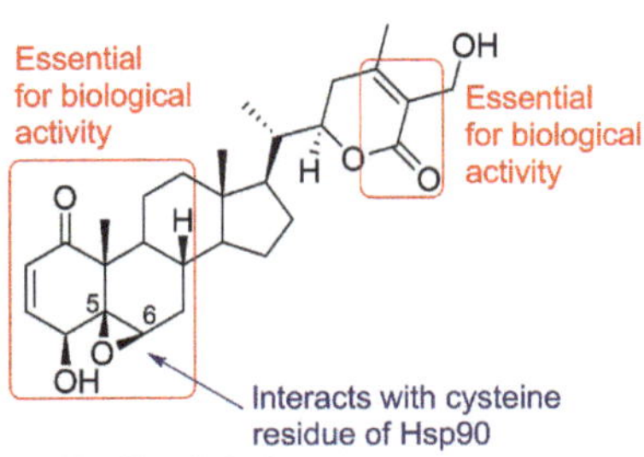

46: Withaferin A
(Panc1, IC_{50} = 1.24 μM)

Figure 13, Gaurav Garg *et al.* (See Page 74 of this volume.)

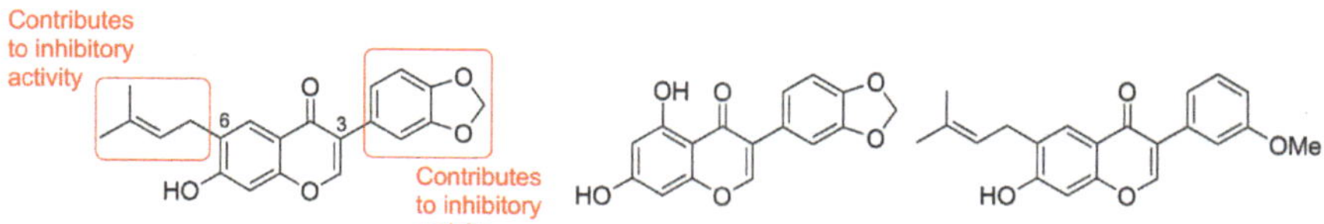

47: Derrubone
(MCF-7, IC_{50} = 11.9 μM)
(HCT116, IC_{50} = 13.7 μM)

48
(MCF-7, IC_{50} > 100 μM)
(HCT116, IC_{50} > 100 μM)

49
(MCF-7, IC_{50} > 100 μM)
(HCT116, IC_{50} > 100 μM)

Figure 14, Gaurav Garg *et al.* (See Page 76 of this volume.)

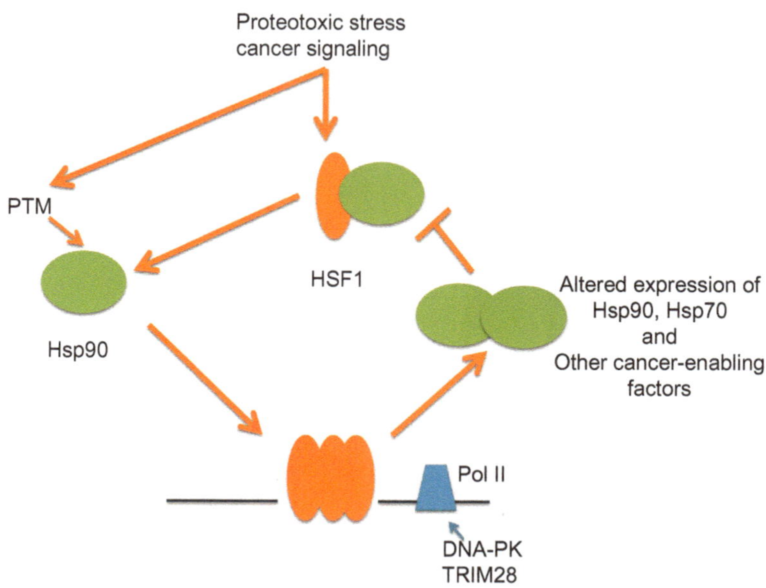

Figure 1, Stuart K. Calderwood and Len Neckers (See Page 92 of this volume.)

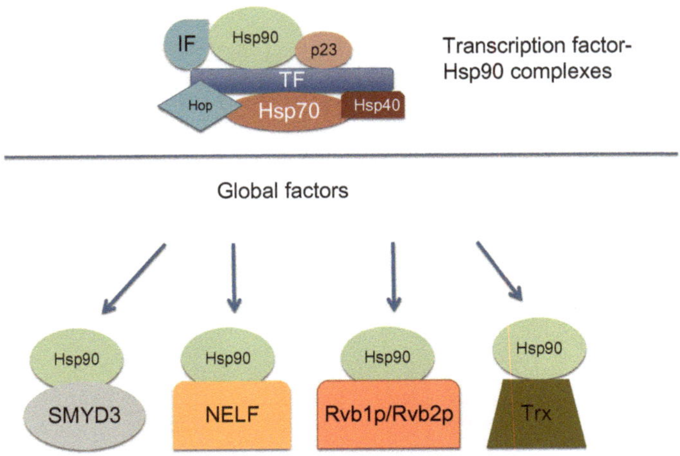

Figure 2, Stuart K. Calderwood and Len Neckers (See Page 93 of this volume.)

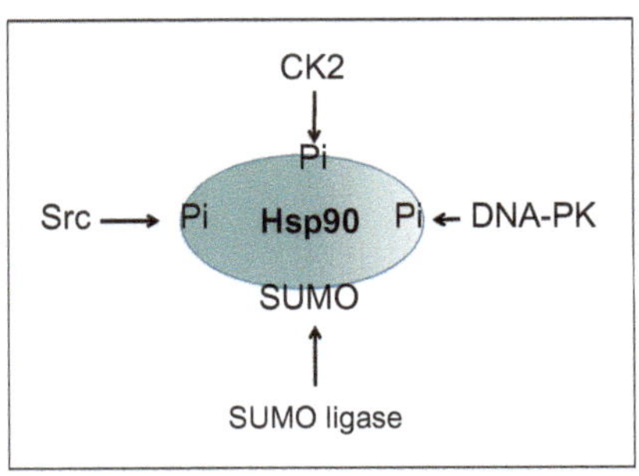

Figure 3, Stuart K. Calderwood and Len Neckers (See Page 98 of this volume.)

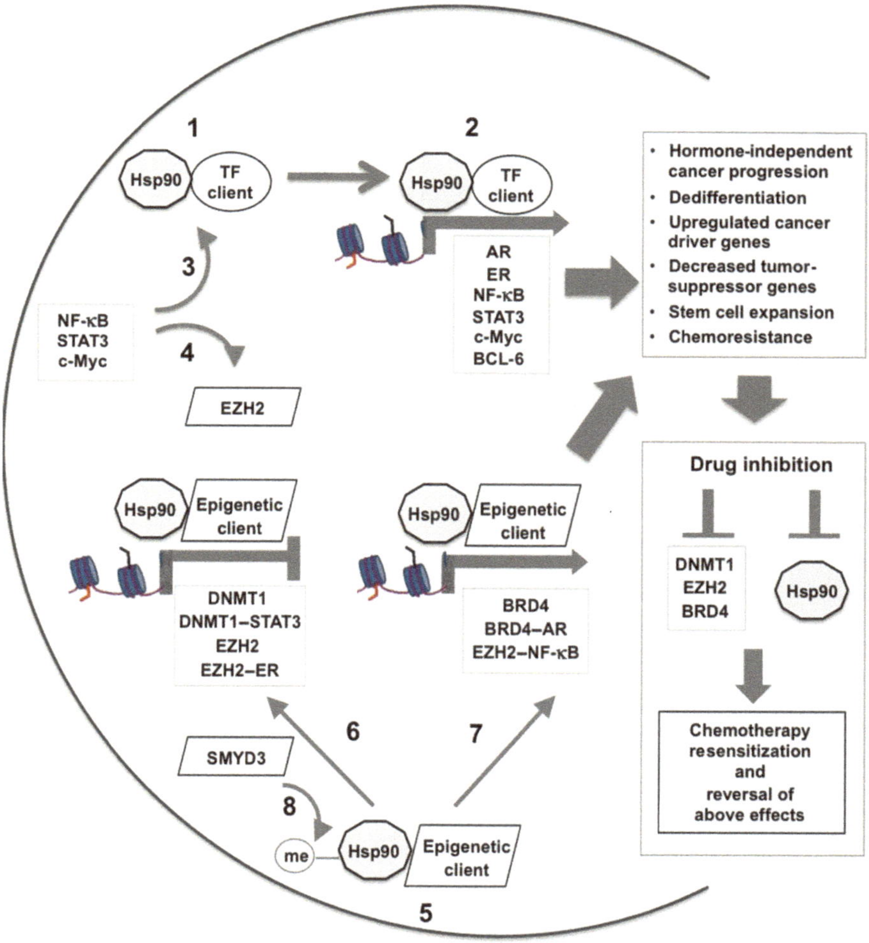

Figure 1, Jennifer S. Isaacs (See Page 110 of this volume.)

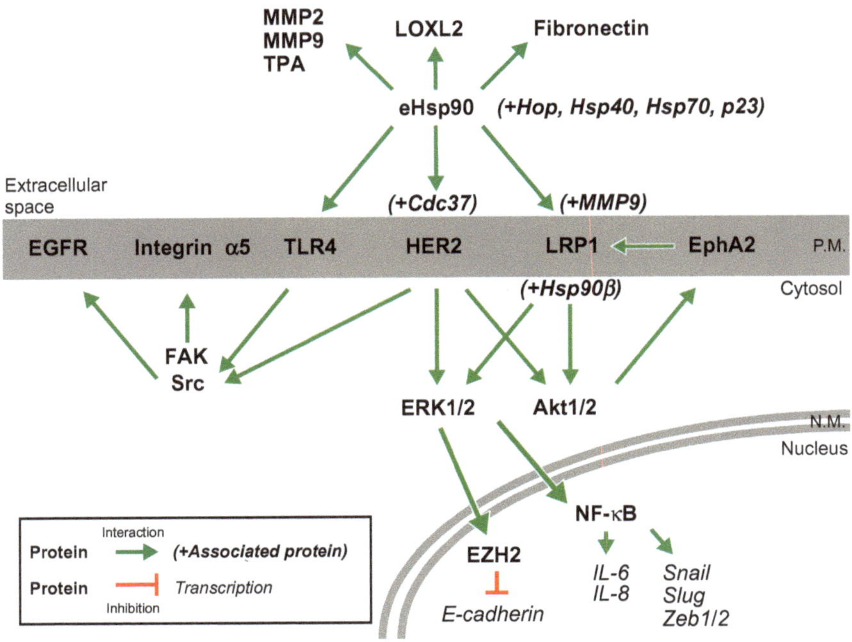

Figure 1, Daniel Senh Wong and Daniel G. Jay (See Page 148 of this volume.)

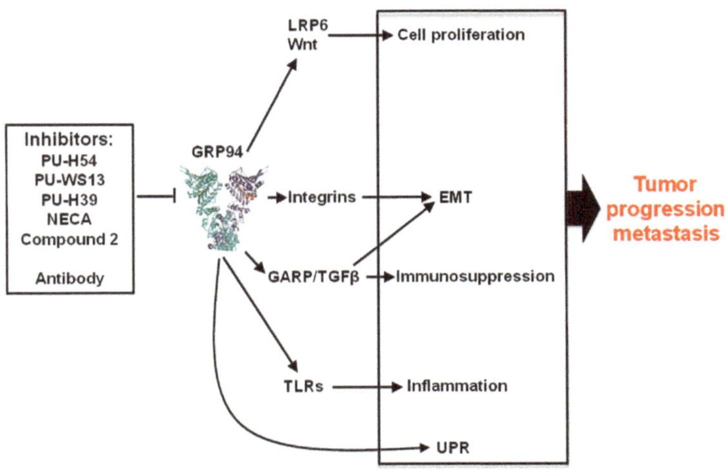

Figure 1, Bill X. Wu *et al.* (See Page 172 of this volume.)

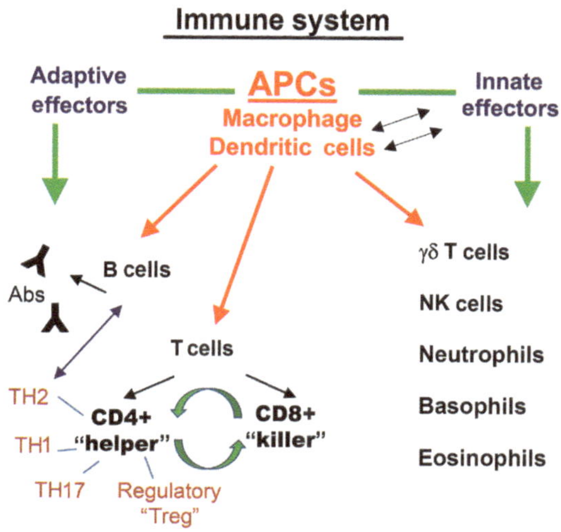

Figure 1, Michael W. Graner (See Page 193 of this volume.)

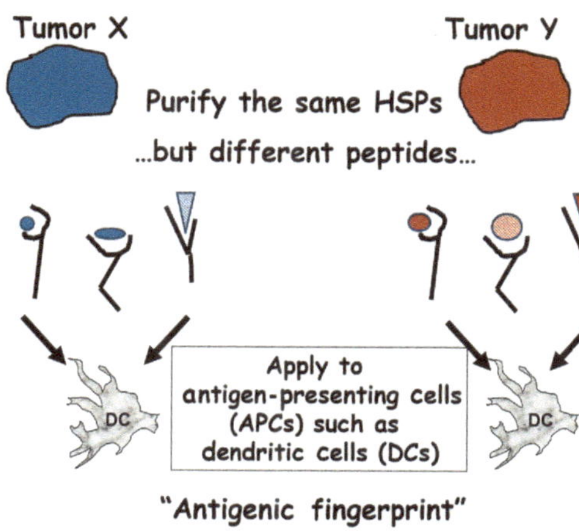

Figure 2, Michael W. Graner (See Page 196 of this volume.)

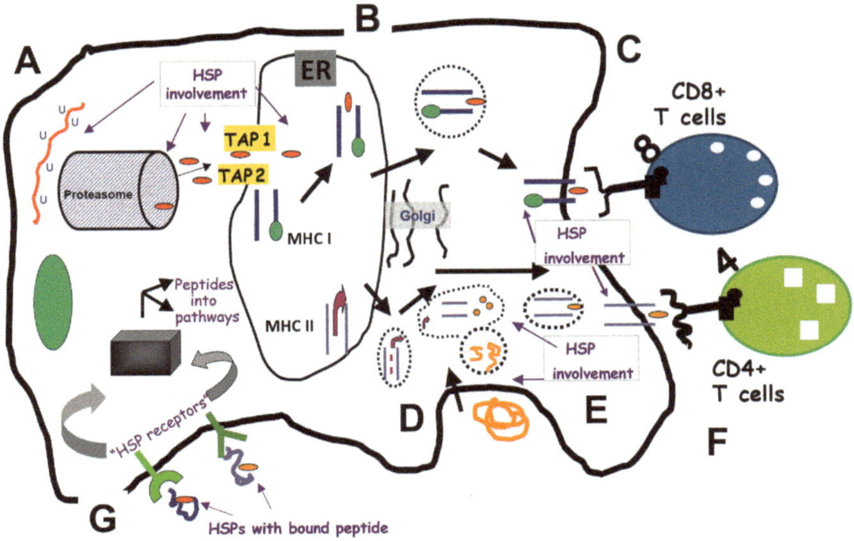

Figure 3, Michael W. Graner (See Page 197 of this volume.)

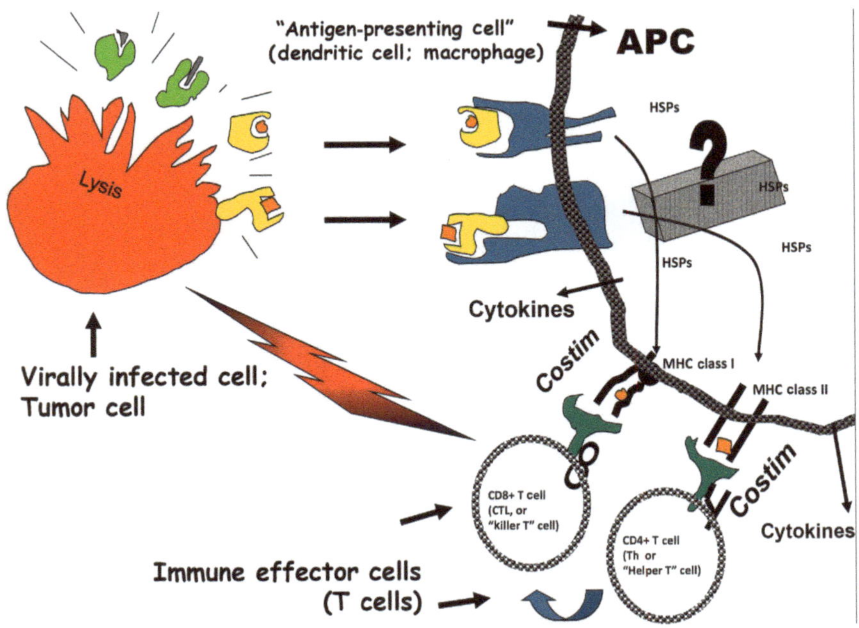

Figure 4, Michael W. Graner (See Page 201 of this volume.)

Kolibri 360 GmbH
Ferdinand-Jühlke-Straße 7
99095 Erfurt, Deutschland
produktsicherheit@kolibri360.de